Build Your Own
Universal Computer
Interface

Build Your Own Universal Computer Interface

Second Edition

Bruce A. Chubb

McGraw-Hill

New York   San Francisco   Washington, D.C.   Auckland   Bogotá
Caracas   Lisbon   London   Madrid   Mexico City   Milan
Montreal   New Delhi   San Juan   Singapore
Sydney   Tokyo   Toronto

Library of Congress Cataloging-in-Publication Data

Chubb, Bruce A.
    Build your own universal computer interface / Bruce A. Chubb.—
2nd ed.
       p.   cm.
      Includes index.
      ISBN 0-07-912638-3 (hc).—ISBN 0-07-912639-1 (pbk.)
      1. Computer interfaces—Design and construction—Amateurs'
manuals. 1. Microcomputers—Amateurs' manuals I. Title.
TK7885.5 C577 1997
004.6'4—dc21
                                 96-49215
                                      CIP

McGraw-Hill

*A Division of The **McGraw·Hill** Companies*

1 2 3 4 5 6 7 8 9 0   DOC/DOC   9 0 2 1 0 9 8 7

P/N 011688-1   P/N 011689-X
PART OF   PART OF
ISBN 0-07-912638-3 (HC)   ISBN 0-07-912639-1 (PBK)

*The sponsoring editor for this book was Scott Grillo, the editing supervisor
was Ruth Mannino, and the production supervisor was Suzanne W.B.
Rapcavage. It was set in ITC Century Light by Joanne Morbit of McGraw-Hill's
Professional Book Group composition unit, Hightstown, N.J.*

Printed and bound by R. R. Donnelley & Sons Company.

McGraw-Hill books are available at special quantity discounts to use as premiums and
sales promotions, or for use in corporate training programs. For more information,
please write to the Director of Special Sales, McGraw-Hill, 11 West 19th Street,
New York, NY 10011. Or contact your local bookstore.

Contents

Introduction

This book provides in-depth coverage of how you can build your own universal computer interface system (UCIS). The book details the complete technology of computer interfacing from theory to step-by-step assembly instructions. Both hardware and software are covered in depth. The UCIS works with everything from the newest Pentium back to the earliest IBM PC, XT, and their compatibles. It works with the early Commodores. It works with the earliest Apple II up through the latest Macintosh. The result is that "You can hook up most any computer to UCIS and have it perform any function you want it to perform."

Nearly 10 years of UCIS experience have been achieved since the original edition was published, and I've worked personally with thousands of UCIS users in a wide variety of applications. This second edition takes advantage of this experience base to help you better meet your interfacing requirements. Where original readers might have had difficulty, the material presented in this second edition has been enhanced for added clarity and completeness.

New circuits and new circuit boards are introduced, including an entirely new internal bus extender card that fits directly into Pentium and x86-class computers. Entirely new general-purpose digital input and output cards are introduced. More coverage is provided for the distributed serial option, enabling I/O devices to be placed closer to the objects they monitor and control. Software examples are expanded, with increased emphasis on using different languages.

It is easy to build a UCIS using the ready-to-assemble circuit cards available from the readily accessible sources provided in this second edition. An order blank is included. For readers desiring to get started in applying the interface without building the components, new sources are provided for purchasing fully assembled and tested interface cards. Alternatively, circuit artwork is provided for those wishing to fabricate their own circuit cards. Recommended source lists are provided for easy location of all electronic parts.

The UCIS does it all by providing direct bus-connected parallel, serial, and distributed serial interfacing for the whole spectrum of popular computers. For the serial case, the UCIS handles multiple standard hookup options, including RS-232, RS-422, and 20-mA current-loop, plus it makes provision for an RS-232 to RS-422 converter.

The UCIS provides tremendous expandability, with up to 1536 separate I/O lines in the parallel and basic serial options and up to 24,576 I/O lines with the distributed serial option.

This second edition provides extensive software coverage including standard serial protocol subroutines, real-time software control, multiple application examples, system test programs, and automated diagnostics. All software is incorporated in a 3.5-in disk enclosed with the book. To fit the backgrounds and needs of different readers, examples are included in an assortment of languages including BASIC, QBasic, QuickBASIC, Visual Basic, Pascal and C++.

Step-by-step instructions are included for building general-purpose digital input, digital output, analog-to-digital (A/D), and digital-to-analog (D/A) cards for connecting to all types of external hardware, such as switches, LEDs, lamps, relays, digital circuits, potentiometers, stepper motors, thermocouples, dc motors, and more. Comprehensive illustrative application examples are provided including model railroading computer cab control, automatic signaling, and train speed/direction control.

There are many different computers, and most every other interfacing book covers either one particular computer, or no specific computer at all (by treating the interface in only a generic sense). As proven with 10 years of UCIS experience, interfaces that work require specific detailed knowledge of the idiosyncrasies of each computer in terms of pinouts, line functions, polarities, and timings. In this book, all the required details for most every popular computer have been worked out. They are spelled out in step-by-step assembly detail for straightforward construction of an interface that really does the job.

The contents of this second edition provide the full complement of required information applicable to the neophyte with no prior electronic/computer experience, or to the professional electrical engineer. Following the step-by-step assembly and test instructions requires no prior experience to build a functional interface that works.

The first edition proved that this book, and the resulting UCIS, have universal appeal to the broadest possible audience. The industry technology trend is toward more effective applications of computers as part of real-time systems. This book shows specifically how a computer can be mated effectively to any type of external hardware for real-time control by software.

Numerous universities, colleges, and technical schools have developed UCIS applications. Instructors, having applied the text, declare that they see the material well-suited for classroom use. It is especially helpful when used as a supplementary text for special lab projects involving a computer connected to hardware, and for college and technical courses in digital computer interfacing and real-time software development.

In addition to the major commercial/industrial applications and the expanding educational interest, this book provides significant insight for the electronic and computer hobbyist. Experience is not required. People who have successfully constructed the UCIS include those with no exposure to any electronic project up to electrical engineers who daily work on digital electronics. In other words, anyone can build the UCIS. If you absolutely had to label this book, it is for the beginner to intermediate hacker/developer/system user; however, the expert can use the material to effectively and quickly build up to any level of interface complexity desired.

The text is set up so the reader can easily skip the theory and functional material and go right to the construction steps if desired. Likewise, if interest is exclusively in parallel interfacing, the reader is instructed where to skip over the serial interface material and vice versa. For those readers wishing to purchase their interface cards fully assembled and tested, they can jump right to the application examples.

Chapter 1 sets the stage for interfacing by whetting the appetite with a look at typical UCIS applications. It defines how the UCIS functions. An overview look at the available interface options is presented, followed by an introduction to software. The procedures for assembling your own UCIS are generally described, as well as a readily accessible source for purchasing the ready-to-assemble circuit cards. Additionally, new sources are provided where the interface boards can be purchased fully assembled and tested. The chapter concludes with discussions on circuit card soldering, system troubleshooting, and system reliability.

Chapter 2 provides added insight into selecting the best set of interface options to meet a particular need. Topics include computer selection and which interface options work with which computers, system cost, and what it takes to switch from one option to another. The attributes provided via direct parallel attachment to the computer's main bus, serial, and distributed serial interfacing are presented. Commonality of the major interface elements is stressed, along with the tailoring necessary to fit specific computer capability and external hardware requirements. System modularity and expandability as well as forward and backward compatibility are presented.

Chapter 3 begins the assembly of the UCIS parallel interface option by constructing the internal bus extender card (IBEC)—a new card designed specifically for interfacing to the Pentium, 486, 386, 286, and AT-class computers. Fundamental bus architecture is explained, along with interfacing highlights for the x86 computer family and their important predecessors. Also provided is a description of the IBEC functions, including signal buffering, address decoding, I/O mode selection, and data line control. Once functionality is understood, touring the schematic presents further insight into how the IBEC handles interface demands. Easy-to-follow, step-by-step assembly instructions are presented in a format that leads to a quality operational system.

Chapter 4 introduces the UCIS serial interface option by describing the universal serial interface card (USIC). Serial interface standards are described, including 20-mA current loop, RS-232, and RS-422, all of which are handled by the USIC. The operation of the USIC is explained first from a functional viewpoint and then via its schematic. This is followed by step-by-step assembly instructions. The chapter also covers assembling an RS-232 to RS-422 conversion card for setting up the distributed serial option.

Chapter 5 covers IBEC installation in Pentium, x86, and AT-class computers, as well as the IBM System/2 Model 30 and the earlier IBM PC, XT, and their compatibles. Initial card testing is followed by how to select different I/O modes, including memory-mapped and port-mapped I/O. The reader is led through proper card configuration procedures and then operational testing with the installed IBEC.

Chapter 6 expands the UCIS parallel interface option by describing the universal bus extender card (UBEC). Also provided is a description of the UBEC functions and a schematic tour. The UBEC assembly instructions are presented, including how to set jumper configurations and part substitutions to effectively mate the card with a multitude of different computers.

Chapter 7 explains how the UBEC is connected to the computer using either memory-mapped or port-mapped I/O. The chapter provides extensive coverage of computer connections and the use of computer bus adapter cards (CBACs) for a wide range of computers including the IBM PC, XT, XT-286, AT, 286, 386, 486, Pentium, SYSTEM/2 Model 30, the TRS Color Computer family, the Commodore family, and the entire Apple II family. Even a hardwiring example shows how the UBEC can be tied directly into computers that do not have an available JLC adapter card. Covering such a broad spectrum makes it easier to figure out how to mate with any particular computer requirement.

Chapter 8 covers building a newly designed set of general-purpose digital I/O circuits and a test card. General understanding of I/O handling is achieved by presenting address decoding and I/O card functional descriptions. Step-by-step assembly instructions are provided for the I/O motherboard, general-purpose digital I/O cards, building a 5-Vdc power supply, a test panel card and a wraparound test cable.

Chapter 9 covers testing of the parallel IBEC-based interface. Simple test software is presented, along with information on how to read the test LEDs built into the IBEC design. Recommended courses of action are exemplified for areas of difficulty. Testing is completed by using a provided output card test computer program followed by a complete wraparound test program guaranteeing correct operation of the complete system.

Chapter 10 explains in an easy-to-understand format the complexities of serial interface software. Topics include general message formats, protocol subroutines, and the initialization of program variables. Complete UCIS standard protocol software is provided for direct reader applications. Program timing considerations are discussed, along with recommended application program format for ease in real-time user software development.

Chapter 11 covers testing of the serial USIC-based interface. The reader is led through the complete testing process starting with visual inspection and simple voltage testing. Specific debugging steps are defined. Testing is completed by using provided serial versions of the output card test computer program followed by a complete wraparound test program.

Chapter 12 covers testing of the parallel UBEC-based interface. The reader is led through the complete testing process starting from simple continuity and voltage testing up through using test programs guaranteeing complete system operation.

Chapter 13 defines how to make I/O card connections to a broad range of external hardware including lamps, LEDs, panel switches, pushbuttons, limit sensors, other digital circuits, relays, stepper motors, and a booster circuit for driving heavy current-demanding loads. Decoding and encoding circuitry is presented to further simplify system wiring along with how to take advantage of current-sourcing output cards and how to drive matrixed outputs. The chapter also explains the design and assembly of an example decoder card, as well as the principles of good ground wiring.

Chapter 14 increases the reader's available repertoire by adding analog interfacing cards to the UCIS. Digital-to-analog converter (DAC) chips are described, and then via step-by-step assembly instructions, a complete three-channel UCIS-compatible DAC card is generated. The same proven process is repeated for understanding and assembling a UCIS-compatible three-channel analog-to-digital converter (ADC) card. Motor speed control is covered as a further example, using a special throttle card followed by the application of pulse power control for improved slow-speed operation.

Chapter 15 covers the important topic of generating I/O-handling software. The basic steps for generating effective I/O software are explained so they are applicable to the full range of application needs. Recommended easy-to-use I/O tables are provided to help establish program organization and correctness. Program format for the real-time loop is presented along with making I/O addressing easy. Important methods define how to read and unpack input bytes and how to pack and write output bytes. To solidify the procedures, an I/O program example is prepared using the IBEC, then the UBEC, and then repeated using the USIC.

Chapter 16 covers building the original I/O circuits and the corresponding test card. Although not as sophisticated as the new card designs, these single-sided versions still have plenty of merit, especially for systems where the reader desires to fabricate his or her own cards.

Chapter 17 starts a very meaningful two-chapter interface application example by defining and building up a set of external hardware to interface a model railroad to a computer using UCIS. This example of UCIS application is designed to go a long way toward helping every reader adapt UCIS to best meet his or her specific interface application. The cards discussed might appear at first to have a specialized purpose; however, in reality the technological ideas presented (as well as the circuits that result) have far-reaching implications applicable to most any interface project.

Chapter 18 completes the comprehensive interface example by hooking up the system and generating the application software. Additional topics include setting up a monitor status display, preparing display subroutines, simulating UCIS inputs for program testing, and program timing. Once the software is complete for IBEC and UBEC, it is then regenerated for USIC users followed by a general discussion of further advanced capabilities provided by automated diagnostics and fully automated operation.

Build Your Own Universal Computer Interface

1
CHAPTER

Basics for Building a Computer Interface

Here is your invitation to join in a major technology advancement: the easy application of a *universal computer interface system* (UCIS) to connect external hardware to your computer. This enhanced and significantly expanded second edition advances the technology several steps, making it easier for the user to obtain an interface optimized to meet any set of needs at the lowest possible cost. Another great advantage of the UCIS is the simplicity with which it can be applied to any setup, from an IBM-compatible to a Macintosh or to any other computer.

As a special feature of the UCIS, I've made it possible for you to obtain ready-to-assemble printed circuit boards, or cards, from JLC Enterprises. Having the JLC cards readily available makes it simple and enjoyable to assemble your own interface system. For readers electing to fabricate their own cards from scratch, I've included circuit artwork in Appendix A.

Also, by popular demand this second edition provides new sources for readers preferring to purchase their interface cards fully assembled and tested. The assembled and tested boards cost about twice as much as if you purchase the bare boards and parts and do your own assembly and testing. Most commercial and industrial UCIS applications find it more affordable to purchase the assembled and tested boards. Many educational and home users do the same, while others find it more affordable (and often more enjoyable) to perform their own assembly and testing using ready-to-assemble boards.

The first 16 chapters show how to select and then (through step-by-step instructions) how to assemble your own UCIS to support whatever applications you desire. The last two chapters illustrate the capability of the UCIS through a detailed example in which the UCIS interfaces to model railroad hardware, with monitoring and control provided by application software.

Obviously a computer can perform many useful computational and paperwork functions. The benefits really expand, however, when you electrically interface the computer with the outside world for a wide variety of monitoring and control functions.

1

Example UCIS Applications

Applications range from a personalized computer-controlled security system to a completely automated home or business, and could include an interface with test equipment to monitoring and control of laboratory experiments, lighting control systems, data acquisition systems, machine control, automated testing and system diagnostics, model railroad operation and display, material handling systems, games and simulators with actual hardware hookup, or even full robotics.

For example, watching a game show on one of the major TV networks, the blinking lights and sound effects may well be controlled by one of several UCISs they assembled for such purposes. A woodworking company in the northeast is using multiple UCISs to control stepper motors to automate wood engraving machines. A utility company is experimenting with a UCIS to monitor and control the switchover of their generator systems. A Midwestern city is doing the same in well water systems. Aerospace companies are using the system in laboratory test apparatuses, and in controlling resin-mixing processes.

A fruit producer in the northwest is applying the system, interfaced with scales, to control the weighing and packaging process. Several aerospace companies are using the system to automate system testing, airfoil design, and aerodynamics analysis. A company in China uses the system for manufacturing safety monitoring in aerosol filling operations. A major computer manufacturer uses the system for training its marketing personnel. A social learning center uses the system in rehabilitation work. A chemical company uses the system in process control.

A major machine tool producer applies the system in robotics. A computer services company makes use of systems in automated diagnostics. A prominent eastern university uses the system in their optics program. A major research hospital is using a UCIS to support advanced work in their Medical Physics department. A California company specializing in custom computer systems uses the UCIS in rapid prototyping. Another corporation uses it in food processing. A forest products company has the system supporting its manufacturing operations, and a major national corporation uses the system in refinery process control.

Numerous research centers, schools, and industries make use of the UCIS. For example, a nationally famous research center used a UCIS to demonstrate advanced capability in material handling systems. Many universities and technical schools around the world use the system as part of their electronics and computer education programs. The world's largest maker of computer chips used UCIS components to interface with a model railroad layout; the purpose was to demonstrate the release of a new real-time software package.

Advanced, technically oriented high schools use the UCIS as part of their classroom and computer science club activities. Students in one school use a UCIS to control a room full of robots.

In addition to the United States and Canada, installations of UCIS systems can be found worldwide, including Australia, Brazil, China, Denmark, England, France, Germany, Italy, Jamaica, Japan, Luxembourg, Mexico, Netherlands, New Zealand, Norway, and Sweden.

The most common UCIS application areas could be summarized as follows:

- Process control systems
- Monitoring and control of laboratory experiments
- Education in developing real-time systems
- Automating museum displays
- Lighting control systems
- Security systems
- Automatic weighing and packaging systems
- Data acquisition systems
- Environmental control systems
- Support of university/industrial research projects
- Machine control of every type
- Software development for real-time systems
- Rapid prototyping where need computer interfacing
- Automated testing and system diagnostics
- Fireworks ignition display systems
- Greenhouse climate-control systems
- Home or office automation
- Model railroad operation
- Material handling systems
- Games and simulators with actual hardware hookup
- Robotics

I'm using a UCIS to connect my 133-MHz Gateway 2000 Pentium computer to support interface hardware development and testing. The Pentium is also connected via UCIS to a 2800-ft^2 model railroad, providing both operator-assisted and fully automated operation. Using the UCIS, the Pentium monitors the location of all trains operating on the system, interfaces with all the control panels, drives hundreds of trackside signals, and can control the trains themselves. Up to 20 or more trains can operate simultaneously on the system. I'm using another UCIS with a Compaq 286 computer to support special laboratory experiments and computerized testing.

I used the original UCIS to connect a Heathkit computer to a model railroad display system. The model railroad example is a great way to demonstrate the real-time hardware-in-the-loop capabilities of the UCIS. Numerous universities and technical schools use a UCIS tied to model railroad hardware as part of their computer interfacing or real-time software development curricula. Student interest in such projects remains high, and they can easily visualize the results of their efforts.

Thousands of model railroaders are also using UCISs, typically known for their application as the *Computer/Model Railroad Interface* (C/MRI), originally published in a 16-part series in *Model Railroader* magazine. These applications cover interfacing to *digital command control* (DCC) and other command control systems, *computer cab control* (CCC), *computer block control* (CBC), prototypical signaling systems, hump yard control, automatic staging, interlocking at junctions and terminals, lighting control, and display-mode operation.

Incorporating the experiences gained from these diverse industrial, commercial, educational, and hobby computer setups, I've expanded this second edition of the UCIS manual, making it even more universal and attachable to the broadest possible range of computers and applications.

How the Universal Computer Interface System Works

To understand the workings of the UCIS, you need to understand how the computer interfaces with itself—that is, how the central processing unit, or CPU, communicates with computer memory and with peripheral devices like the display monitor or cathode ray tube (CRT), keyboard, disk drive, and printer. We'll be looking at a simplification in terms of the full capability of modern computers, but it covers the fundamentals and that is all you need to understand to grasp how the interface works. Most CPUs (or *microprocessors*) in personal computers communicate using a 16- or 32-wire internal address bus, an 8-, 16-, or 32-wire data bus, and a few control wires; these together are referred to as the *internal* or *system bus* (Fig. 1–1).

To send to or receive data from one of the units on the internal bus, the CPU first transmits an address on its address bus. In this way it defines the memory location or the I/O interface to be accessed.

Once the address is set, a signal on the appropriate control lines causes the message (data) to be transmitted on the data bus, either as input to the CPU or output from the CPU to the memory or peripheral. Every transmission over the data bus goes to every device on the internal bus, but the only device that responds is the one whose address matches the setting of the address bus. Note that, while the data bus is bidirectional, the address bus (at least as far as the UCIS is concerned) is unidirectional.

A straightforward way to interface the computer with your external hardware is to use a parallel I/O interface card. Such cards are available as add-ons for many popular computers. They plug into spare slots on the computer's bus, and provide from 8 to 24 separate input/output lines, each of which can be directly or indirectly connected to external hardware devices. At 8 or even 24 lines each, the cards needed for all but the very smallest of external hardware applications would soon fill the spare slots in any computer, as well as rapidly drain your pocketbook.

The UCIS is a more efficient and economical way to move data, and its parallel interface form using the internal bus extender card (IBEC) is also illustrated in Fig. 1–1. The IBEC extends the computer's internal bus to an external set of UCIS-provided, general-purpose input and output (I/O) cards. Each one of these I/O cards handles 24 separate lines to or from your external hardware. These general-purpose digital I/O cards are covered in detail in Chaps. 8 and 16. Their analog counterparts are detailed in Chap. 14.

With large amounts of I/O, it's a good idea (and usually necessary) to have an electrical buffer between the multiple I/O cards and the computer's internal bus. The IBEC performs this function, too. As a result, you can use a separate power source for the I/O cards, plus keep the UCIS from loading down the computer's internal bus.

The number of address and data lines have gradually increased as newer machines incorporate increased memory size and performance capability. Because the function of these lines remains essentially the same regardless of design, however, it's easy for our UCIS to handle the variations. The more significant differences among the various machines are in the control lines; these vary in number and func-

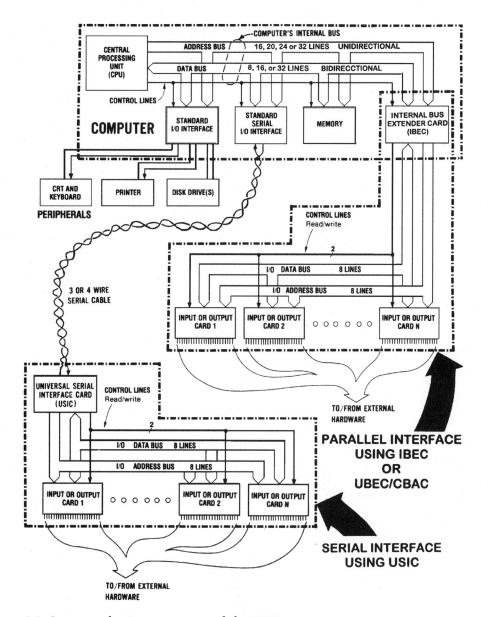

1-1 Computer busing structure and the UCIS.

tion. Also, the connector type and pinout arrangement of all the lines vary between different computer families. As a result, you would need to make a unique bus extender card to fit into each different type of computer.

The ISA bus structure found in the Pentium, 486, 386, and 286 or AT-class computers is thoroughly standardized. Considering these computers' popularity, it is practical to provide a bus extender card that fits right into this class of machine. This is exactly what JLC Enterprises did with their offering of the internal bus extender card. The card works in the earlier IBM PC, XT, and their compatibles, as well as in the IBM System/2 Model 30. The IBEC plugs right into the ISA bus 62-pin connector and functionally matches the address, data, and control-line structure found in all these machines.

You could make other special bus extender cards, with each one specifically designed to fit right into each particular computer that incorporates expansion slots. However, you would need a specific design for each of the Apple II family, the Macintosh, IBM's Micro Channel Architecture computers, and S100 bus computers, and the list goes on and on. Putting the bus extender card right inside the computer is a very neat way to go, so long as you have space for it. However, except for the IBEC, the alternative approach I've taken for UCIS is not to build up a special card for each computer type or family. There are just too many different computers on the market, and more coming out every year.

The alternative to the IBEC approach effectively replaces the IBEC's function with two separate cards. First a *computer bus adapter card* (CBAC) is used to mate with each computer's connector type. This card then connects via a short cable to a *universal bus extender card* (UBEC) placed outside the computer. Because the CBAC must vary from computer type to computer type, its design is kept as simple as possible; its primary purpose is to rearrange the computer connections to make them fit a standard cable configuration at the UBEC. The more complex functions such as address decoding, data and address line buffering, and control line manipulation are placed in the UBEC. By altering jumper locations and implementing a limited number of part substitutions on the UBEC, it is possible for the CBAC/UBEC combination to handle the widest possible range of computers.

To further complicate the situation, many smaller computers do not have spare card slots. To keep size and cost down, they are designed with the bus coming out to an external connector. This device goes by a variety of names, including *expansion connector*, *ROM connector*, *cartridge connector*, and *game connector*. By design, however, the truly universal capability of the UCIS typically handles these situations. Because the UBEC is designed to sit just outside your computer, it can plug into the expansion connector just as easily as it reaches inside your computer via the CBAC to plug into a spare expansion slot.

A computer without either spare expansion slots or a compatible external bus expansion connector is not a good candidate for parallel interfacing with either the IBEC or the UBEC. However, UCIS is readily attachable to these computers as well, through the serial interfacing capability provided by the *universal serial interface card* (USIC).

Looking back at Fig. 1–1 illustrates how the UCIS serial interface setup is achieved using the USIC. With the USIC, in contrast to the IBEC and UBEC, the group of general-purpose I/O cards can be set up at quite a distance from your computer. A simple three- or four-wire serial cable runs from the USIC to connect to a standard serial I/O interface port on your computer.

It is important to note that the UCIS components on the external hardware side of the interface (that is, the local UCIS I/O bus consisting of eight lines of address, eight lines of data, and two lines of control plus the bank of general-purpose I/O cards) remain identical regardless of whether you elect to use the IBEC, UBEC, or USIC.

System Options

Figure 1–2 illustrates what you'll need to build the IBEC-parallel form of UCIS. The IBEC plugs directly into one of the ISA expansion slots in your Pentium or *x*86-class computer, and also works with IBM PCs, XTs, and their compatibles. On the external hardware side of the IBEC, everything is the same no matter which computer or form of UCIS you choose. A standard 40-wire ribbon cable connects the IBEC to the I/O motherboard, which in turn holds the general-purpose I/O cards.

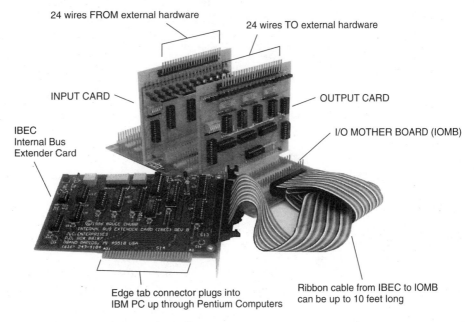

24 wires FROM external hardware

24 wires TO external hardware

INPUT CARD

OUTPUT CARD

IBEC
Internal Bus
Extender Card

I/O MOTHER BOARD (IOMB)

Edge tab connector plugs into
IBM PC up through Pentium Computers

Ribbon cable from IBEC to IOMB
can be up to 10 feet long

1-2 UCIS-parallel (IBEC-based) system.

Figure 1–3 illustrates what you'll need to build the UBEC parallel form of UCIS. The UBEC and its computer connection are the only parts of the system that need to be tailored to specific machines. Everything else is the same. The identical-to-IBEC 40-wire ribbon cable connects the UBEC to the I/O motherboard, which in turn holds the general-purpose I/O cards. I'll explain the various *computer bus adapter cards* (CBACs) and their connections to fit a multitude of different computers in Chap. 7.

Figure 1–4 illustrates what you'll need to build the serial form of UCIS. The USIC plugs directly into the I/O motherboard. A three- or four-wire cable connects the USIC to the computer's serial port, which can be RS-232, RS-422, or 20-mA current loop. For applications in which you desire more distance between the USIC and your computer than is typically provided by RS-232, or for cases where you desire multiple USICs connected to the same serial port, you can employ the optional RS-232-to-RS-422 conversion card. I'll cover its use and construction in Chap. 4.

In all three approaches—parallel with IBEC, parallel with UBEC, and serial with USIC—the input and output cards remain identical, and each has 24 lines, three 8-bit ports, which can be directly connected to a varied assortment of electrical

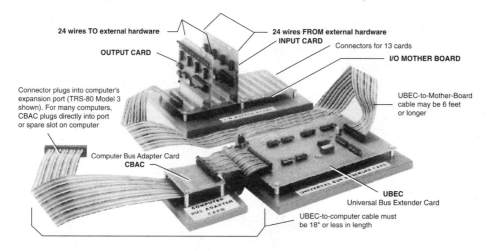

1-3 UCIS-parallel (UBEC-based) system.

devices associated with your external hardware. Such devices include lamps, LEDs, panel switches, pushbuttons, limit sensors, other digital circuits, and relays. I'll show you how to connect these and many other devices to the I/O cards and how to prepare your own computer programs to operate them.

In designing the UCIS, I have selected components and laid out the circuit cards so that construction and testing is a straightforward matter. I also have built and tested IBECs, UBECs, and USICs with most popular computers. Thousands have successfully built UBEC and USIC systems from the first edition.

We'll look at how to select the best UCIS configuration to meet your particular needs in Chap. 2. In the remaining chapters you can delve into the UCIS to whatever level you find necessary to make your system totally operational.

Introduction to Software

Software refers to the instructions placed in a computer memory to tell the computer what action to take. These instructions can be mathematical or logical. A group of instructions is referred to as a *program,* and a program loaded into a computer defines, step-by-step, what kind of a machine the computer will be.

When we load a check-balancing program, the computer helps us balance the checkbook. Load a space-war program and the same computer becomes a space-war game. I'll be showing you interface programs you can use to communicate with your external hardware using UCIS. These programs will include test software to make sure your system is operating correctly, as well as sample application programs.

The software examples presented in this second edition are written with primary emphasis being ease of understanding and simplicity of adaptation. A 3.5-in disk, enclosed with this second edition, includes all the software presented in the book, plus numerous other examples; the code is written in several different languages, including BASIC, QuickBASIC, QBasic, Visual Basic, Pascal, and C++. Be sure to start with the README file for a more detailed explanation of the disk's con-

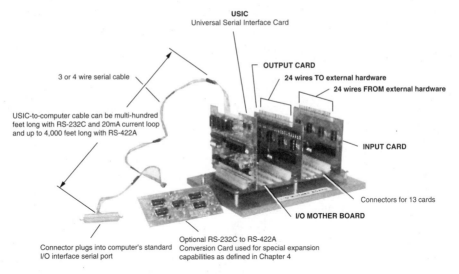

USIC
Universal Serial Interface Card

3 or 4 wire serial cable

OUTPUT CARD
24 wires TO external hardware
24 wires FROM external hardware

USIC-to-computer cable can be multi-hundred
feet long with RS-232C and 20mA current loop
and up to 4,000 feet long with RS-422A

INPUT CARD

Connectors for 13 cards

I/O MOTHER BOARD

Connector plugs into computer's standard
I/O interface serial port

Optional RS-232C to RS-422A
Conversion Card used for special expansion
capabilities as defined in Chapter 4

1-4 UCIS-serial (USIC-based) system.

tent. Also, make a working copy of the disk, because you will want to be tailoring the programs to best fit your own personal needs.

Programs can be written in many different programming languages, but the most common one for personal computers is BASIC. It is easy to learn, and for years a BASIC interpreter was included as an integral part of nearly every personal computer. However, BASIC has limitations for *real-time* operation with external hardware, that is, interacting with external hardware as it runs. Because it is an interpreted language (not separately compiled into digital machine code), BASIC is very slow in executing instructions. When we get to applications that demand higher program operational speed, it is best to move beyond interpreted BASIC; but that language is more than adequate for getting started.

To greatly speed up program operation, you can use one of the compiled versions of BASIC. The one I like especially well is Microsoft's QuickBASIC for the IBM PC and compatibles. The latest version number as of this writing is version 4.5, and it works great with UCIS. QuickBASIC has a user-friendly editor that makes programming easier, it is better suited to structured programming techniques, no statement numbers are required, and meaningful variable names can be applied. Fundamentally, I find that it works very well for our real-time UCIS programming requirements.

QuickBASIC should not be confused with QBasic, an interpreted BASIC that comes bundled with DOS 5.0 and higher, for example. QBasic can be used for operating UCIS, but it lacks the programming power and speed provided by the higher-level compiler-based programs such as full QuickBASIC, Visual Basic, Pascal, Turbo Pascal, C, Turbo C, and C++, to name just a few. All will work with the UCIS.

If you have a favorite programming language, then use it with UCIS. If you are not familiar with programming, then BASIC is a good place to start. If BASICA or QBasic is installed on your machine, then use it. If not, then I would recommend you purchase a copy of Microsoft QuickBASIC Version 4.5. A similar version is available for the Macintosh.

As with any programming language, interpretive BASIC instructions vary a bit from one manufacturer to another. The BASIC I'll use in this book is Microsoft interpreted BASIC. Dialects vary from machine to machine, but Microsoft BASIC is the most widely used version. For every example presented in this second edition I will, at a minimum, include programming compatible with Microsoft interpreted BASIC and Microsoft QuickBASIC. My programs include extensive commenting, so that reading the resulting code is much like reading plain English; therefore, these programs should be easy to convert to your favorite language if this has not already been accomplished for you via the enclosed disk.

I can't take the space to teach BASIC (or any other programming language) in this book, but most statements we'll need are easy to understand. Any language you purchase comes with a manual, and computer stores have a wide assortment of books on most every programming language. One title I've always liked is *BASIC BASIC*; you can hardly get more basic than that!

If you desire more information on Microsoft QuickBASIC, I highly recommend The Waite Group's, *MICROSOFT QuickBASIC BIBLE*, published by Microsoft Press. Also, *BASICs for DOS* (Gary Cornell; McGraw-Hill) provides excellent coverage of GW-BASIC, BASICA, and Microsoft's QBasic

A few software examples included on the disk enclosed with this second edition are written to operate with UCIS under a Windows environment; however, for the most part, real-time software applications demand fast-response, dedicated-type processing, and work best when their performance is closely linked to the computer hardware. Therefore, while developing your software under a Windows environment (including Windows 95) is fine, you will typically obtain your best real-time software operation by performing all interface software execution under a DOS environment or its equivalent for your particular computer. Therefore, the majority of all material presented in this book assumes a dedicated-processor DOS-type environment.

Binary Number System

When we talk to a computer, we use the *decimal* number system for the most part: The digits 0, 1, 2, 3, 4, 5, 6, 7, 8, and 9. The computer, however, uses a *binary* number system with only two digits, 0 and 1, to do everything we do with 0 through 9.

Computers use the binary system because it is very easy to handle this system electronically. Inside a computer are millions of electronic switches, called *transistors*, that store and control the flow of electrical signals. For example, the Pentium chip itself has 3.1 million transistors, and that's only a small fraction of the total in a Pentium-based system. You can think of these transistors as if they were switches, like those in Fig. 1–5.

Each switch is separately controlled and can be either on or off, representing binary digit 1 when on and 0 when off. A single switch can send only two codes, but a pair can send four, equivalent to decimal numbers 0 through 3. Each time a switch is added, the number of codes that can be sent doubles. Three switches can send 8 binary codes, four switches can send 16, five switches can send 32, six switches can send 64, and so on.

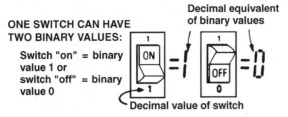

ONE SWITCH CAN HAVE
TWO BINARY VALUES:

Decimal equivalent
of binary values

Switch "on" = binary
value 1 or
switch "off" = binary
value 0

Decimal value of switch

TWO SWITCHES CAN HAVE FOUR BINARY VALUES:

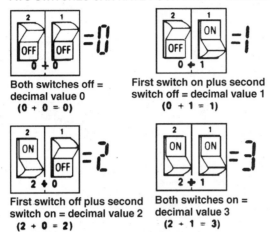

Both switches off =
decimal value 0
(0 + 0 = 0)

First switch on plus second
switch off = decimal value 1
(0 + 1 = 1)

First switch off plus second
switch on = decimal value 2
(2 + 0 = 2)

Both switches on =
decimal value 3
(2 + 1 = 3)

EIGHT SWITCHES CAN HAVE 256 BINARY VALUES:

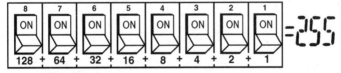

**All eight switches on (the sum total of each switches decimal
value) equals 255; plus all switches off (decimal value 0) yields
a total of 256 binary codes (or decimal numbers). The decimal
equivalent value of each bit (switch) starts with 1 on the right
and doubles with each position to the left.**

1-5 Binary codes.

The abbreviation for a single binary digit is *bit*. A bit can have only two states, but they may be referred to as on or off, high or low, true or false, +5 V or 0 V, as well as 1 or 0. The personal computers you will want to devote to UCIS applications will be 8-, 16-, 32-, or 64-bit machines, which means that their electronic switches are arranged in groups of 8, 16, 32, or 64. A group of eight contiguous bits is called a *byte*, and as Fig. 1–5 shows, it takes eight binary switches to send or store one byte of information.

If all eight switches are on, they indicate binary 11111111, which is equivalent to decimal 255. The decimal equivalent is the sum of the decimal equivalent values of each switch position that is turned on, and from right to left each switch doubles in decimal equivalent value. If all the switches are off, the indicated value is zero, so one byte can represent all decimal numbers between 0 and 255, a total of 256 binary codes.

Table 1–1 shows the relationship between the number of address bits and the size of memory that a computer can address. The entries are simply powers of two, for example, a 16-bit bus supports an address space of 2^{16} or 65,536 decimal locations that can be individually addressed. In computer jargon, 65,536 is referred to as 64 kbytes, or 64K for short. In the general sciences, the symbol k typically stands for *kilo*, Greek for 1000; however, in the computer industry K stands for 2^{10}, or 1024. This difference can be confusing, but in this book K will always be 1024.

When dealing with computers the 1024 interpretation is handiest, because computers "think" in powers of two, that is, in the binary number system. Likewise, in the computer field, M (standing for *mega* or million), is equated with 2^{20} or 1,048,576 locations, and G (standing for *giga*, meaning "giant" in Greek), is equated to 2^{30} or 1,073,741,824. Note that M = K × K, and G = K × K × K, or more simply M × K. It's also interesting that each time you add a single bit, you double the address space.

Also note that the Table 1–1 numbers, for instance, 8, 16, 32, 64, 128, 256, 512, and 1024, coincide with numbers we'll be using throughout this second edition. Virtually every number we use with computers has its roots in this powers-of-two table. We'll use these relationships extensively as we proceed through the book, so you might find it handy to refer to Table 1–1 from time to time.

Binary is good for computers but cumbersome for humans, so I won't deal in binary any more than is absolutely necessary. One reason binary is cumbersome is that it takes a combination of eight 1s and 0s to represent a byte. A decimal number from 0 to 255 can also define the contents of a byte, but it's difficult to see the correspondence between the decimal number and the binary bit pattern. This is where the hexadecimal number system comes to the rescue.

Hexadecimal Number System

In the hexadecimal number system, there are 16 *digits*, or characters: 0 through 9, plus A through F. The letters are symbols used as digits; A = 10, B = 11, C = 12, D = 13, E = 14, and F = 15. These 16 digits, 0 through F, can represent all possible bit combinations for a 4-bit group, sometimes called a *nibble;* that is, half a byte. Figure 1–6 shows corresponding hexadecimal, decimal, and 8-bit binary-coded numbers. A complete table would take 256 rows, 0 through 255, so I've eliminated many higher-order rows.

Being able to eliminate these rows, yet still have all the information we need, is one reason hexadecimal is so handy. It takes only two hexadecimal characters to represent any eight-bit binary code, a byte. That's convenient, but the main advantage of hexadecimal is the ease of converting binary to hex and hex to binary—much easier than with decimal numbers.

You can ignore the shaded portion of Fig. 1–6. The small unshaded portion is the only part we need to understand, because everything else can be derived from it. For example, a hexadecimal 11 is binary 0001/0001, a hex BB is 1011/1011, and a hex 1B is 0001/1011. The left hex character defines the left four binary bits, and the right hex character defines the right four bits. This will come in very handy when you

Table 1-1. Address Bits and Memory Size (Powers of Two)

Number of Address Bits	Number of Addresses	Range in Bytes
1	2	2
2	4	4
3	8	8
4	16	16
5	32	32
6	64	64
7	128	128
8	256	256
9	512	512
10	1,024	1Kb
11	2,048	2Kb
12	4,096	4Kb
13	8,192	8Kb
14	16,384	16Kb
15	32,768	32Kb
16	65,536	64Kb
17	131,072	128Kb
18	262,144	256Kb
19	524,288	512Kb
20	1,048,576	1Mb
21	2,097,152	2 Mb
22	4,194,304	4 Mb
23	8,388,608	8 Mb
24	16,777,216	16 Mb
25	33,554,432	32 Mb
26	67,108,864	64 Mb
27	134,217,728	128 Mb
28	268,435,456	256 Mb
29	536,870,912	512 Mb
30	1,073,741,824	1Gb
31	2,147,483,648	2Gb
32	4,294,967,296	4Gb

HEXA-DECIMAL	DECIMAL	BINARY
00	0	00000000
01	1	00000001
02	2	00000010
03	3	00000011
04	4	00000100
05	5	00000101
06	6	00000110
07	7	00000111
08	8	00001000
09	9	00001001
0A	10	00001010
0B	11	00001011
0C	12	00001100
0D	13	00001101
0E	14	00001110
0F	15	00001111
10	16	00010000
11	17	00010001
12	18	00010010
13	19	00010011
14	20	00010100
15	21	00010101
16	22	00010110
17	23	00010111
18	24	00011000
19	25	00011001
1A	26	00011010
1B	27	00011011
1C	28	00011100
1D	29	00011101
1E	30	00011110
1F	31	00011111
::	::	
2A	42	00101010
::	::	
6B	107	01101011
::	::	
8B	139	10001011
::	::	
BB	187	10111011
BC	188	10111100
::	::	
D5	213	11010101
::	::	
FD	253	11111101
FE	254	11111110
FF	255	11111111

1-6
Hexadecimal, decimal, and binary numbers.

write application software to use with your UCIS. When I use hexadecimal numbers in this book, I'll append a lowercase h, so that hex number D9 will appear as D9h.

Assembling Your Own Interface

Please don't worry if all this sounds too complicated. I'm going to lead you through each step of the construction and installation process, and in the course of building the hardware and programming the software, you'll find yourself learning what you need to know about computers and computer interfacing technology. If you dig in and follow along with me through the essential chapters applicable to your setup, you'll enjoy the new experiences and be very proud of the result.

For readers who feel comfortable with electronics, I will be presenting complete circuit schematics, along with block diagrams to illustrate circuit functions. You won't need to study them unless you want to understand more about how the UCIS works. The assembly instructions are written with the novice in mind, and don't even refer to the schematics or block diagrams.

If you want to learn more about computers and electronics than I have room to explain in this book, there are other books that can help you. I especially recommend:

- *Understanding Electronics,* 3d ed. (R. H. Warring and G. R. Slone; McGraw-Hill, 1987).
- *Mastering IC Electronics,* (Joseph J. Carr; McGraw-Hill, 1991).
- *TTL Cookbook,* (Donald Lancaster; Howard W. Sams & Co, 1974).
- *Computer Technician's Handbook,* 3d ed. (Art Margolis; McGraw-Hill, 1989).
- *Rescued by Upgrading Your PC* (Kris Jamsa; Jamsa Press, 1994).

The last book is very basic and quite pictorial in nature, but for readers new to computers it is a great place to start. I'll reference additional books as we move through other chapters.

If you desire to stay simpler yet, you can purchase the necessary UCIS boards fully assembled and tested. Using this approach you can skip all the assembly and test sections of this book and simply move right into making the interface work for you.

To build the parallel form of UCIS, you will need to assemble either five or six different kinds of circuit cards: an IBEC (or UBEC and CBAC), an I/O motherboard, as many input and output cards as your interface to external hardware requires, and a computer output test card. You will also need a 5-Vdc regulated power supply for the I/O cards; I'll cover that in Chap. 8 when I explain how to build the cards.

Building the serial form of UCIS is identical, except that you substitute the USIC for the IBEC (or UBEC and CBAC). All the rest of the interface is identical, including the optional *digital-to-analog* (DAC) and *analog-to-digital* (ADC) cards I'll cover in Chap. 14.

I hope you'll build a UCIS for your computer. You will find building your own interface both enjoyable and rewarding. Working one step at a time, you will find the steps are meaningful and easy to understand. Before you realize it you'll have a working system.

UCIS Parts

Ready-to-assemble printed circuit boards, or PC boards or cards, are readily available from JLC Enterprises, P. 0. Box 88187, Grand Rapids, MI 49518-0187, telephone 616-243-4184. For your convenience, card description information and a blank order form are included in Appendix A. Circuit artwork is also provided in Appendix A in case you want to etch and drill your own UCIS cards.

Because of the number and size of the UCIS cards, and because many are double-sided with plated-through holes and thus difficult to etch on your own, I encourage you to buy the ready-to-assemble cards from JLC Enterprises. They are top-quality cards of epoxy-glass construction with holes drilled and all traces solder-plated for easy assembly. The feed-through holes on double-sided cards are plated through for circuit continuity, and all tab connector surfaces are gold-plated for high contact reliability. Quantity discounts are available.

If you are interested in the fully assembled and tested boards, they are available from ECO Works, P.O. Box 9361, Wyoming, MI 49509-0361, (616) 243-2893; EASEE Interfaces, P.O. Box 92260, Lakeland, FL 33804, (941) 858-8702; and Automation In Action (AIA), 1313 16 Mile Rd., Kent City, MI 49330, (616) 887-0817. Please send ECO Works, EASEE Interfaces, or AIA a business-sized envelope to receive a copy of its current price list. An alternative "best-way" to reach AIA is via its worldwide web site HTTP://WWW.ISERV.NET/~AIA.

During the generation of the material for this book, I put in countless hours searching through catalogs, price lists, and part specification sheets to come up with the lowest priced good-quality parts. The goal is to make the UCIS affordable and very easy to assemble. These are the specific part sources I recommend:

- Digi-Key Corp., P.O. Box 677, Thief River Falls, MN 56701; (800) 344-4539.
- Jameco Electronics, 1355 Shoreway Rd., Belmont, CA 94002; (800) 831-4242.
- JDR Micro Devices, 1850 South 10th St., San Jose CA; (800) 538-5000
- Mouser Electronics, 11433 Woodside Ave., Santee, CA 92071; (800) 346-6873.

All four provide free catalogs, an 800 number for ordering as noted above, and excellent service. For example, both Digi-Key and Jameco advertise fill rates well over 99 percent, and with Digi-Key your order is typically in UPS's hands within four hours after receipt of your call. You just can't beat such service, and it sure beats shopping around town for substitute parts that might or might not work in the defined application. Ordering the specific recommended parts from the specifically recommended sources, you can be confident that you have the correct parts. Such an approach goes a long way to help make your system operate correctly the first time you turn it on.

There are several different family choices for the *integrated circuits* (ICs), and new versions are periodically introduced. Throughout the book, I've primarily listed what is still currently the most popular family, the LS (*low-power Schottky*) versions. In almost every case you can substitute chips from any other TTL compatible family. For example, in place of a 74LS00 you can use a 7400, 74S00, 74F00, 74ALS00, or a 74HCT00. I chose the LS versions because they work great in the UCIS, they provide a good balance between power consumption, speed, and drive capability, plus they are typically better-stocked by suppliers and usually at the lowest price.

In the I/O card area, where typically you need multiple identical cards, I do list HCT and F parts in a few cases where in current catalogs I found them to be less expensive than the equivalent LS part.

A few of the ICs are static sensitive, so I'll give you special instructions for them. The general procedure for installing and extracting ICs is shown in Fig. 1–7. Always make sure that you have the correct part installed with the correct orientation. Figure 1–8 illustrates many of the parts we'll be installing and points out their correct orientation.

It's hard to believe, but probably the most confusing parts in the whole UCIS are DIP (*dual in-line package*) switches. They are easy to install and use, take up little board real estate, and are inexpensive compared to any other type of switch. However, on the negative side, their labeling can be confusing. Some manufacturers label some of their switches ON when the switch circuit is closed, and others are labeled ON when the switch circuit is open. Some switches have their segments numbered from right to left, while others are numbered left to right. It's easy for us to totally circumvent all this potential confusion, if you simply follow this procedure:

- Always ignore *all* the labeling printed on the switches!
- Use your VOM (*volt-ohm meter*—we'll discuss it later) to test which way is the closed-circuit position, and then install the switch exactly per the assembly instructions printed in this book.
- Then, when looking at each card in the same orientation as drawn up in this book, the switch ON position will always be the switch thrown toward the top of the card, and switch OFF will always be the switch thrown toward the bottom of the card. *These are the ON and OFF position designations we will always use*, even though they may be entirely different from the manufacturer's switch labels.

The side-actuated DIP switches used on the IBEC are a bit different, and we'll treat them accordingly in Chaps. 3 and 5.

Capacitor values also need a little clarification. The basic unit of measurement for capacitance is the *farad*, with typical unit sizes being defined in either µF for microfarad or pF for picofarad. Because µ denotes 10^{-6} and p 10^{-12}, to convert between one and the other, you simply shift the decimal point six places. For example, a .1-µF capacitor is the same as a 100,000-pF unit, and a .01-µF equals a 10,000-pF, and so on. Also, these capacitors are sometimes marked with code numbers rather than capacitance values. For the values we'll be using, these codes are 102 for .001 µF, 103 for .01 µF, and 104 for .1 µF.

INSERTING

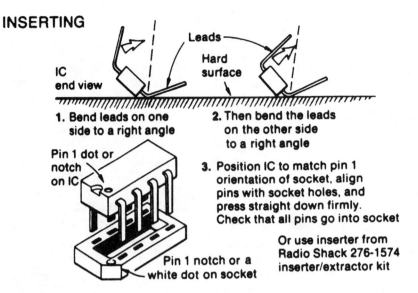

1. Bend leads on one side to a right angle

2. Then bend the leads on the other side to a right angle

3. Position IC to match pin 1 orientation of socket, align pins with socket holes, and press straight down firmly. Check that all pins go into socket

Or use inserter from Radio Shack 276-1574 inserter/extractor kit

Pin 1 dot or notch on IC

Pin 1 notch or a white dot on socket

EXTRACTING

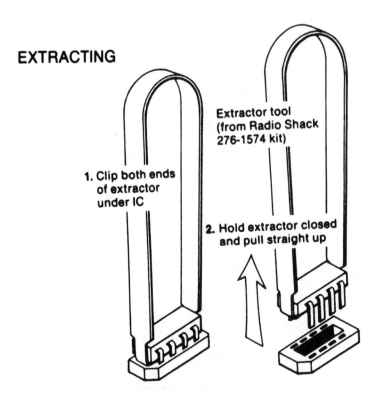

Extractor tool (from Radio Shack 276-1574 kit)

1. Clip both ends of extractor under IC

2. Hold extractor closed and pull straight up

1-7 Inserting and extracting ICs.

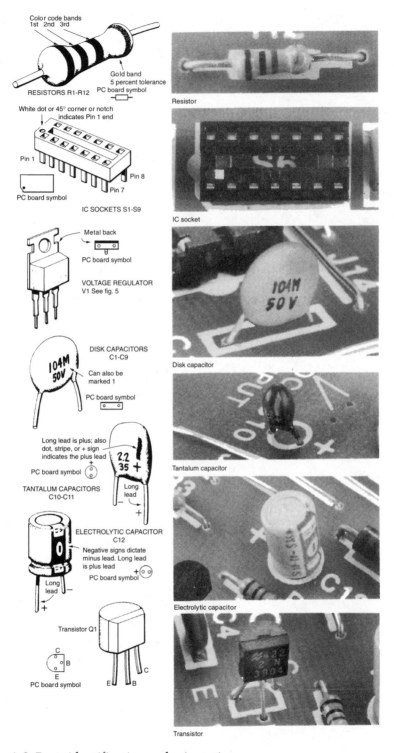

Color code bands
1st 2nd 3rd

Gold band
5 percent tolerance
PC board symbol

RESISTORS R1-R12

Resistor

White dot or 45° corner or notch
indicates Pin 1 end

Pin 1

Pin 8

Pin 7

PC board symbol

IC SOCKETS S1-S9

IC socket

Metal back

PC board symbol

VOLTAGE REGULATOR
V1 See fig. 5

Disk capacitor

DISK CAPACITORS
C1-C9

104M
50V

Can also be
marked 1

PC board symbol

Long lead is plus; also
dot, stripe, or + sign
indicates the plus lead

PC board symbol

2.2
35 +

TANTALUM CAPACITORS
C10-C11

Long
lead

+

Tantalum capacitor

ELECTROLYTIC CAPACITOR
C12

Negative signs dictate
minus lead. Long lead
is plus lead

PC board symbol

Long
lead

+

Electrolytic capacitor

Transistor Q1

C
B
E

PC board symbol

E   B

Transistor

1-8 Parts identification and orientation.

Catalogs sometimes list part dimensions in inches, and sometimes in millimeters. To convert from millimeters to inches, simply divide by 25.4. For example, a capacitor with a 2.5-mm lead spacing is equivalent to .098-in spacing, which is what is required to fit the new IBEC design we'll cover in Chap. 3 and the new digital I/O card designs in Chap. 8.

PC Card Soldering

Good soldering is the most important requirement for building a successful UCIS, and that starts with the right equipment. Soldering guns or large irons are absolute killers on circuit cards, but so is an iron with too little wattage. To do a good job and do it easily, what you need is a "just right" soldering iron.

I use a Weller WTCP temperature-controlled soldering station, as well as a newer station with a digital temperature readout. Such items can be expensive tools (each costs about $100), but the results they provide make for a wise investment. Such an investment is especially worthwhile if you plan to build up a large number of cards and circuits.

The primary advantage of the higher-priced soldering stations is their ability to electrically monitor tip temperature and adjust the heat-flow required to maintain optimum soldering temperature. This way your iron is always at the optimum temperature independent of whether you are soldering a tiny transistor lead to a small circuit board pad, or a joint between two heavy wires, or one of the 4–40 nuts to the large circuit traces found on the UCIS motherboard. Such action tends to make quality joints easier and quicker to accomplish. Also, your iron continues to maintain exactly the right temperature, even when sitting idle.

Conventional irons tend to get too hot when sitting idle. Consequently, when you first apply them to a pad and component lead, they tend to overstress the electronic part, as well as possibly loosen the adhesion between the copper pad and circuit board base material.

In spite of their nonoptimality, there is still a great amount of quality soldering accomplished with conventional soldering irons. You'll need the pencil hand-grip type, with an approximate $\frac{1}{32}$ in tip diameter and a 25- to 33-W heating element. Make sure the iron is specifically designed for light work such as PC board soldering, but not too light. For example, I find the smaller irons with 20-W capacity result in too low a tip temperature for proper PC card soldering. Also, do not use an iron with much more than the recommended maximum 33-W element, because they can easily damage the PC card. The iron-clad nickel-plated tips are much preferred over the bare copper tips, which require constant tinning for good soldering. Also, make sure you have a soldering iron stand and a sponge holder. Digi-Key, Jameco, JDR, and others have a variety of irons, as does your local Radio Shack store.

You'll also need good solder, a 60/40 (or 63/37) mix of tin/lead with either rosin-core flux or the newer water-soluble-core flux. You shouldn't use additional flux, as quality solder has all you need. Never use acid-core solder for electronic projects. It's important to use small-diameter solder in PC card work, .031 in diameter or less. JDR SLDR-031 fits this description for the rosin core, and SLDR6403 for the water-soluble. The latter advertises less harmful fumes, reduced environmental harm, and easier board cleaning. I'm personally enjoying the easier board cleaning provided by the water-soluble flux.

Figure 1–9 illustrates the steps in making a good solder joint on a PC card. Both the soldering tip and the parts have to be clean for efficient heat transfer. Wipe the tip on a damp sponge after soldering every few joints, to keep the tip bright and shiny. If you find a component lead that has oxidized and is no longer shiny, scrape it lightly with a knife to remove the oxidation before soldering.

Most importantly, hold the tip *firmly* against *both* the lead and the pad. *Most poor solder joints come from either the pad or the lead not getting enough heat.* One gets to the proper temperature, but the other does not. I can't overstress the point too much. *The lead and the pad must both be at the correct temperature in order for you to achieve a quality soldered joint.*

For pads that are joined, heat each separately. Don't feed a blob of solder to one pad and hope it will get to the other before cooling. Heating each pad-and-lead combination separately also helps prevent heat damage to the components.

Learn to solder quickly and correctly, but don't rush either. When you complete a group of joints, look them over with a magnifying glass under a bright light and from different angles. Figures 1–9 and 1–10 show what to look for in good and bad joints. If you find any bad traits, reheat the joint, pad, and lead, and apply a bit more solder to get a good connection.

I've made soldering the UCIS cards as easy as I could by designing, wherever possible, good-sized cards with large pads and ample space between traces. Still, there is no substitute for careful soldering with the right equipment.

Figure 1–11 shows a close-up of the solder connections for IC socket S1 on UBEC. This is the kind of area where you have to be especially careful to avoid solder bridges. Study the pads before you start soldering to note which ones are joined— like the second and third from the left in the bottom row—and which are separate; then make sure they're still that way when you're done.

If you need to remove an incorrectly installed or faulty part, heat each pin with desoldering braid (Radio Shack 64-2090) or solder wick (Jameco 41081) between the iron and pin to wick up the solder until you can remove the part.

Cleanup and Inspection

How well your interface performs the first time you turn it on is highly dependent upon how well you perform the cleanup and inspection step at the end of each circuit card assembly. Do it well, and you are likely to have a system that performs perfectly the first time you turn it on. Each time you come to this step during a card's assembly, make certain you seriously follow the procedures covered in this section.

Prior to a serious inspection you must clean the solder side of each board to remove solder flux residue. If using rosin core solder, spray on a flux remover such as Jameco 118041, and for the water-soluble use a mixture of water and ammonia, or plain water with a trace of detergent or general-purpose cleaner. Use a brush with very stiff bristles (brass bristles work well) to thoroughly scrub the board. Any stubborn flux can be further loosened with an X-Acto knife. Then re-scrub, and lightly rinse, until all traces of flux are removed.

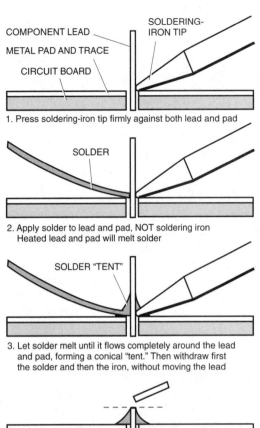

COMPONENT LEAD

METAL PAD AND TRACE

CIRCUIT BOARD

SOLDERING-
IRON TIP

1. Press soldering-iron tip firmly against both lead and pad

SOLDER

2. Apply solder to lead and pad, NOT soldering iron
Heated lead and pad will melt solder

SOLDER "TENT"

3. Let solder melt until it flows completely around the lead
and pad, forming a conical "tent." Then withdraw first
the solder and then the iron, without moving the lead

4. Once the solder has cooled, clip off the excess lead
flush with the top of the solder tent

1-9
Soldering techniques.

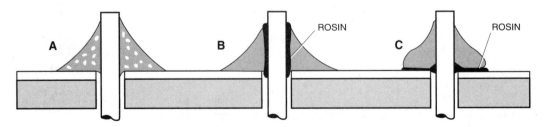

ROSIN

ROSIN

A

B

C

A. Solder tent has a frosted or stippled look: component lead was moved before solder completely cooled

B. Solder fails to flow on lead, and a dark ring of rosin insulates the lead from the pad: soldering iron did not heat lead

C. Solder appears to flow <u>inward</u> and blobs on top of pad: soldering iron did not heat pad

1-10. Common soldering faults.

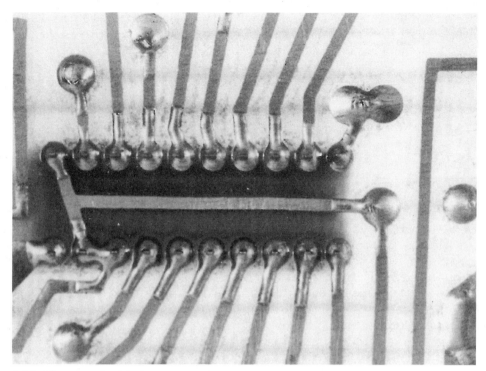

1-11 Closeup of soldering connections.

I cannot overstress the importance of giving each final board a thorough cleaning, to remove every trace of the solder flux, and then giving both sides of the board a detailed visual inspection. After cleaning, make sure every joint is correctly soldered and that there are no solder bridges. Even hairline traces of solder between places where they do not belong can cause problems.

Always use a bright light and a magnifying glass to perform every visual inspection. Look over every single joint from different angles to make sure there is a nice conical buildup of solder around each component lead. The tip of the lead should be visible sticking out the top of the conical tent. There should be no traces of solder flux between the lead and the solder or between the solder and the pad. There should be no void or unsoldered region around the circuit board hole.

If the slightest question arises concerning how any joint looks, reheat it and apply a bit more solder. Make sure each joint is *perfect* before proceeding to the next. Remember, most every problem with UCIS can be traced to poor soldering. Poor soldering far outweighs the problems resulting from a part being inserted incorrectly, or the wrong part installed.

Use your same bright light and magnifying glass to make sure every IC has the correct part number, that it is installed with the correct pin orientation, and that all are correctly inserted into the socket. Study every pin of every part before moving on to the next.

Make sure that all polarity-sensitive capacitors are installed with the correct polarity orientation. Make sure that all LEDs (*light-emitting diodes*) are installed with the flat portion of their lip facing away from the + sign marking. Make sure all DIP switches have the correct closed-circuit orientation. Make sure each resistor network has its pin 1 marking oriented as in the parts layout diagram, and that all separate resistors have the specified color code markings.

When everything looks perfect on both sides of the board, spray the bottom of the card with a protective coat of clear acrylic, making sure the spray doesn't get on any connector contact surface. I like Krylon #1303, available at most hardware stores.

Test Meters

To build and test the UCIS, you will need a volt-ohm meter (VOM). Either an analog or a digital meter will do fine, and no high precision equipment is required. A digital meter is more expensive, but it is also easier to read and more accurate because you have less chance of misinterpreting its measurements.

When using an analog meter, or a non-auto-ranging digital meter, you must pay careful attention to setting the scale so that you don't damage the meter or make an incorrect reading. If you don't get the prescribed voltage when testing with an analog meter, the first thing to do is make sure you're reading the correct scale.

We'll be using the VOM to perform two basic kinds of tests: continuity/resistance tests, and voltage tests.

Continuity/Resistance Testing

First make sure that all power is turned off in the circuit being tested. Set the meter on the ohms scale (an ohm is a unit of electrical resistance), then connect one lead of the meter to one point of the circuit and the other lead to another point. The meter will measure the resistance of the circuit paths between the two points under test.

If you read zero ohms, or very near zero, you have a solid conduction path (or a *short circuit*) joining the two points. If the meter doesn't move, that is, it shows infinite resistance, you have an open circuit. Readings in between define the actual resistance of the circuit, components, and so forth under test. Get to know your meter by using it to measure the resistance of a few resistors. Compare your readings to the resistor values to make sure you are doing it correctly.

Voltage Testing

You'll typically be making all voltage measurements with respect to a common point, called the *ground*. To do so, first verify that you have selected the correct range, then connect one of the meter probes to the ground and the other to the point where the voltage is to be measured.

Whenever possible, use an alligator clip to attach the ground probe firmly to the circuit card. If you try to hold both test leads and the card at the same time while looking at the meter, you are asking for trouble, because it's just too easy to slip and short something out.

For all but a very few select tests, which I'll call out in the instructions, the negative (−) probe connects to the ground and the positive (+) probe touches the point to be tested. Make sure that your probe touches only the point to be tested, and that it doesn't bridge between closely spaced leads such as IC pins. Don't let the probe slip and touch another component lead accidentally. When you measure a dc voltage with an analog meter and the needle tries to go the wrong way, reverse the two leads.

System Troubleshooting

"I built up the UCIS but I can't get it to work. What do I do?" This might be a question that many may feel I frequently receive, but it's probably less than one out of every 100 letters or phone calls. The system is really easy to assemble, and very few readers have reported a problem that they don't seem to be able to solve themselves. Of those who have written or called for help, I have been able to help most quite quickly by mail or over the phone; even they are then elated on how easy the UCIS is to debug.

I find that most readers of the first edition, and especially those with little or no experience with electronic assembly, seem to perform the assemble steps with excellent results. Those with some experience tend to skip over details in the instructions, which can lead to troubles later. Thus, if you have zero experience you are almost *better* equipped to jump in and start building the interface!

In essence, if you use the JLC provided boards, take it slowly, carefully follow the specific step-by-step instructions provided in this book, and use the parts recommended from the sources recommended, you will have no trouble getting the interface to work. Test and debug procedures are included in the book to help ensure success. As a result, a large number of first-edition users report that the system worked the first time it was turned on.

However, if you do have extended trouble getting your system to work, and if you purchased your circuit boards from JLC Enterprises, I do offer to physically help find any assembly problem for a nominal hourly charge plus parts and return shipping and handling. I'm extremely knowledgeable at the system debug, so it usually doesn't take me very long to locate problems and repair the system to working order—typically less than 4 h, which I would say is an outside maximum for a system that is really in bad shape. I find the solution to the majority of problems by performing a very detailed visual inspection of the boards. If you perform your examination under a bright light, with a thoroughly cleaned board and a magnifying glass, then you can achieve the same high-class results.

Along with the returned cards, I include a list to tell you what I found wrong and what I repaired so that you can improve your skills. However, please call for instructions and authorization of such service before shipping your system in for repair. Also, if you desire contracted system application and/or software programming assistance, please contact Dr. Bruce A. Chubb, 3902 Wilton Dr. SE, Grand Rapids, MI 49508, (616) 243-4184.

If you are in doubt of your capabilities, give consideration to having ECO Works, EASEE Interfaces, or AIA assemble and test at least your first few boards.

System Reliability

I've had my first UCIS components installed now for over 14 years, and I don't recall ever replacing a part that failed during operation. Solid-state systems that are properly designed and assembled, once operational, tend to operate forever. One of the best demonstrations of reliability is the extensive UCIS setup at the Chicago Museum of Science and Industry, where the system has operated day in and day out 365 days per year for years without UCIS problems.

If you ever have trouble, the plug-in modular features provided by the JLC Cards and the diagnostic software make locating and correcting the problem a straightforward task. In summary, I couldn't be happier with the ease of assembly that people have reported and the superreliability of the system.

In the next chapter we'll look at selecting the best UCIS configuration to meet your interfacing needs—whether it be parallel with IBEC, with UBEC, serial with USIC, or even distributed serial with multiple USICs. Once this is decided, you are ready to start building a system.

2
CHAPTER

Selecting the Best Interface Option

One of the greatest attributes of the UCIS is its versatility and adaptability to meet different needs. For example, you can pick different parallel and serial options, different I/O configurations (including analog-to-digital and digital-to-analog conversion capability), and you can select from an abundance of auxiliary cards for attaching all types of special devices to the I/O cards. Let's explore some of these features.

Versatility and Adaptability

Fundamentally, you can think of the UCIS as a collection of over 30 different circuit boards, a shopping list from which you can pick whatever cards you need (and how many of each) and then combine them in different ways to best meet your specific defined need. In addition, numerous boards can be assembled and/or configured with different options to further tailor the system.

The UCIS provides a building-block approach to computer interfacing, giving you (the user) complete freedom to select and tailor the blocks you need, and only those blocks, and then assemble them in the manner you choose to best meet your requirements. You only need pay for and assemble the specific pieces you require. This makes the UCIS system very affordable and easy to apply. You can also elect to benefit from the system's modularity and start very small and gradually expand the UCIS as your needs and capabilities grow.

And to top it off, you can purchase each of the circuit boards in ready-to-assemble format, you can purchase the boards fully assembled and tested, or you can fabricate boards from scratch using the artwork provided in Appendix A.

New material presented in the second edition is based on experiences gained from thousands of applications that took advantage of the first edition. The success of those applications speaks well to the soundness of the UCIS designs and the quality of the resulting circuit boards. When selecting any version of UCIS, you are signing up for a well-proven technology with a solid record of doing what it says it can do.

The UCIS has an outstanding track record, and with its breadth of coverage, it truly is a universal computer interface system.

The selections and options available with this second edition have significantly expanded over those available with the first edition. That's great, and certainly opens the door for the interface user to obtain the best possible system for a given need at the lowest possible cost.

However, with such a broad selection capability, some confusion can develop. This was true with the first edition. Consequently, with the much-expanded choices in the second edition we definitely need to help each reader answer the question, "How do I select the UCIS options to best meet my need?" This is the reason for this added chapter, and for the extensive added clarification presented throughout the second edition each time I explain the available options.

Computer Selection

I could easily write a whole book on selecting the best computer to match a set of needs. My main advice is to start with a system that provides good expansion capabilities. We'll start out by looking at using the interface with the Pentium, $x86$, and AT-class computers. These newer machines (and especially the Pentium) offer tremendous capability to the interface user. With a Pentium tied to the interfacing capability of UCIS, you have true power at your fingertips. With this power you can very effectively and efficiently monitor and control any external hardware you desire. The modern programming languages with their built-in user-friendly editors and mouse interface, coupled with the software examples presented in this book, make programming for your interface a straightforward process.

We'll cover interfacing to the Macintosh family of computers and see how using their readily available RS-422 serial port you can directly attach the UCIS. Also, we'll cover interfacing to the older IBM PC and XT computers, and their compatibles, as these machines remain excellent candidates for interfacing to hardware applications not requiring superfast computer power.

For certain real-time, controller kinds of UCIS applications in which you desire to dedicate a computer to a specific interface task, and the task isn't too demanding, and (most importantly) you don't want to spend a lot of money, give consideration to older, smaller, hobby-type computers. You can obtain them for little or nothing. Such examples include several Tandy/Radio Shack models, the entire Apple II series, and the complete Commodore family.

In addition to the great news we constantly hear about the newer Pentium computers and the newest P6 Pentium Pro processor, there is still an interest in these older computers. As this second edition is being published, JLC continues to receive calls concerning interface cards for these older computers. I'll show you how to interface the UCIS with each of them. Maybe you have one of these computers tucked up in your closet, and the UCIS may just be the ticket you need to put that machine back to work satisfying your smaller interfacing needs.

The original of all home/hobby computers, the Heathkit H-8, falls in this class and has one neat interfacing advantage: its alphanumeric front panel display digits can illuminate any memory location or register variable in real time as the program operates. This feature comes in handy for debugging real-time systems, so I'll cover it as well.

The main reason, though, for keeping the H-8 example in the second edition is that it demonstrates how different computers can be *hardwired* into the UCIS even when there is no available JLC adapter card. Also, by studying the broadest possible example range of how UCIS is attached to different computers, you learn how it might be attached to your particular computer even if that machine isn't specifically covered in this second edition.

For many applications, there is no need to dedicate a computer to the real-time task, or you might well desire a more capable machine. Most new operating systems (for example, Windows95) support multitasking, so you can (with proper application software) operate your interface while you are performing other tasks with your computer. In other cases, your interfacing requirements might be so demanding that they take up the capacity of your machine; however, for the times when you are actively using the UCIS, you can afford to dedicate your computer to this cause. For example, when you are operating a large model railroad or a group of wood engraving machines with your computer via UCIS, you typically do not have much time for doing word processing.

For more ambitious applications, or for where you want to time-share your interfacing with other computer activity, larger-capacity systems make excellent choices for the UCIS. Software flexibility and capability provided by the newer computers is a big plus to justify their selection. That's the reason I'm now using the Pentium tied into my railroad and hardware development systems. Fundamentally, I've designed the UCIS as an easily assembled interface that mates with almost every popular personal computer—past, present, and future! That's the reason I call the interfacing system a *Universal Computer Interface System.*

Even if your machine isn't mentioned in this book, I'll be showing you enough different ways to connect into both a computer's internal bus and its serial port that, with the help of your computer manuals, you should be able to figure out what you need to do.

For the parallel interface option, and for the newer Pentium to AT class computers, I'll show you how to assemble the UCIS using the newly developed Internal Bus Extender Card (IBEC). This card combines many features of the UCIS into a single board that directly plugs into one of your computer's bus expansion slots. A built-in switch selects between memory-mapped I/O and port-mapped I/O.

As an alternative method to parallel interfacing, I'll show you how to assemble the UCIS using a universal bus extender card (UBEC) coupled with a computer bus adapter card (CBAC). By simply selecting the correct CBAC to fit each particular computer, this approach yields tremendous flexibility to the broadest possible range of computer applications.

For the serial interface option, I'll show how you can assemble the UCIS using the Universal Serial Interface Card (USIC) for connecting your interface to any computer having a 20mA current-loop, RS-232, or RS-422 port. The latter works great for Macintosh applications. Using these methods, the UCIS is easily attachable to a large spectrum of computers, including laptops and portables that do not have built-in expansion slots. Upward compatibility is another advantage of the serial option, because your interface is not subject to changes in internal computer bus architecture.

An expansion feature provided with the serial option is the capability to use *distributed serial* I/O. This way you can distribute your I/O over a large area (involving distances up to 4000 ft) using only a single 4-wire cable. This approach greatly reduces the amount of wiring required for large systems.

Building the IBEC is covered in Chap. 3, the serial USIC in Chap. 4, the UBEC in Chap. 6, and the CBACs in Chap. 7. By using one or more of these cards, you can truly adapt the UCIS to nearly every type of computer and meet nearly every possible interface requirement.

Looking at the Options

Should I go parallel with IBEC, parallel with UBEC, or go serial with USIC? If serial, should I go with RS-232 or RS-422? Or should I go with distributed serial and use multiple USICs? How many I/O cards do I need? Which I/O cards should I use? What about analog conversion capability? Which auxiliary cards do I need, and how many of each? What set of cards should I order to start applying the system? How much is it going to cost? These are some questions asked by potential users as they first become acquainted with the broad scope of UCIS material. In this chapter I'll work for you, to help provide insight into the UCIS system, so that you will be in a better position to answer these questions. Let's look at the cost issue first.

System Cost

The cost of your UCIS depends on whether you build up your own system (and the price you have to pay for parts) or if you purchase the interface cards assembled and tested; it also depends on the number of interface lines you need and the functions they perform. If you elect to assemble your own cards, I have provided detailed parts lists and recommended sources.

The UCIS is very cost-effective. Assuming a basic assemble-yourself interface with, say, 48 I/O lines, the cost should be about $300. (Obviously, this price does not include the cost of the computer.) Adding more I/O capacity will cost in the neighborhood of $2 per line. Going the fully assembled-and-tested route, the costs will be approximately two times higher.

Costwise there typically isn't much difference going the IBEC, UBEC, or USIC route. The basic card is cheaper going with USIC, but when you add the cost of the onboard computer chip it can come out a bit more expensive than parallel. However, by the time you add in ribbon cables, the costs are nearly identical.

Adding in the cost of the I/O motherboard and some I/O cards that are identical for all three approaches, any slight cost difference between the IBEC, UBEC/CBAC, and USIC becomes less significant. Fortunately, therefore, the parallel versus serial decision can be achieved based on the interface's technical need, the computer you have available, and your own personal desires.

Parallel or Serial?

One of several important features contributing to the universal capability of UCIS is the choice between implementing: (1) a parallel bus approach using the internal bus extender card (IBEC); (2) a parallel bus approach using the universal bus extender card (UBEC); (3) a serial approach using the universal serial interface card (USIC); or (4) a distributed serial approach with multiple USICs.

The two parallel approaches (IBEC and UBEC), with their direct expansion of the computer's address and data bus, provide the fastest I/O time between your external hardware and the computer. The IBEC (presented in Chap. 3), the newest of the options, is designed specifically for optimizing the parallel interface to Pentium and x86-class computers. It also works with the IBM PC, XT, and their compatibles.

As newer-than-Pentium computers come on the market, they should also continue to work with IBEC as long as they include compatible expansion slots. Because of the popularity of these machines, I expect the IBEC to become more prevalent than the more universal UBEC option. It can be easy, however, to rule out the IBEC approach, for it only works with the above listed family of computers. If your computer doesn't fit in this category, or if it doesn't have spare ISA slots, then the selection of IBEC is not for you. Thus enters UBEC and USIC.

The UBEC, presented in Chap. 6, is the forerunner to IBEC. Based on the broad spectrum of computers that interface easily with UBEC, it is the most universal parallel option. For example the UBEC is easily mated with the IBM PC and XT and their compatibles, the entire Apple II family including the II, II+, IIe, and IIGS, the entire Commodore family, and so on. In fact, as I'll illustrate in Chap. 7, the UBEC also works with the Pentium, 486, 386, 286, or AT class of IBM-compatible computers. Although for these specific machines, the IBEC offers more readily available options.

To attach the UBEC to each different class or computer type, you need a computer bus adapter card (CBAC). The CBAC mates the unique connector type and pinout arrangements found on different computers to required connections on UBEC. Currently there are seven different CBACs available, and they each typically support several different computers. As new computers emerge, new corresponding CBACs might also become available from JLC Enterprises. If a CBAC isn't available for a specific machine, it's frequently possible to *hard-wire* UBEC directly to a computer. We'll look at that option as well in Chap. 7.

As a third major option, the UCIS can be assembled as a serial interface between the computer and the bank of I/O cards connecting to your external hardware. With the serial UCIS, you also can select a further distributed option, whereby a four-wire bus distributes your I/O over distances up to 4000 ft to reach dispersed groups of I/O cards.

Figure 2–1 illustrates these three methods of interfacing with UCIS. In the UCIS parallel interface (Fig. 2–1a), the IBEC (or the UBEC with CBAC) performs as an extension to the computer's address, data, and control lines. Up to 64 I/O cards can be attached to the parallel interface, providing a total of 1536 I/O lines connected to your external hardware (64 × 24). In Chap. 13, I'll show how you can expand the number of effective I/O lines even further by using encoding and decoding circuitry at your external hardware.

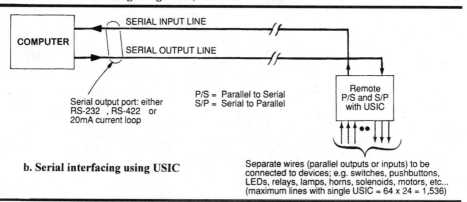

A = Address lines
D = Data lines
C = Control lines

Separate wires (parallel outputs or inputs) to be connected to devices; e.g. switches, pushbuttons, LEDs, relays, lamps, horns, solenoids, motors, etc...(maximum lines with IBEC or UBEC = 64 x 24 = 1,536)

a. Parallel interfacing using IBEC, or UBEC and CBAC

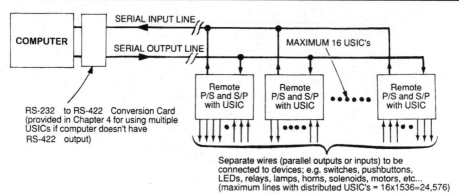

Serial output port: either RS-232 , RS-422 or 20mA current loop

P/S = Parallel to Serial
S/P = Serial to Parallel

b. Serial interfacing using USIC

Separate wires (parallel outputs or inputs) to be connected to devices; e.g. switches, pushbuttons, LEDs, relays, lamps, horns, solenoids, motors, etc... (maximum lines with single USIC = 64 x 24 = 1,536)

RS-232 to RS-422 Conversion Card (provided in Chapter 4 for using multiple USICs if computer doesn't have RS-422 output)

Separate wires (parallel outputs or inputs) to be connected to devices; e.g. switches, pushbuttons, LEDs, relays, lamps, horns, solenoids, motors, etc... (maximum lines with distributed USIC's = 16x1536=24,576)

c. Distributed serial interfacing using multiple USICs

2-1 Parallel, serial, and distributed serial interfacing.

Figure 2–1b illustrates serial interfacing using a single USIC. As with the parallel versions, you can attach up to 64 I/O cards to USIC for a total I/O line count of 1536.

Figure 2–1c illustrates how, using RS-422, you can connect up to 16 USICs to a single serial port. This approach provides a distributed capacity to interface with up to 16 × 64 × 24, or 24,576, separate I/O lines connected to your external hardware.

The serial and distributed-serial interfaces typically have the advantage of using less wire, because the I/O cards can be located close to the devices they read or control. For most I/O lines, however, you can use low-cost, small-diameter wire to offset the expense of running long cables. In fact, surplus multiconductor telephone cable is ideal for many applications in which you desire to keep the cost low.

The parallel interface has a significant advantage in speed because it receives or transmits on all data lines at once, with the execution of a single computer instruction. A serial interface transmits each digital data bit individually. That technique is slower, and because the computer itself processes data in a parallel format, the data must be converted into serial format before they can be sent. The receiving device, the USIC in our case, then must collect the data and reconstruct them into a parallel format for monitoring or controlling our separate external hardware devices.

Addresses also have to be transmitted over the serial bus to identify which device is being accessed, and they too must be converted from parallel to serial and back again at each end. A serial bus system also needs a communications protocol, which, as we will see in Chap. 10, involves extra bits to separate address from data and to determine the direction of data flow.

It's not hard to see the time penalty paid for using a serial bus. However with the superfast processing speeds of today's computers, serial bus systems can give us sufficiently fast responses to meet most interfacing needs. Also, there are plenty of other factors that are important considerations in selecting the best interface approach. Serial provides a definite advantage in many applications.

In the next two sections I'll list numerous attributes associated with selecting the parallel (IBEC- or UBEC-based) interface and then the same for the serial (USIC-) interface. Seeing which group of characteristics better fits your personal desires and your interfacing needs should help you select the best approach.

Attributes for Selecting the Parallel (IBEC or UBEC) Approach

I want the fastest possible interfacing speed. If speed is your ultimate driving factor, meaning that you need your interface operating in the range of one to a few hundred microseconds, then either the IBEC or UBEC solutions provide a definite advantage. But with the speed comes possible limitations, such as portability, closeness requirements of external hardware, and so forth, so don't stop here with your decision. Review all the attributes associated with both approaches.

I can mount the interface close to my computer. Parallel interfaces must be close to the computer for direct connection to its high-speed internal bus. For example, the I/O motherboard should be 10 ft or less from your IBEC or UBEC. The wiring from the I/O cards plugged into the motherboard to your external hardware can be of whatever length your application requires. This can result in substantial wiring requirements if you require a large number of I/O lines and your external hardware is quite distant from your computer. However, when external devices and the computer are in close proximity, the parallel approaches work great with wiring requirements about equal to serial.

My computer doesn't have any available serial ports. Many older computers do not have RS-232 or RS-422 serial ports, and newer machines may very well have their ports quickly filled up with a mouse, fax/modem, and an abundance of other features. Without any available form of serial port, the serial USIC option is out and your interfacing options become IBEC or UBEC.

My computer has available bus expansion slots. That's good, because having some method of attaching to the computer's bus is essential for IBEC and UBEC. Rather than internal slots, some computers provide equivalent external sockets typically referred to as expansion ports, game ports, and so on. We'll look at several such examples in Chap. 7.

I feel comfortable opening my computer and plugging boards into the computer bus. This process is routine for many people, but others shy away from the requirement.

I don't mind system configuration challenges. With the two parallel approaches, you need to locate a free address space, that is, a block of addresses not occupied by memory or other devices attached to the computer's bus. This can be taxing at times, and even after your system is operating wonderfully, at some future time you add another new XYZ device, it can cause conflict with the UCIS. If and when this happens, you'll need to reconfigure your system by, for example, changing the DIP switch settings on your IBEC to possibly select a different I/O mode and/or a different UCIS starting address, and then correspondingly change the starting address in your software.

The somewhat easier parallel-system programming approach sounds better to me. Because I provide all the serial protocol subroutines for you as part of this second edition, there really is little difference in the programming requirements between parallel and serial. However, the little difference that remains does make the parallel versions a bit easier to program.

Attributes for Selecting the Serial (USIC) Approach

My computer has a serial port available for interfacing. If so, then you are all set to plug in a serial UCIS. Even if you don't have a serial port free, or later you find you want to add another serial device, consider the possibility of adding a switch box to your serial port so that you can switch between operating the interface and other devices. Or possibly installing an add-on expansion card to make a serial port available. The port can be RS-232, RS-422, or even the older current-loop standard, but one of these is essential to make the serial version of UCIS operational.

My computer doesn't have bus expansion slot capability. Many computers don't provide internal expansion slots, nor do the manufacturers make available an expansion connector giving access to the address, data, and control lines needed for the IBEC and UBEC parallel approach. The TI/99, highly produced in its day; the Apple IIc; most of the Macintosh family; and many portable computers and laptops are prime examples in this category. You could bring out the required bus lines by soldering leads to appropriate traces on the computer's internal circuit cards, but a serial UCIS interface is much easier to implement.

My computer must be remote from the external hardware. Parallel interfaces must be close to the computer for direct connection to its high-speed bus. With a serial interface you can make long-distance connections. For example, with an RS-422 four-wire cable, the computer can be as far as 4,000 feet from your external hardware. With three-wire RS-232 the specification is 50 feet maximum, but typically distances of 100 to 200 feet and more are successfully operated.

Interface portability is very important to me. I want to be able to hook up the interface to laptops and be able to take it to different locations and hook it up to different computers, and maybe even to different external hardware. Fast changeover between my desktop or tower case system to my laptop or portable is important. It's hard to beat moving one serial cable and having your interface switch between different computers.

Upward compatibility is very important to me. I'm seeking maximum assurance that future computers will be easily attachable to my interface with minimum modification. With all the serial peripherals on the market, it's hard to imagine a new emerging computer that wouldn't provide serial port compatibility. The parallel bus used by the IBEC and UBEC is also pretty well entrenched, but probably not as solidly as the RS-232 and RS-422 standards are for serial. Going with the serial USIC, you should have maximum upward compatibility.

Maximum possible I/O speed is not the driving factor. Yes, I might need response in the ten- to a few-hundred-millisecond range, but I don't need microseconds. Actual interface speed is highly dependent upon the computer used, the amount of computer processing required, how well the software is written for fast operations and in what language, and the number of I/O lines being monitored and controlled. However, all else being equal, the serial will be considerably slower than the parallel; how fast do you need it? Not requiring response times in microseconds can make the serial option very attractive.

I would like to avoid opening my computer to plug in the interface card. This comment is certainly understood, as there is always the risk of causing problems each time you get inside any type of electronic box. Something inserted wrong, a defective IC or a solder bridge across bus lines on the card you are inserting, accidentally dropping something into the computer, and the like, can cause the computer to hang up when turned back on and possibly even do damage to the computer. However, with a little caution and experience, you'll soon see there's little reason not to perform these functions.

I want minimum interface wiring. A serial interface lets you distribute I/O cards nearer to the external hardware devices they monitor and control. For large systems, or even as my system grows, I like the idea of being able to expand to distributed serial I/O using multiple USICs. Such factors are of little importance on small-to-average-sized interfaces, but on large systems, going with serial can be a big advantage that greatly reduces wiring requirements.

I don't want to be concerned about changes in my computer's configuration. If in the future you add a new serial card or change a serial peripheral device, you still might need to reconfigure your system a bit. Typically, though, if any changes are required they are easier to accomplish with serial than with parallel.

Narrowing the Choice

If you still experience difficulty in making the choice between serial and parallel, you can read further in this book to refine your understanding of each approach. As you gain additional insight, your best approach should become clear. As a further refinement, many first-edition readers pursued building up both the parallel and serial systems. There isn't really much duplication, because the motherboard, all the I/O cards, and any other special-purpose cards you might require are identical and totally independent of whether you go with the IBEC, UBEC, or USIC. Building both, you learn the maximum about both parallel and serial and the universal nature of the interface. Such knowledge and experience makes you uniquely qualified to tailor a specific system to ideally meet every interfacing need.

Assuming you have decided parallel, then going IBEC or UBEC is pretty much a decision based upon the computer you plan to use. If its a Pentium, $x86$, or AT class machine, then the IBEC is your best choice. If you already have built up a UBEC from an earlier application, you can still go ahead and use it with the newer machines, including the Pentium. You will not have the flexibility provided by IBEC, but it should handle most requirements.

If your computer is not one of the above, then for parallel I/O the UBEC is your best choice. You'll need to select a CBAC that matches the UBEC to your particular computer, or hardwire it following the example in Chap. 7, if a CBAC is not available.

Assuming you elect to go with serial I/O, then you replace the IBEC (or the UBEC/CBAC combination) with the universal serial interface card (USIC). The next decision is whether to use the USIC's RS-232 or RS-422 option. This again depends to a large extent on which ports your computer supports. If you have a Macintosh with RS-422 then the decision is easy—go with the USIC configured for RS-422. Likewise, if you have an IBM compatible with RS-232, then you can go with the USIC configured for RS-232. However, in the latter case you might well want to invoke the RS-232-to-RS-422 conversion card between your computer's RS-232 port and the USIC, thereby keeping your USIC configured for RS-422.

Why use RS-422 if your computer has RS-232? Well, RS-422 does offer advantages, including: (1) It isolates your external hardware ground from your computer ground; (2) You can communicate over longer distances; (3) The system has greater noise immunity; and (4) Building in the conversion to RS-422 up front makes your system easily expandable to the distributed serial option using multiple USICs.

However, like most all UCIS advantages, initial selections do not tie you to a specific interfacing path. It is very easy to make changes later and with very little time or cost penalty. For example, you can switch the USIC back and forth between RS-232 and RS-422 simply by substituting two ICs back and forth. By having your ICs in sockets, such swaps take very little effort. Converting back and forth between serial and parallel is also quite easy to accomplish.

Table 2-1 summarizes the major features available for connecting the interface. We've discussed most of them, but having them collected into one easy-to-compare spot can be effective in helping seek out the best approach for a given application. Before we move on though, we should briefly look a bit deeper into two interface approaches; namely distributed serial systems and combined parallel/serial systems.

Table 2-1. Options Available for Connecting UCIS to Computer

	Interface Card(s)	Primary Features of Selected Option
Parallel	IBEC	Optimized specifically for plugging directly into IBM and IBM compatibles with an 8-bit or 16-bit ISA bus slot including PC, XT, AT, 286, 386, 486, Pentiums, etc. Includes a built-in switch for selecting between Memory-Mapped I/O and Port-Mapped I/O. Built-in-Test LEDs and easily accessible switches provide assistance in system setup and testing. Up to 1,536 devices can be monitored/controlled using this option.
Parallel	UBEC with CBAC	Original design optimized to fit a wide range of computers. Using different Computer Adapter Cards (CBACs) with the Universal Bus Extender Card (UBEC) you can tailor the design to fit earlier computers including Apple II, II+, IIGS, etc., Commodore 64, 64C & 128, TRS Color Computers, etc. Using the IBPC adapter card works for the IBM PC, XT, AT or 286, 386, 486 and Pentium applications. Up to 1,536 devices can be monitored/controlled using this option.
Serial	USIC with RS-232	Optimized to attach the interface to any computer that has a RS-232 Serial Port. Extremely useful for portable/laptop computers that do not have expansion slots. Using the commonly available RS-232 standard, this option has the advantage of being upward compatible to most every new computer. Also enables the interface to be placed remote from the computer (up to at least 50 ft per the RS-232 specification and frequently implemented several hundred feet) using a 3-wire cable. Up to 1,536 devices can be monitored/controlled using this option.
Serial	USIC with RS-422	Especially attractive for Macintosh applications that have a built in RS-422 port. Uses a 4-wire cable, 2 wires for communication in each direction, to achieve a greater ground isolation between the computer and the interface hardware and greater noise immunity with distances up to 4000. Computers that do not have a built-in RS-422 port can still take advantage of this option by using a JLC provided RS-232 to RS-422 conversion card. Up to 1,536 devices can be monitored/controlled using this option.
Distributed Serial	Multiple USICs	Up to 16 USICs can be connected to the same RS-422 serial port. Each USIC becomes a node supporting up to 1,536 devices providing a total expansion capability to monitor/control up to 24,576 devices. Especially attractive for application where the devices you want to monitor/control are spread out over a large area. A single 4-wire cable, up to 4000 feet long, runs from the computer to node 1, then to node 2, etc. up to 16 nodes. Local wiring, to devices being monitored/controlled, need only connect to the nearest node. Overall wiring is minimized.

Distributed Serial Systems

Frequently computer interfacing requires connecting external hardware, located at different locations throughout a facility, to one computer. For example, you might have five different test stations, or three robots, or several engraving machines, or a number of mixing tanks, or a large material-handling system, or for that matter a combination of these items, that you desire to tie into one PC for monitoring and control. In a model railroad example it might be a large terminal yard, a staging area, the main line, and a central-dispatcher control panel. For any of these situations, the distributed serial option of UCIS provides a definite advantage.

Figure 2-2 illustrates the setup with five UCIS serial nodes connected to one computer. Each node consists of a USIC plugged into an I/O motherboard (IOMB).

Each node is individually configured with the desired compliment of input and output features by simply plugging in the appropriate I/O cards. Each inserted card can be a general-purpose digital input or digital output card, as we'll cover in Chaps. 8 and 16, or an analog-to-digital or digital-to-analog card as covered in Chap. 14. Each digital I/O card provides 24 I/O lines and each of the analog cards has three channels. Each node can hold up to 64 I/O cards. Assuming all discrete I/O, that's a capacity to interface with 1,536 separate I/O lines per node.

The UCIS is capable of handling 16 nodes, for a total capacity of 16 times 1,536 or 24,576 devices being monitored/controlled by the computer. The computer can be

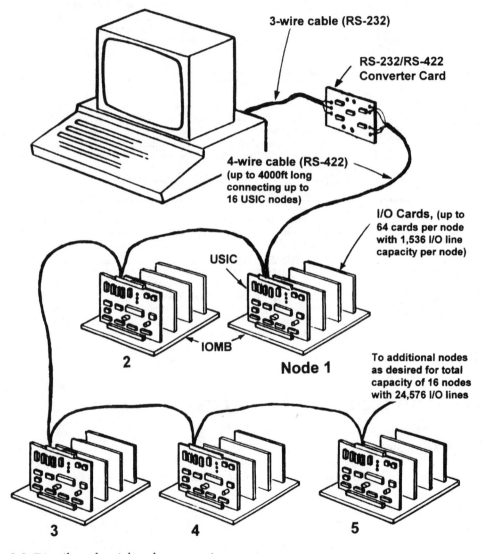

2-2 Distributed serial node connections.

a Pentium, x86, XT, PC, Apple II, Power Macintosh, or whatever you choose as long as it has an available RS-232 or RS-422 serial port. If the port is RS-232, then you need to incorporate the converter card covered in Chap. 4 to match the computer to the RS-422 requirement for multiple USICs.

Only a single four-wire cable is required to run from node to node, and the distance between the computer and the furthermost node can be up to 4000 feet. The only wiring required between the external hardware and the I/O cards is that associated within the immediate vicinity of each node. Total system wiring requirements are substantially minimized.

For interfacing requirements where the I/O need is distributed over a large area, or otherwise grouped into clusters, the distributed serial option of the UCIS provides an excellent solution.

Combined Parallel/Serial Systems

As I mentioned, many first edition readers obtained a USIC and a UBEC for the benefits of learning about and trying out both approaches. Also, it can be beneficial to use the USIC and IBEC (or UBEC/CBAC) together on the same system. I've personally done this on several systems, and I expect numerous others have combined their features as well.

Maximum benefit is achieved when the parallel system is tied to the subset of your external hardware that requires superfast response, and the serial system connects to areas needing an abundance of I/O distributed over a significantly large area, but with a somewhat slower response. This way you reap the benefits provided by overall minimized wiring consistent with the system's speed requirement. Under this combined configuration, your IBEC's I/O might be in the few-microseconds to hundreds-of-microseconds range, thereby satisfying your most demanding time-critical need; while the USIC I/O, handling a much larger number of I/O lines, would be in the tens- to hundreds-of-milliseconds range. These times can be changed substantially, of course, depending upon what computer is used, how the software is programmed, and in what language it is programmed.

In the remaining chapters, I'll cover both the parallel and serial design approaches in detail so you will be able to answer remaining questions as well as assemble the system that best meets your need. However, if you already have decided to just assemble the parallel version using an IBEC, then as you read through the book you can skip Chaps. 4, 6, 7, 10, 11, and 12. If you desire only the parallel version using a UBEC, then you can skip Chaps. 3, 4, 5, 9, 10, and 11. If you desire only the serial version using a USIC, or multiple USICs, then you can skip Chaps. 3, 5, 6, 7, 9, and 12. If you aren't sure which way you want to go or if you just really want to develop a good overall understanding of computer interfacing, both parallel and serial, then all chapters are applicable.

With the basics covered, you are now ready to begin building the universal computer interface system. In the next chapter, we'll start by building the internal bus extender card. If you have decided that you only want to build the serial approach using the USIC or the parallel approach using the UBEC, you can skip ahead to Chaps. 4 and 6, respectively.

3
CHAPTER

The Internal Bus Extender Card

In this chapter we'll start building the UCIS. We'll start with step-by-step instructions for building the internal bus extender card shown in Fig. 3–1. I assembled this IBEC on a ready-to-assemble PC card made by JLC Enterprises, which comes with the parts layout printed on its component side. If you don't care to assemble the card yourself, they are available assembled and tested from ECO Works, EASEE Interfaces, or Automation In Action (AIA). See Appendix A for details.

The IBEC is designed to plug into a spare ISA card slot in a Pentium, x86, IBM PC, or their compatibles. If you don't have one of these computers, or if your computer doesn't have a spare expansion slot, then you can't use the IBEC (or the UBEC discussed in Chap. 6). In these cases you should focus on the serial USIC approach presented in Chap. 4.

You don't need to understand electronic circuits to build the UCIS. The circuit cards are readily available, and all you have to do is follow my instructions. To explain how the IBEC works with different computers, and to help in debugging in case you have any trouble, I'll explain some fundamentals about computer bus architecture, followed by a description of applicable computers. Then we'll look at IBEC functions and I'll give you a tour of its schematic. If you really just want to get started assembling the card, skip ahead now to IBEC Parts.

Bus Fundamentals and the x86 Computer Family

Understanding bus fundamentals is core to understanding how a UCIS interfaces with IBM's PC series of computers. The x86 nomenclature refers to the entire computer family beyond the original PC, where the x takes on values 2 through 5. Knowledge of how these computers evolved helps explain the various UCIS options

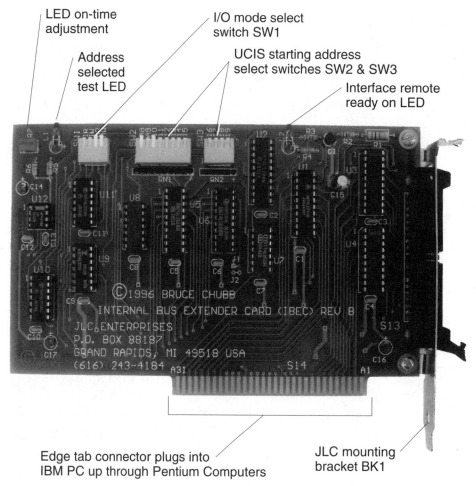

LED on-time
adjustment

I/O mode select
switch SW1

Address
selected
test LED

UCIS starting address
select switches SW2 & SW3

Interface remote
ready on LED

Edge tab connector plugs into
IBM PC up through Pentium Computers

JLC mounting
bracket BK1

3-1 Internal bus extender card (IBEC).

and how to tailor them to each computer. Even if your computer isn't of this family, you should find the material helpful to better grasp how UCIS works.

The microprocessor chip inside a computer executes step-by-step instructions defined by the software programs loaded into the computer. These include instructions to interface with memory, the keyboard, disk drives, a printer, and so forth. Each time the microprocessor needs to read data from or write data to an external device, whether it be a sound card, printer, or our UCIS, the microprocessor uses three sets of signal lines to perform the read or write operation. Each set is referred to as a *bus,* and the computer's three buses are:

- Address bus
- Data bus
- Control bus

The general terms *internal bus* and *system bus* refer to the group of all three buses. The microprocessor uses the address bus to identify the external device, and in most cases the exact location within the device, with which it wishes to communicate. It accomplishes the addressing by placing a pattern of ones and zeros on the address bus. Newer computers typically have more address lines, and therefore support a greater range of addresses.

The data bus provides a communications path for the transmission of data between two devices, such as the microprocessor and memory, or the microprocessor and our UCIS. The control bus synchronizes (controls) all operations taking place on the system bus. Its structure includes all the signal lines that are not address or data lines. The microprocessor communicates with external devices under the following circumstances:

- When it needs to fetch (read) the next instruction from memory
- When the currently executing instruction directs the microprocessor to read data from an external device
- When the currently executing instruction directs the microprocessor to write data to an external device

For UCIS applications we are concerned with the last two items. The term *external device* refers to everything outside of the microprocessor such as the memory, keyboard, monitor, printer, disk drives, and the UCIS.

Each external device attached to the system bus includes an address decoder that constantly monitors the address bus to see if the address currently on the bus is an address assigned to the device. If it is, the address decoder informs the device that the microprocessor is communicating with it. Further decoding may then occur to identify the exact location within the device where the data are to be exchanged. The control lines then identify the type of data transaction such as a read or write operation and then synchronize the typically fast microprocessor to the typically slower external device.

Using the system bus, the address is set up and the data are transferred during what is called a *bus cycle*. Although there are other direct memory access (DMA) devices that can take control of the system bus, and thereby generate bus cycles, in this book we devote our attention to bus cycles that are generated by the computer's main microprocessor. When the microprocessor generates a bus cycle, it drives the system bus with an address of a memory location or an I/O port, controls the direction of data flow, and is either the sink or source for the data. The latter depends on whether the microprocessor is reading or writing the data.

You can think of the bus cycle as having two distinct time slots. The first (which we'll call address time) is where the microprocessor sets the address values on the address bus. The second time slot (which we'll call data time) is where the data are transferred from the microprocessor to the external device or vice versa. The bus operations we discuss will always be from the viewpoint of the microprocessor; therefore, data going toward the microprocessor is a read (or input) operation, and data going away from the microprocessor is a write (or output) operation. In general, there are four basic bus cycles that can be used by our UCIS design: memory read, memory write, I/O port read, and I/O port write.

Extended Memory-Mapped I/O

Starting with the IBM AT class computer up through the latest Pentium system, the memory read and write bus cycles are further subdivided into two types: bus cycles addressing below 1 Mbyte, referred to as conventional memory, and those addressing above 1 Mbyte, referred to as extended memory.

As we'll see in Chap. 5, all machines from the first PC up to the newest Pentium have plenty of space available for fitting UCIS into their conventional memory space. Even if your Pentium has 16 Mbytes, 32 Mbytes, or more installed memory, there still remains an abundance of unused address space in the region between 640 and 896K. This 262,144-byte range includes reserved space for devices, such as our UCIS, that attach to the computer's ISA bus.

We only need a contiguous block of 256 bytes to support a fully expanded UCIS. Therefore, there is little if any advantage for an IBEC to directly make use of extended memory. Even if IBEC provided an extended-memory-mapped option—and I constructed prototypes to use this option—the option becomes invalid for computers having 16 Mbytes or greater of installed memory. This is because only the lower 24 address lines are brought out on the ISA bus. Thus, any adapter cards (including an IBEC) can only decode addresses up to 2^{24}, or 16 Mbytes. Also, direct interaction by the user with addresses above 1 Mbyte is more difficult with most readily available software packages, including DOS. Coupling these facts with the fact that we have more than we need available in the conventional memory space, plus for AT and newer computers we also have the option to use the 64-K port-mapped address space, there isn't any need for implementing extended memory-mapped I/O.

Predecessor Computers

Let's put the above described fundamentals to work for us by viewing the properties of various computers leading up to the $x86$ family. With this information it is easier to see how the IBEC fits each application.

Intel 8080

The Intel 8080 1-MHz chip is the original seed for the entire $x86$ family. Consequently, many $x86$ characteristics stem from this design. The 8080 uses 16 address lines on which the microprocessor can place any address pattern, from all zeros (0000h) to all ones (FFFFh). Note that we need four hex characters to represent the 16 lines, because each hex character represents four bits. If this is unclear, you might want to go back and review the hexadecimal number system in Chap. 1.

The 8080's data bus is eight lines wide, or one byte wide, which is also the effective width of each memory location. Therefore, we say the 8080 has a memory capacity of 64 kbytes, or 64K for short. In addition to using this space for addressing memory devices, the microprocessor must also be able to use a number of locations to address I/O devices. In simple terms, any device that is not a memory device (used to store programs and data) is considered to be an I/O device. The addresses associated with I/O devices are referred to as ports.

Rather than directly tie up part of the limited 64K address space for I/O ports, Intel's 8080 design includes two special programming instructions, namely IN and OUT, for communicating to I/O ports. When these instructions are executed, the 8080 notifies external logic hardware that the address bus is addressing an I/O port and not conventional memory.

Intel 8085

The 8085, the next step up from the 8080, incorporates the same 16-bit address bus and eight-bit data bus. It uses a pin labeled *IO/M**, standing for input/output or memory. The asterisk on the *M* signifies that if this line is low the multiprocessor is talking to a memory address, and if high it is talking to an I/O port address. This action creates two separate address spaces, frequently referred to as maps; one is for memory and one for I/O ports. This fundamental two-address space concept is used extensively as we apply the IBEC to the IBM PC and later machines in that line. It also applies when using the UBEC/CBAC for an assortment of other computers, as we'll discuss in Chaps. 6 and 7.

With the limited number of I/O devices available when the 8085 was introduced, Intel decided to implement only 256 I/O ports. Consequently, only the lower eight bits are sent out on the address bus when I/O ports are being addressed, versus the use of all 16 when addressing memory. Many port locations are dedicated to functions like keyboard interfacing, serial ports, and disk drives. Coupling this with the fact that our parallel UCIS, when fully expanded with 64 I/O cards, requires the full 256-byte address space, it's essential for UCIS to use memory-mapped I/O with the 8085.

The IBM PC, XT, and Their Compatibles

Intel's introduction of their 8088 microprocessor set the stage for the tremendous growth in today's world computer market. Its outstanding success in the IBM PC and its many clones accelerated the development of ever-faster and more powerful systems. Let's look at the key UCIS interfacing-related features of the PC.

IBM PC

Original IBM PCs used the Intel 8088 microprocessor running at 4.77 MHz, with 32K of RAM on the system board and five expansion slots. Shortly after its introduction, the system board was upgraded to 64K. The 8088 is a 16-bit microprocessor, meaning that its instructions can manipulate 16-bit data. However, the data and instructions are read from and written to memory eight bits at a time—the width of the data bus.

The 8088's address bus contains 20 lines; this permits a physical memory size of 2^{20} or 1,048,576 bytes, or 1MB. Each expansion slot incorporates a 62-pin female edge-tab connector. Matching adapter cards plug right into the computer's bus, with

31 tabs on each card side and a spacing between tab centerlines of .1 in. Throughout the computer industry, the resulting 62-pin bus is typically referred to as *The PC bus,* and it is the standard for many, many clones.

For the 8088 chip itself, the size of the I/O map address space was increased from the previous 256 bytes up to 64K, thereby supporting 65,536 unique port addresses. This sounds great; however, PC designers using the 8088 decided to only activate the lower ten address bits on the PC bus when addressing I/O devices. This limits the usable I/O port space to 2^{10} or 1024 addresses. The lower 512 bytes are reserved for system board addresses, and the upper 512 allocated for use by adapter cards plugged into the computer's expansion slots. We'll see in Chap. 7 that this limitation makes I/O mapping with a UCIS not very attractive for PC applications.

The 4.77-MHz clock results in a system clock time of approximately 210 nanoseconds (ns), or 210 billionths of a second. Clock time, the period between clock pulses, is calculated as one divided by the clock frequency. Memory read and write bus cycles in the PC design take four clock cycles; therefore, a typical PC memory bus cycle is 840 ns, or .84 µs.

I/O port read and write bus cycles include an additional wait state; that is, an idle clock time. This wait state, automatically inserted by the computer, compensates for the typical situation where external hardware operates more slowly than computer memory. Correspondingly, I/O port read and write bus cycles take five clock cycles, requiring a completion time of 1.05 µs. These baseline times continue to be very important as we look at interfacing to the complete *x*86 family.

IBM XT

The original XT retains the same 8088 microprocessor operating at 4.77 MHz with the same busing structure, but increases the number of expansion slots to eight. The XT's cabinet size was kept the same, and, to accommodate the added slots, the spacing between slots is reduced from 1 to .75 in. Our UCIS cards fit in either slot spacing. Memory on the original XT system board was increased to 256K. Later XT versions, sold by a number of manufacturers, incorporated higher clock speeds and higher-performance microprocessors. For example, a number of XT compatibles use the Intel 8086 microprocessor. The 8086 is identical to the 8088, except that it supports a larger internal instruction queue and a 16-bit data bus. An 8086-based XT runs about 1.6 times faster than one with an 8088 at the same clock speed.

The x86 Series

The accelerated development pace continues to gain momentum with each new release of the *x*86 computer family. Interfacing with the Pentium seems ultimately powerful, making one even more excited about the interfacing power afforded by the new P6 microprocessor, Pentium Pro.

Let's not get too far ahead though, for there are plenty of interfacing applications ideally suited to AT- (or 286-) class machines.

PC AT and 286

The PC AT introduces a new generation Intel 286 microprocessor with a 16-bit data bus and a 24-bit address bus. Backward compatibility is maintained so that the AT/286 can make use of the abundance of bus adapter cards already on the market for the PC and XT. This is accomplished by keeping the 62-pin PC bus connector, while adding a second connector with 36 pins to most card slots. The added pins support the increase in address lines from 20 to 24, the increase in data lines from eight to 16, and numerous increases in the control bus structure.

Slots that have the two connectors, the PC slot plus the AT extension connector, are referred to as 16-bit adapter or I/O slots because they support the 16-bit data bus. Likewise, slots with only the 62-pin PC bus connector are called 8-bit adapter slots. The resulting dual-connector bus, meaning the 62-pin plus the 36-pin connectors, is now well established as the 16-bit ISA (Industry Standard Architecture) bus or simply the *AT bus*. This bus is the standard upon which the IBEC card is based.

Considerable confusion in bus naming conventions results from the fact that the original PC 62-pin bus is also frequently referred to as the (8-bit) ISA bus, as well as the *IBM or PC bus*. However, in this book we will use what has become the more common nomenclature, which is to refer to the 62-pin bus as the PC bus and the 62-pin plus the 36-pin as the ISA bus.

The original IBM AT uses a clock rate of 6 MHz. This was quickly increased to 8 MHz. Many manufacturers of AT compatibles increased the clock rate significantly above these values. Besides improved performance resulting from higher clock speed, the 286 microprocessor introduced more capable instructions while requiring fewer clock cycles per instruction, and cut in half the required clock cycles per bus cycle. These factors, coupled with a 16-bit data bus, give later AT systems a tenfold performance improvement over the 4.77-MHz PC and XT.

The 286 requires only two clock cycles per bus cycle, versus four required by the 8088. However, as with every new x86 design, system boards introduced added wait states into ISA bus cycles to maintain compatibility with already existing adapter cards. ISA bus speeds have increased slightly, but nothing like the increase in internal speeds of the microprocessors.

With the 24-bit address bus the AT is capable of addressing 2^{24} or 16,384,000 bits, or 16MB. Sixteen of the 24 lines are activated during I/O operations, providing 2^{16} or 65,536, or 64K unique port addresses. This 64-fold increase over the PC and XT makes port-mapped I/O very attractive for use with the UCIS.

386 Systems

IBM first introduced the Intel 386 microprocessor into their PS/2 family of computers using a new proprietary bus architecture called the Micro Channel Architecture. The remainder of the industry departed from their heretofore path of closely cloning the IBM machines, and continued using the ISA bus on their 386 systems. The 386 supports 32-bit address and data buses, even though only 24 and 16 lines respectively are brought out to the ISA bus.

By placing the system memory (located on the system board) on a separate local bus, the 386 supports a split-bus architecture as illustrated in Fig. 3–2. This sets up a separately optimized 32-bit bus for high-speed memory, while keeping the AT bus for lower-speed I/O that is fully compatible with the ISA specification. The advantages provided by this split-bus architecture are even more fully exploited with each new system design, including the Pentium.

The 386 offered a quantum jump in performance while retaining software and bus compatibility with the 8088- and 286-based system. Typically 386 systems offer a 12–50-fold performance improvement over the original PC, and 2–10-fold over an 8-MHz 286. The 32-bit address lines provide an address space of 2^{32} or 4,294,967,296 bytes, or 4096 Mbytes, which is 4 Gbytes. The 24 lines brought out on the ISA bus retain the same 16-Mbyte address range for expansion slot cards, and I/O port-mapped space is retained at 64K.

486 Systems

The 486 introduced the Intel 486 microprocessor. One neat attribute of the 486 chip is that it incorporates the 387 math coprocessor as an integral part of the basic chip design. Not having to transmit between chips results in higher-speed mathematical operations. This makes for faster handling of math in our real-time software. Also, bus speeds are increased to the point where many bus transactions take place in just over 1 clock cycle. Typically for the same clock speed, the 486 offers a twofold performance improvement over the 386. The address and data buses remain at 32 lines each, with the lower 24 address and 16 data lines brought out to the ISA bus. I/O port space remains at 64K.

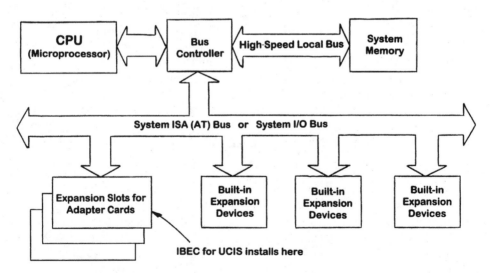

3-2 Dual- or split-bus architecture.

Pentium Systems

The next logical "name" in the progression of Intel CPUs would have been 586. However, mainly to improve their patent position, Intel introduced the Pentium. Some people still refer to it as the 586. Actually, in Greek *pente* means five. As with each new chip introduction, the computing power of the Pentium is a giant leap ahead of the 486. It has a 64-bit data bus and two separate instruction execution units. With these and numerous other enhancements, the Pentium is capable of executing four instructions per clock cycle—four times the capability of the 486. Bus transactions include transferring eight bytes (64 bits) in a single clock cycle.

Pentium clock speeds have surpassed 200 MHz, which corresponds to a clock time of 5 ns, or 5 billionths of a second. Five nanoseconds is about three times shorter than the propagation delay of a single chip on IBEC. That's superfast. To give you an idea of how fast, the speed of light is 1 ft/ns. To support these superfast speeds, voltage levels are reduced to 3.3 Vdc in the Pentium for less heating and faster circuit response.

The Pentium is a fascinating processor, and if you would like to learn more about Pentium-based systems, I recommend the book *Build Your Own Pentium Processor PC and Save a Bundle* by Aubrey Pilgrim (McGraw-Hill). It's a great book to read, even if you have no desire to build your own. It beautifully presents what is inside the computer's case.

From the UCIS viewpoint, the Pentium gives tremendous real-time software processing speed while maintaining adapter card I/O bus compatibility with standard ISA bus connector slots and a typical ISA bus clock frequency of 8–12 MHz.

With this understanding about how the different computers operate with regard to their bus structure, it's time to look at the IBEC's interfacing capability.

IBEC Functions

Figure 3–3 is a functional look at what the IBEC does. The IBEC's four functions are:

- Signal buffering
- Address decoding
- I/O mode selection
- Line enabling/data direction control

The arrows indicate the direction of signal flow; double lines with a slash and number indicate that multiple parallel wires are involved, 4 through 12 in this example. All signals going between the computer and the external hardware are buffered within IBEC, as illustrated by the triangular IC representations in Fig. 3–3. These buffers provide a degree of isolation between interface I/O and the computer, and serve as line-driver amplifiers to keep the pulse inputs strong and sharp enough to drive a significant number of remotely located I/O cards. Without this buffering,

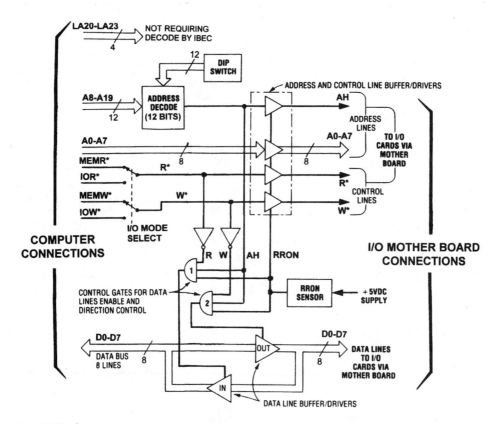

3-3 IBEC functions.

external hardware electrical demands would so overload the computer's internal bus that it couldn't work properly. Also, when the interface is turned off, all the buffer/drivers transform to open circuits, effectively disconnecting the interface from the computer. The RRON (*remote-ready-on*) sensor monitors the 5-Vdc supply that powers the interface's I/O cards and, when turned off, sends the signal to set all buffers to open circuits.

The computer's low-order address lines, A0 through A7, pass directly from the computer side of IBEC through the buffers to the interface I/O motherboard connections. These lines are decoded by the bank of I/O cards attached to IBEC.

The 16 high-order address lines, lines 8 through 23 from the computer, are of two different types. A8–A19 behave exactly like lines A0–A7, which is to say that they are latched and held valid during the complete bus cycle. In contrast, the uppermost four lines, LA20–LA23 (referred to as *latchable address* lines), are not latched by the computer, thus enabling faster address decoding under certain conditions. Our IBEC doesn't need this feature and, because we achieve full interfacing potential in the conventional memory region, we can ignore the LA20–LA23 lines. The fact that we can bypass decoding these four lines is also a direct result of the neat manner in which the ISA bus designers expanded the control bus to include two-each memory read and memory write control lines. We'll look at these shortly.

Consequently, the IBEC needs to decode only lines A8–A19 to see if the computer's CPU is addressing the interface. Decoding means that when each of 12 input lines match the external hardware UCIS starting address code set on the IBEC's DIP switches, decoding output line AH (address lines high-order bits) goes high. I'll explain how to define and set the starting address in Chap. 5.

At any point in time, AH (and every other wire connecting to or from the IBEC) conveys a single bit of information. A bit is a single numeral in the binary code of digital logic, either a 0 or a 1. To say that any digital line or device goes high means that it is carrying or putting out +5 Vdc, the signal for the numeral 1 in conventional digital logic. The other possible condition is low, or 0 V, the signal for logic 0. When AH is high the computer is set to communicate with the external hardware; when low the computer is not communicating with the external hardware.

IBM computers, starting with the AT class and including machines up through the Pentium, make use of the following six control lines for directing bus traffic:

SMEMR*

This conventional memory read signal, available on both the PC and AT bus, is activated on memory read cycles where the computer is addressing memory below the 1-Mbyte boundary. The * after the acronym signifies active low, meaning that the line is pulled low by the computer during the read operation and otherwise left high. (On the PC and XT computers this signal is simply called MEMR*, because these machines can only address 1MB.)

SMEMW*

This signal performs exactly like the SMEMR* except that it is activated low for writes rather than reads.

IOR*

This I/O read signal, available on both the AT and PC bus, is activated low on a read cycle when the computer is addressing an I/O port address rather than memory.

IOW*

This signal performs exactly like the IOR* except that it is activated low for writes rather than reads.

MEMR*

This memory read signal, available on the AT bus but not on the PC bus, is identical to the SMEMR*, except that it is activated low on all memory read cycles, including those above and below the 1MB boundary. For example, if MEMR* is low and SMEMR* is also low, then the computer is set to read from convention memory. If MEMR* is low and SMEMR* is high, then the computer is set to read from extended memory, meaning above the 1-Mbyte boundary.

MEMW*

This signal performs exactly like the MEMR* except that it is activated low for writes rather than reads.

The fact that ISA bus designers retained the PC bus read MEMR* and MEMW* signals (renaming them SMEMR* and SMEMW*) is of vital importance to ISA bus users like UCIS. If all we had available was the AT bus MEMR* and MEMW* signals, then we would need to decode all the higher-order address lines, even when we were using only the conventional memory region.

Why? Consider an example case where the UCIS is set to use a starting address of 640K, which is A0000h, the higher four bits being a binary 1010. We would set the IBEC's DIP switch segments to A19 = logic 1 and A17 = logic 1, and all others logic 0.

Now let's say that operating software had a need to write to location 1664K, which is 1A0000h. The corresponding address bits placed on the address bus would be A20 = 1, A19 = 1, A17 = 1, and all others zero. Because we aren't decoding above A19, and all the bits A19 and below match our DIP switch settings, we would incorrectly determine that the computer was communicating with our interface. This same decode problem would occur for all addresses above 1 Mbyte that had their lower-order bits defined by A0000h.

All these memory-overlapping problems are circumvented by the ISA bus providing the SMEMR* and SMEMW* signals. Using these as IBEC's control lines, our UCIS ignores all AH high decodes except those that occur when either SMEMR* or SMEMW* are active low, indicating that the computer is addressing the conventional memory space.

Because the SMEMR* for the AT and later machines is on the same pinout as MEMR* is for the PC and XT, and likewise the same for SMEMW* and MEMW*, our IBEC works great on the whole spectrum of machines from the earliest IBM PC up through the Pentium using the ISA bus.

To try to avoid confusion in the remainder of this second edition, I'm going to drop the use of SMEMR* and SMEMW* and simply use MEMR* and MEMW* as our designations for input to the UCIS, independent of whether we are connected to an AT or a PC bus. A switch on IBEC sets the desired I/O mode as either memory-mapped I/O or port-mapped I/O by attaching IBEC to the appropriate computer control lines.

During the computer's operation, each of the control lines plus all the address and the data lines change state continuously, multimillions of times a second, as the computer talks to memory and every other device hanging on the bus. Only when the UCIS interface address is selected (AH is high) is it important that the UCIS take action in response to the control lines for which it is programmed by the I/O mode select switch.

For each selected I/O mode, the R* (read active low) output of I/O mode select switch will be low if the computer is executing a read operation and the W* (write active low) output will be low if the computer is executing a write operation. These signals are buffered and sent out to the interface I/O cards via the motherboard connector, and also are used within IBEC to enable and set the direction of the data line buffers.

Data is sent back and forth on lines D0-D7, the data bus. Because data flow must be bidirectional, the data lines are double-buffered. One group of eight buffers (labeled IN in Fig. 3–3) handles data flow into the computer, and the group labeled OUT handles data out to the external hardware.

Control gates 1 and 2 logically determine which set of data buffers is active by combining the AH address, the R and W control signals, and the RRON signal. The control gate logic is summarized in Table 3–1. Note that if the external hardware is turned off, the data lines—just like the address lines and the control lines—cannot be enabled, so the interfaced hardware is effectively disconnected from the computer.

Table 3-1. Control Gate Logic

IF:	THEN:
External hardware is turned on (RRON high) and External hardware is being addressed (AH high) and Computer is executing a write command (W high)	Data bus is enabled for data flow from computer to external hardware
External hardware is turned on (RRON high) and External hardware is being addressed (AH high) and Computer is executing a read command (R high)	Data bus is enabled for data flow from external hardware to computer
Neither of above sets of statements has all three conditions true	Data bus is disconnected

IBEC Schematic

The schematic in Fig. 3–4 is simply an expansion of the Fig. 3–3 functional diagram. Digital circuits can appear complicated because there are a lot of parallel lines doing basically the same thing. Once you understand the function of a few of the separate lines, however, you quickly pick up the understanding of the companion lines, and soon the overall diagram becomes understood.

Integrated circuits, especially the TTL types shown here, are easy to work with because you can think of them as building blocks. Each block performs a particular function, and each is designed to be plug-compatible with the next. By connecting them in different patterns, you can build any of several functions, such as a lamp-flashing circuit, a signal decoder, or (say for example) a computer bus extender card.

With that in mind, let's take a closer look at the IBEC schematic, starting with the address decoding function. In every case, the address lines A0 and A1 feed directly through the IBEC's low-order address line buffer/drivers to determine the internal port status on the interface I/O cards. Lines A2 through A7 carry six-bit addresses for the individual I/O cards. Six bits can address 2^6 or 64 I/O cards; that provides direct access to 1536 lines (64 × 24 lines/card) connected to your external hardware.

IBEC takes the high-order address lines A8–A19 and decodes these bits to see if they match the interface starting address placed on the DIP switches SW2 and SW3. This is accomplished by feeding the 12 input lines into ICs U5 and U6, which are eight-bit identity comparators. For example, U5 looks at the A8–A15 inputs from the computer and sees if they compare to those set on the address DIP switch, SW2. If all eight input pairs match, then the comparator's pin 19 output is set logic low;

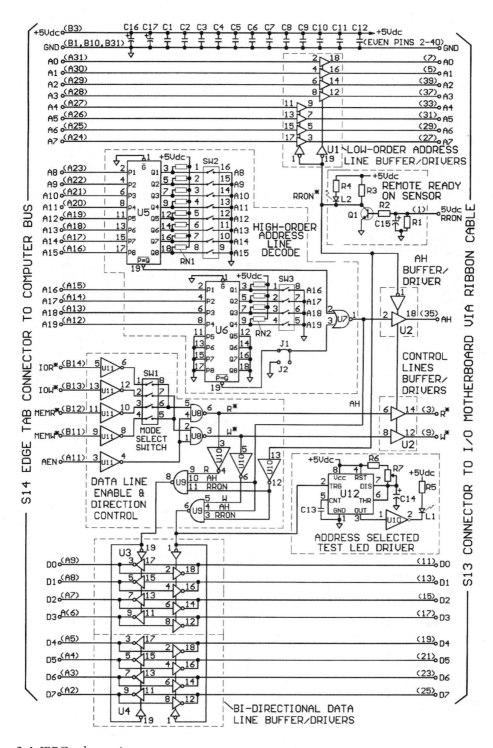

3-4 IBEC schematic.

otherwise pin 19's output is high. (The small circle on the output or input of an IC symbol indicates that the corresponding pin's function is active low.)

U6 performs the same function on A16–A19, but because that's only four bits, the unused pins 11 through 18 are grounded. This way U6's pin 19 is low when the four address lines match the four settings on SW3.

As a result, the computer is addressing our interface hardware only when U5 pin 19 and U6 pin 19 are both low. These signals are fed into one section of U7, a 74LS02 quad NOR gate. The pin 1 output of U7 goes high only for the condition when both inputs are low, meaning that each of the 12 address inputs exactly match the corresponding setting on SW2 and SW3. This output is then further buffered by U2 and sent out to the I/O cards as AH.

The important point is that AH is high only if the computer is addressing the external hardware via UCIS. With AH high, the specific I/O card that is being addressed is defined by A2–A7, and the specific I/O port on the card is defined by A0 and A1. I'll provide more on this later when I talk about the specifics of the general purpose I/O cards in Chap. 8.

Only address lines A0–A15 are active when AT and newer computers use port-mapped I/O. The computer sets the higher-order unused lines to logic zero. Therefore, when using this mode SW3 is set to all logic 0.

Now let's look at the IBEC's next function, buffering the address, control, and data bus lines. Each of the four ICs U1–U4 is a 74LS244 octal buffer/line driver. Octal simply means that each IC can handle eight lines. These ICs provide tri-state logic; that means that in addition to the usual two digital states, high or low, their output capability also includes a high impedance state, which for these purposes is effectively an open circuit. You can think of the output of tri-state devices as being a single-pole three-position rotary switch, as illustrated in Fig. 3–5, with the center pole being the device output. Set in position 1 the output is a +5 Vdc (known as logic high or logic 1); in position 2 the device output is at ground (know as logic low or logic 0), and in position 3 the output is open circuited, meaning no connection (N.C.).

Tri-state devices are used extensively to attach devices like our interface to a computer bus. In contrast, if two regular devices (non-tri-state devices) had their outputs connected to a computer bus line, we would have bus contention. For example, one

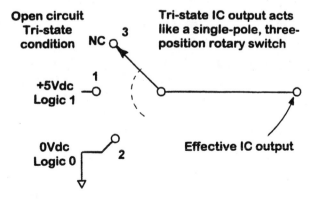

3-5 Tri-state switch.

device could be trying to pull the bus low while the other one was trying to pull it high. The devices would be at war with each other, and nothing would work. With tri-state components, many output devices can be attached to a bus and can be arranged so that only one component is sending data over the bus at a time. This tri-state capability makes it possible to have up to 64 I/O cards connected to our IBEC's bus output.

The IBEC's address bus and control lines are unidirectional, as are their buffers. The data bus must be bidirectional, so two buffers are hooked up back-to-back for each of the eight data lines. If the rightmost vertical column of data buffers in Fig. 3–4 is enabled (pin 1s low), data can be transmitted from computer to external hardware with the buffers facing the opposite direction in tri-state open-circuit condition. Likewise, if pin 19s are low, data can be transmitted from external hardware to the computer, and the opposite direction buffers are open-circuited. If all of the pins 1 and 19 are high simultaneously, then both sets of buffers are in their tri-state, open circuit condition and there is no data connection between the computer and external hardware. Pretty straightforward, isn't it?

The next function of our circuit is generating the enable inputs to control the direction of data flow, either to or from the external hardware. Remember, the CPU controls data flow on the computer data bus with the read and write control lines MEMR* and MEMW*, or IOR* and IOW*. Which set of two lines you want to use is selected by closing the corresponding two segments of DIP switch SW1, the I/O mode select switch. At any point in time only segment pair 1 and 2 *or* 3 and 4 should ever be closed, and the other two segments should be open. A 74LS04 hex inverter in each of the control input lines prevents incorrect switch settings from shorting control lines together back in the computer. The correct procedure for setting SW1 is to always first open all segments, then set the desired two segments closed.

Devices other than the CPU, such as DMA (direct memory access) devices, can control the address and control lines. To distinguish when the CPU is in charge, the PC pulls the AEN (address enable) line low. Bus socket pin A11 is that line, and to keep other devices from inadvertently communicating with our UCIS, we gate AEN with the read and write control lines selected by SW1.

For the selected I/O mode, the pin 6 output of U8 (a 74LS00 quad NAND gate) is low only if the CPU is executing a read operation, and correspondingly pin 3 is low only if the CPU is executing a write operation. If a DMA is executing a read or write, both control lines used by IBEC remain high.

Depending on which set of control lines you select via SW1, the selected set controls the data line buffers through the two U9 control gates. Each U9 section is a three-input NAND gate, meaning that the output (pin 8 for the read case and pin 6 for the write case) are low only if all three of the corresponding gate inputs are high. For this to happen the interface must be turned on (meaning that RRON is high), the interface must be addressed (meaning that AH is high), and either a CPU read or write operation must be taking place (meaning that R or W is high). The read and write control lines are also buffered by U2 for extending to the I/O cards as control lines R* and W*.

The card's next function is to switch the whole bus extender to the open-circuit state when the external hardware is turned off, in effect disconnecting the hardware while the computer is used for other tasks. This is accomplished by the RRON sensor. If the external hardware power is on then transistor Q1 conducts, causing Q1's collector to be low. This signal, active low, is designated as RRON*.

RRON* is used directly to enable U1, the low-order address line buffer/drivers, the AH buffer/driver, and the control line buffer/drivers. It also turns on an IBEC indicator LED, namely L2, to signify that the IBEC is sensing that the external hardware is powered up. The RRON* signal is then inverted back to RRON via pins 13 and 12 of U10 as required to drive the U9 data line direction control gates.

This conversion from RRON to RRON* and back to RRON demonstrates an important concept for understanding digital circuits: that of regular or *noninverted logic* versus *inverted logic*. The pin 12 output of U10 is regular logic in that a high (+5-Vdc) output corresponds to a high input to the RRON sensor; in other words, the external hardware's +5 Vdc is turned on. Pin 13 of U10 is the inverse of pin 12, and follows inverted logic; a low on pin 13 corresponds to the external hardware being on. The small circles at the output of buffer triangle symbols indicate that the output is inverted from the input.

In place of using the asterisk notation for inverted logic (active low), it is common practice to use a bar or macron over a variable symbol, as follows: $\overline{\text{RRON}}$. I'll use the asterisk notion when symbols are discussed in text, and both notations will be used in figures. Note in Fig. 3–4 that all of the IBEC outputs to the interface I/O cards are noninverted except for W* and R*. Thus, the interface I/O cards work with noninverted signals for the most part.

During system operation, and especially to help out when setting addresses and making sure that your hardware DIP switch settings and software are working in harmony, I've included an address-selected test LED, namely L1 on IBEC. The LED's on-state indicates that the computer, as directed by software, has selected the UCIS starting address set by IBEC's address DIP switches.

The LED could be connected to AH directly, but this wouldn't help much for two reasons. First, since AH pulsed-high times are typically around a microsecond, the LED's on-time would be so short it couldn't be seen. Secondly, AH can go high when the CPU or DMA is addressing other devices, for example the memory if IBEC is set for port-mapped I/O, or an I/O port if IBEC is using memory-mapped I/O. To eliminate this possibility, I've triggered the LED with the output of AH gated with a write operation to IBEC, which is found at the pin 6 output of U9.

To lengthen the LED on-time, I've added a "one-shot" pulse extender circuit using U12, an NE555 timer IC. By adjusting potentiometer R7, you can stretch the LED on-time from near zero to just over 1 s. As long as software is periodically setting AH high and writing to IBEC, you will see the LED blink on or stay on, depending on how frequently your software is addressing the interface and the R7 adjustment. If AH is not set high (meaning that your software address doesn't match the DIP switch settings) or if the decode circuit is not working correctly, the LED will stay off. Observing this LED can significantly help with getting the interface up and running quickly.

IBEC Parts

A key advantage to UCIS is that ready-to-assemble printed circuit boards are available from JLC Enterprises. If you elect to buy them assembled and tested, they

Table 3-2. IBEC Parts List

Qnty.	Symbol	Description
1	J1 **or** J2	Jumper, make from no. 24 uninsulated bus wire (Belden 8022). Install in J1 position unless not installing U7, then install in J2 position (see text).
1	S13	40-pin protected, right-angle header(JDR IDH40SR)
1	BK1	Card mounting bracket (JLC BK177)
2	—	4-40 x ¼" pan-head machine screws (Digi-Key H142)
1	R1	110Ω ½W resistor [brown-brown-brown]
1	R2	1.0KΩ resistor [brown-black-red]
1	R3	2.2KΩ resistor [red-red-red]
2	R4,R5	330Ω resistors [orange-orange-brown]
1	R6	10kΩ resistor [brown-black-orange]
1	R7	1MΩ potentiometer (Digi-Key 3362W-105)
6	S1-S6	20-pin DIP sockets (Digi-Key A9320)
5	S7-S11	14-pin DIP sockets (Digi-Key A9314)
1	S12	8-pin DIP socket (Digi-Key A9308)
1	RN1	2.2KΩ 9-element resistor network (Digi-Key Q9222)
1	RN2	2.2KΩ 5-element resistor network (Digi-Key Q5222)
2	SW1, SW3	4-segment side actuated DIP switches (Digi-Key CT1944MST)
1	SW2	8-segment side actuated DIP switch (Digi-Key CT1948MST)
12	C1-C12*	.1μF, 50V monolithic capacitors (JDR .1UF-MONO)
1	C13*	.01μF, 50V monolithic capacitor (JDR .01UF-MONO)
1	C14*	1μF, 35V tantalum capacitor (JDR T1.0-35)
1	C15*	100μF, 16V radial lead electrolytic capacitor (JDR T100R16)
2	C16,C17*	2.2μF, 35V tantalum capacitors (JDR 2.2-35)
1	Q1	2N3904 NPN small signal transistor (JDR)
1	L1	Diffused red LED, T1 size (Digi-Key P363)
1	L2	Diffused green LED, T1 size (Digi-Key P364)
4	U1-U4	74LS244 octal buffer/drivers, non-inverting (JDR)
2	U5,U6	74LS688 8-bit magnitude comparators (JDR)
1	U7	74LS02 quad 2-input NOR gate (JDR)
1	U8	74LS00 quad 2-input NAND gate (JDR)
1	U9	74LS10 triple 3-input NAND gate (JDR)
1	U10	7404 hex inverter (JDR)
1	U11	74LS04 hex inverter (JDR)
1	U12	NE555 or LM555 timer (JDR LM555)

Author's recommendations for suppliers given in parentheses above, with part numbers where applicable. Equivalent parts may be substituted, however if you substitute capacitors, make sure they have .098in (or 2.5 mm) lead spacing (as denoted by asterisk above). Resistors are ¼ W, 5 percent except R1 is ½ W and color codes are given in brackets.

are available from ECO Works, EASEE Interfaces, and AIA; See Appendix A for details. The parts layout for the IBEC is shown in Fig. 3–6, and the parts are listed in Table 3–2. Follow this table to assemble your IBEC.

Figure 3–6 shows the parts locations on the top or component side of the IBEC; you should refer to this diagram as you select and mount parts. Keep the card oriented the same way as the drawing while you work, to help you place each part in its proper holes with correct orientation.

You should also take care to see that you are installing the right components. Sometimes their markings are so small that you need a magnifying glass to read them, but extra effort at this stage is worthwhile. Check resistors against the color

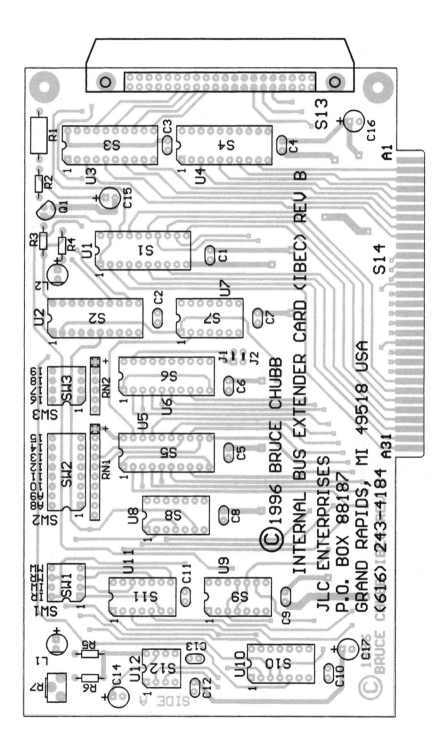

3-6 IBEC parts layout.

code included in the parts list. In addition to the three color bands, there should also be a gold band, indicating a tolerance of 5 percent.

For the following parts, the correct orientation on the card is very important: capacitors C14–C17; transistor Q1; DIP switches SW1–SW3; resistor networks RN1 and RN2, LEDs L1 and L2, and both IC sockets S1–S12 and the ICs themselves, U1–U12. To remind you to check this as you build your IBEC, I've marked the orientation-critical parts with a plus sign in brackets [+] in the following instructions. As an added reminder, the pin-1 pad on the circuit board for each of these parts is square. Figure 1–8 provides pointers on proper parts identification and orientation.

Double-check your selection and location of each component before soldering it in place. It is a lot easier to remove an incorrect component before you've soldered it to the card! This is especially true with plated-through-holes since as you solder the part in place, the solder flows all the way through the hole.

If you plan to drive more than 30 I/O cards with your IBEC, then it's recommended that you substitute 74S244 chips in place of the 74LS244. The pure Schottky (the S version) costs more, but they have a substantially stronger drive capability. I've specified the standard 7404 version for U10 because either it or a 74S04 alternative is the best choice for driving the LED, while keeping U11 a 74LS04 provides less loading of the control bus.

One item you are going to need quite quickly is a 40-conductor ribbon cable to connect the IBEC to your I/O motherboard. This cable should not be over 10 ft long, and the shorter the better. Female socket connectors are required on both ends. You can purchase the connectors and cable separately to make your own assembly, but I recommend you purchase an assembled cable. The connectors are IDC (insulation displacement connector) types requiring crimping right through the insulation, and a special crimping tool provides the best results.

If you do assemble your own, be sure to use your VOM to "ring-out" the connections from one end of the cable to the other for each of the 40 wires. Digi-Key part numbers for 36-in assembled cables are: A3AAA–4036M (for the multicolored cable) and A3AAA–4036G (for the all gray cable). I personally like the multicolor rainbow effect, but it costs about a dollar more and either will work just fine. If you talk to the customer service department, Digi-Key will also assemble longer cables for a nominal extra charge.

IBEC Assembly

The basic skill for building the UCIS is PC-card soldering. If that's a new one for you, make doubly sure that you have thoroughly digested the information on PC card soldering in Chap. 1. You will need to know the tools and techniques of good soldering. Do all soldering on the back or bottom side of the card—the surface without the part markings.

If you are always going to use port-mapped I/O, you can skip installing U6, C6, RN2, and SW3, being sure to place the jumper in the J2 position. Otherwise, install jumper in the J1 position.

The order of parts assembly on the IBEC is not critical, but for the sake of having a plan, follow the steps in order and check off the boxes as you complete each one.

- ☐ *S13[+].* Install this 40-pin header with the connector opening off the right end of the board and with all 40 pins properly in their respective holes. With the header held firmly against the board, solder only one pin on each end.
- ☐ *BK1[+].* Install the mounting bracket with the header housing protruding through the square opening. Secure it using two 4-40 × ¼-in pan-head machine screws. The bracket flanges should be on the back (the solder side) of the board, and the screw heads should be on the component side.
- ☐ *Check the bracket and protective header mounting.* Plug the ribbon cable into the IBEC header so that the cable is perpendicular to the card and extending further away from the component side; by that I mean not going back toward the solder side. Make sure the header housing protrudes far enough outside the bracket so that the cable connector goes in to its full depth. You may also want to temporarily plug the card into your computer, with the computer turned off, to make sure everything fits perfectly. On some computers you may find it necessary to cut off the header's top cable-lock ejector hook.
- ☐ *Complete S13 soldering.* Once you have confirmed a perfect fit between the card and bracket, your ribbon cable, and the computer, and with S13 mounted firmly against the board with all 40 pins protruding through the holes, then solder the remaining pins.
- ☐ *R1-R6.* Make 90-degree bends in the leads of each resistor so that they fit nicely in the two holes. Insert and solder them while holding the part flat against the card, then trim the leads.
- ☐ *R7[+].* Install this variable resistor while holding it firmly against the board, and solder its three leads. The adjustment screw should face the top edge of the card. If it doesn't, you have an incorrect potentiometer.
- ☐ *S1-S12[+].* To avoid mixing them up, install 20-pin IC sockets S1-S6 first, followed by the 14-pin sockets and lastly the eight-pin socket S12. Make certain that pin-1 orientation is correct for each. Solder only a couple pins first on opposite corners of each socket. Reheat as necessary to make certain that the socket is firmly against the board, then solder the remaining pins.
- ☐ *RN1,RN2[+].* Install these SIP (single in-line package) resistor networks, making sure you have the lead common to each resistor (the common lead is typically marked with a white dot or small vertical line) in the hole marked with a plus sign. You can also use your VOM to locate the common lead. Solder the two ends first. Once you have checked that the part is firmly against the board, solder the remaining pins.
- ☐ *SW1-SW3[+].* Mount each DIP switch with the side-actuated levers facing the top card edge. Use your VOM to be absolutely certain that the contacts are closed when the switch levers are thrown down toward the board surface, and open when thrown up away from the board. Hold the switch securely against the card as you solder it.

- □ *C1-C13.* Insert with capacitor standing perpendicular to the card, solder, and trim. Make sure that C13, having a different value, stays separate from the others.
- □ *C14-C17[+].* Make sure that the positive leads go into the (+) holes as shown in Fig. 3–6. Incorrect polarity will damage these capacitors. Solder and trim the leads.
- □ *Q1[+].* Spread the leads of this transistor slightly to fit the three holes, making sure the center (base) lead goes into the hole closest to R2, and that the flat side of Q1 faces the direction shown on the card's printed legend and in Fig. 3–6. Push it in only far enough for it to fit snugly without stressing the leads. Solder and trim the leads.
- □ *L1 and L2[+].* Carefully bend the leads at right angles, and spread them just a bit, as shown in Fig. 3–7, so they fit into the board holes with the LED facing the top board edge. Make sure the + lead is in the + hole. Solder and trim the leads.

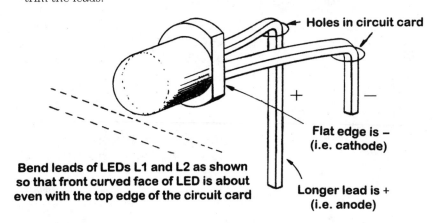

Holes in circuit card

+   **—**

**Flat edge is –
(i.e. cathode)**

**Bend leads of LEDs L1 and L2 as shown
so that front curved face of LED is about
even with the top edge of the circuit card**

**Longer lead is +
(i.e. anode)**

3-7 IBEC LED installation.

- □ *U1-U12[+].* See Fig. 1–7 for IC insertion and extraction procedures. Be sure you have the ICs specified, that you put them in the right sockets with the correct pin-1 orientation, and that all pins go into the socket.
- □ *Cleanup and inspection.* For a professional-looking job and to help ensure that your card functions properly, follow the specific steps covered in Chap. 1 regarding cleanup and inspection. This is a most vital step, so don't cut it short!

With your internal bus extender card complete, you've taken the first big step in building your own UCIS. Skip ahead to Chap. 5, where we'll configure the card's DIP switches and install it in your computer. In Chap. 8 we'll build up the cards for the I/O end of the UCIS, so in Chap. 9 you can put the whole system together.

4
CHAPTER

Universal Serial Interface Card

In Chaps. 1 and 2, I explained how the computer interface could be parallel or serial, and then in Chap. 3 I covered the parallel case with an IBEC. Later we'll look at an alternative parallel approach using a UBEC and CBAC. This chapter is devoted to show how to set up your UCIS as a serial interface, including step-by-step instructions for building the universal serial interface card (USIC).

For a serial UCIS, the USIC replaces the IBEC (or alternatively the UBEC and CBAC), but the rest of the hardware stays identical. Keeping the I/O motherboard and all the I/O cards the same is an important attribute of UCIS. It enables the user to switch between serial and parallel, and vice versa, with very little effort and cost; most of the interface remains unchanged.

Serial Interface Standards

The USIC can use any one of three interface standards: 20-mA current-loop, RS-232, or RS-422. In Fig. 4–1, note that data out from the computer are data in to the USIC, and vice versa. The primary characteristics of each method are presented below.

20-mA Current-Loop Standard

Figure 4–1a shows the hookup for the 20-mA current loop. It is one of the oldest forms of computer interfacing, and is based on teletype equipment. Many low-cost one-card computers use it for simplicity. Current switched on and off transmits binary 1s and 0s, and coding follows teletype practice where Mark (logic 1) equals a 20-mA current, and Space (logic 0) is no current.

Current-loop interfaces use low-impedance circuits and are highly immune to electrical noise, making them well suited to high-electrical-noise environments where remote hardware is switching high currents. The distance between computer

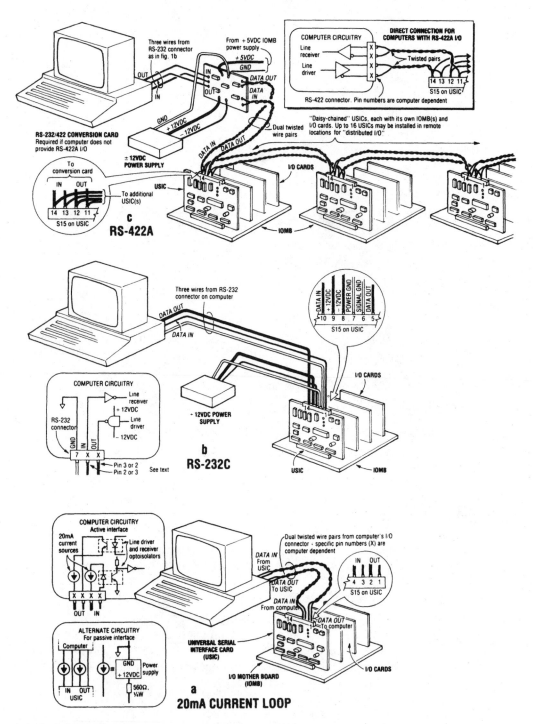

4-1 Serial interface systems.

and USIC may be 1000 ft or more; data transmission speeds, called baud rates, are typically low.

RS-232 Standard

The RS-232 standard, illustrated by Fig. 4–1b, is the most common serial interface. It uses voltage to define the signal with a Mark (logic 1) being –12 Vdc, and a Space (logic 0) being +12 Vdc. Typically, RS-232 ports on computers use either a standard 9- or 25-pin D-type male plug connector, called a DB9P or DB25P, respectively. Thus, to match them you need a female socket connector, a DB25S or DB9S. Pins 2 and 3 are the data wires, and are called the Transmit Data and Receive Data lines. Which is which depends on whether a device is configured as *data terminal equipment* (DTE) or *data communication equipment* (DCE). DTE transmits on pin 2 and receives on pin 3, but DCE transmits on pin 3 and receives on pin 2. Some computers have DTE ports, while others have DCE, so the first step in setting up any RS-232 interface is determining the direction of data flow on pins 2 and 3. The typical RS-232 connections used on most computer ports are:

Function	Connector type and pin number	
	DB25P	DB9P
Transmit data	2	3
Receive data	3	2
Signal ground	7	5

The transmission and reception are from the perspective of the computer. These three wires (pins 2, 3, and 7 or 5) suffice for bidirectional communication between two devices. The other pins, 22 in number for the DB25, are for handshaking between the computer and the peripheral device. For example, pin 6, Data Set Ready on the DB25, is one that a printer might use to tell the computer it is ready to accept the next character.

There is little standardization in how computers and peripherals use the handshaking lines. Along with the interchange of pin 2 and 3 functions, and the use in some terminals of a male DB25P connector and in others of a female DB25S, there are good reasons why the RS-232 is sometimes called "the most nonstandard standard." On the positive side, most computers provide RS-232 ports either as standard equipment or as an option, possibly from an aftermarket supplier.

As Fig. 4–1b shows, the RS-232 is limited to one receiver and one transmitter. That's fine for many applications, but if you want distributed I/O with multiple USICs, the RS-232 won't do the job. You'll need to go to RS-422, and later I'll show you how to build an RS-232/422 conversion card.

Length of cable also affects RS-232 transmission capability and limits the maximum baud rate. The specified maximum cable length with RS-232 is 50 ft without an audio modem (modulator/demodulator) at each end. In practice, however, RS-232 signals are routinely run for 100–200 ft and operated at high baud rates without problems. Such installations aren't standard, but they do work. For more reliable, error-free transmission over long distances, use RS-422.

RS-422 Standard

Few personal computers besides the Macintosh provide a direct RS-422 connection, but its advantages make it worth consideration. Compared to RS-232 it improves noise immunity, lengthens range, and drives multiple receivers from a single transmitter. These features arise from the RS-422's use of twisted-pair, balanced transmission lines (see Fig. 4–1c). The difference between the voltages on the two wires signals a Mark or a Space.

If the difference is positive more than 200 mV, the receiver reads a Mark; if negative more than 200 mV, the receiver reads a Space. A secondary advantage arising from these levels is that the transmitters and receivers can operate with the normal +5-Vdc logic power supply.

Figure 4–1c shows an RS-422 system with multiple USICs and two twisted-pair transmission lines, one pair for signals in each direction. The USICs are daisy-chained, with the same transmission lines running from the computer to the first USIC, on to the second, and so on.

Both RS-232 and RS-422 transmit signals differentially, but RS-232 suffers from using the difference between the signal-wire and ground-wire voltages. This is inferior because each end of the ground wire is locally grounded. Any difference in potential at the ends of the link causes current to flow through the ground wire, and the wire's resistance ensures a difference in ground potential at each end that can cause errors in the data.

RS-422 grounding requirements are much less critical, because local ground potential doesn't affect the Mark or Space signals. With the ground-potential problem reduced, the RS-422 can send its Mark and Space signals closer together, thus operating reliably at higher baud rates and over greater distances than RS-232.

Because typical computer interfacing distances don't push the 4000-ft limit of RS-422, the type of cable used isn't overly important. Regular four-wire telephone cable generally works just fine.

Baud Rate Settings

It's important to understand that all three serial interface standards allow different baud or data-transmission rates. This is significant, because a serial interface transmits data sequentially, one bit after another, whereas our parallel interface transmits one byte (eight bits) at a time.

In a serial interface, the receiving and transmitting devices *must* operate at the same baud rate. The USIC is designed to operate with standard baud rates of 150, 300, 600, 1200, 2400, 4800, 9600, and 19,200 bits per second. Remember that the faster the transmission, the greater the susceptibility to errors from electronic noise.

Universal Serial Interface Card

Figure 4–2 shows a universal serial interface card (USIC) plugged into the I/O motherboard. The motherboard is identical to that used with IBEC and UBEC, except that with USIC the 40-pin header S1 isn't required. The general-purpose I/O cards plug into the motherboard in identically the same way with IBEC, UBEC, or USIC.

The serial connections between computer and USIC are made with the 14-pin, right-angle header at the top of the USIC. Which pins you use depend on whether you're using 20-mA current loop, RS-232, or RS-422 for communication. With RS-232 you must also supply ±12-Vdc power through this same connector. I'll show you how to build such a supply in Chap. 11.

The heart of the USIC is part U1, an MC68701S microcomputer unit (MCU), making it a smart interface. Having the S at the end of the part number is not mandatory, but simply indicates that at the time of this writing, this is the newest available version from Motorola. The very capable MC68701S is a single 40-pin DIP containing a clock oscillator, a microprocessor unit (MPU), 2048 bytes of ultraviolet-erasable programmable read-only memory (EPROM), 128 bytes of random access memory (RAM), a serial communications interface (SCI), a programmable timer, and an abundance of I/O pins. It's a self-contained computer system on one chip, thus giving the USIC lots of power.

The main advantage of a smart serial interface with built-in MPU and memory is great flexibility. The same design works for 20-mA current loop, RS-232, and RS-422 I/O. For RS-232 the interface uses only the pin 2 and pin 3 data lines, along with the pin 5 or pin 7 ground, so it's independent of the sometimes unpredictable RS-232 handshaking lines.

The MC68701S performs all the bookkeeping as to which particular bit the computer is sending and whether it's part of address or data. When the computer is to read data from input cards on the motherboard, it simply tells the MC68701S, which gathers the data from the cards, sets it up in the prescribed format, and transmits it bit-by-bit back to the computer.

Programming the MC68701S EPROM requires special hardware, such as an EPROM programmer with an adapter for handling EPROMs embedded in microcomputer units and/or an extensive software development system. The MC68701S is programmed by first applying high-intensity ultraviolet light to a window on top of the chip, erasing it to all 0s (most other EPROMs erase to the opposite state, logic 1). Then the programmer is used to enter 1s in the desired bit locations. To avoid inadvertent erasure of a programmed chip, the window should be covered with an opaque material, such as a piece of black electrical tape, to protect it against accidental exposure to ultraviolet light.

Don't worry about having to do any EPROM programming yourself. I've arranged to have MC68701S chips programmed for the USIC and tested in a working card by

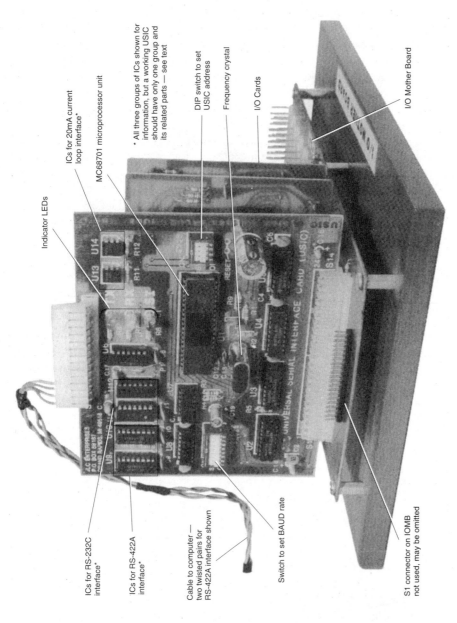

Indicator LEDs

ICs for 20mA current loop interface*

MC68701 microprocessor unit

* All three groups of ICs shown for information, but a working USIC should have only one group and its related parts — see text

DIP switch to set USIC address

Frequency crystal

I/O Cards

I/O Mother Board

ICs for RS-232C interface*

ICs for RS-422A interface*

Cable to computer — two twisted pairs for RS-422A interface shown

Switch to set BAUD rate

S1 connector on IOMB not used, may be omitted

4-2 Universal serial interface card (USIC).

The Chesapeake Computer Group Incorporated, 3903 17th St., Chesapeake Beach, MD 20732. The cost is $65 each, and there is a $3.50 shipping and handling charge per order. (Maryland residents must add appropriate sales tax.)

You don't need to understand electronics to build the USIC. Circuit cards and programmed MC68701S chips are readily available; all you have to do is follow my instructions. If you are interested in how the USIC works, and in debugging in case of trouble, I'll explain what the USIC does and give you a tour of its schematic. If you really just want to get started, skip ahead to USIC parts.

USIC Functions

Figure 4–3 is a functional look at what the USIC does. Its main hardware functions are unique I/O line selection, address decoding, baud rate generation, parallel-to-serial and serial-to-parallel conversion, and signal buffering. The MC68701S also performs software protocols for handling address, data, and control signals, which I'll cover in Chap. 10.

In Fig. 4–3 the arrows indicate the direction of signal flow, and double lines with a slash and number indicate multiple parallel wires—four or eight in our case. The connections to the I/O motherboard are just like with the IBEC and UBEC. Because the USIC is powered from the same supply as the I/O cards, the RRON sensing line isn't needed; so, it's open-circuited. The address high-order bits line, AH, is hard-wired on the USIC to +5 Vdc.

The I/O address lines A0–A7, the two read and write control lines R* and W*, and the eight data lines D0–D7 all pass through buffers before going to the motherboard. Because data must flow both ways, the data lines are double-buffered. One group of eight buffers, labeled IN in Fig. 4–3, handles data flow into the MC68701S, and the group labeled OUT handles data going to the external hardware. The R* and W* lines control which set of buffers is active.

The buffers provide a degree of isolation between the MC68701S and the I/O cards, as well as the necessary drive capability to handle up to 64 I/O cards. Without this buffering the external hardware electrical demands would so overload the MC68701S that it couldn't work properly.

One of three special I/O circuits couples the MC68701S to the serial I/O lines coming to USIC from the main computer. Which circuit is used depends on whether you are using 20-mA current loop, RS-232, or RS-422 I/O. The basic function is the same, however, to buffer and convert the serial signal levels to be compatible with the +5 Vdc and 0 V logic levels used by the MC68701S's transmit (TX) and receive (RX) lines.

To place multiple USICs on the same RS-422 four-wire cable, we must be able to set a unique USIC address for each card. This is handled by a four-segment DIP switch feeding directly into the MC68701S. The four-segment switch can set 16 unique addresses, allowing up to 16 USICs in an RS-422 system.

An eight-segment DIP switch sets the baud rate generator to the desired serial transmission frequency, which in turn is fed into the MC68701S. The MC68701S provides all the parallel-to-serial and serial-to-parallel conversions and determines the operational timings for both the serial and parallel lines. It also keeps track of which

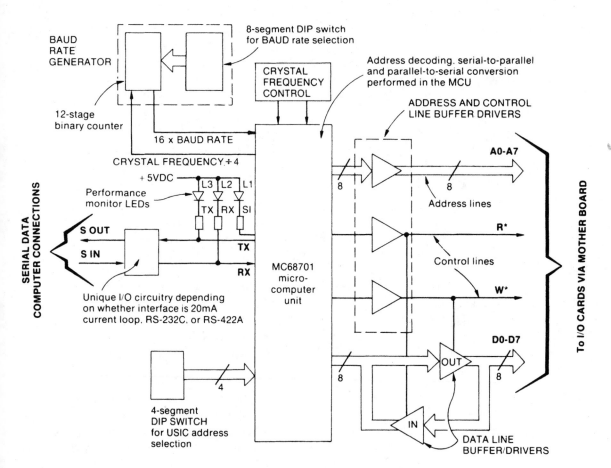

4-3 USIC functions.

information is data and which is address, and determines when to bring the R* and W* lines low for communication with the I/O cards. I'll explain the software for all this in Chap. 10.

USIC Schematic

The schematic in Fig. 4–4 is simply an expansion of the function diagram in Fig. 4–3. Everything focuses on U1, the MC68701S, so we'll talk about it first. Of its 40 pins, 29 are I/O lines, configured in three eight-bit ports and one five-bit port. Port 1, lines P10 through P17, connects to the eight data lines going to and from the I/O cards, while port 3, lines P30 through P37, connects to the eight address lines going to the I/O cards.

That's pretty straightforward. When U1 is to communicate with the I/O cards, it sets the desired address—which port of which I/O card—on port 3, and sends or receives the data through port 1.

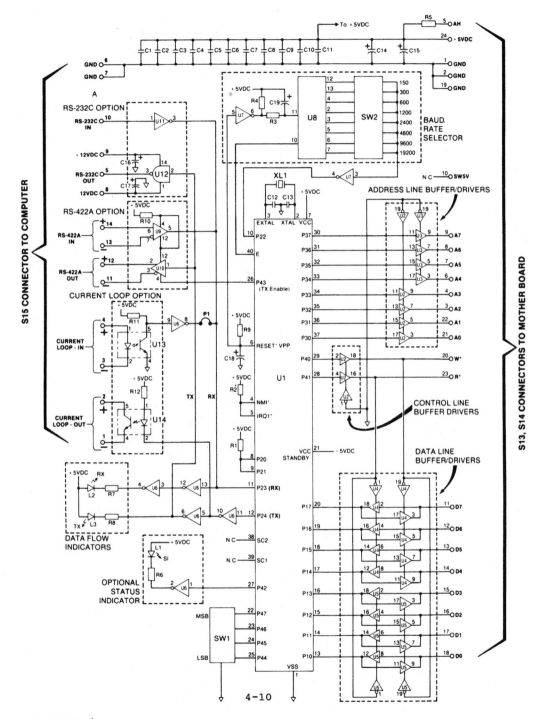

4-4 USIC schematic.

Port 4, the remaining eight-bit port, is configured with lines P40 through P43 as output and lines P44 through P47 as input. Line P40 is the write control line W*, P41 the read control line R*, and P42 drives a USIC operational status indicator LED that helps us see how the USIC is operating.

The four input lines, P44 through P47, are connected to DIP switch SW1, which is set for the USIC address. These input lines have pull-up resistors built into the MC68701S so each line is held high at +5 Vdc unless the corresponding segment of SW1 is closed, thus switching the line to ground.

Port 2 is configured as four input lines and one output. The state of lines P20 through P22 during power-up sequence, defined by the rising edge of RESET*/VPP pin 6 input, sets the mode of operation for the MC68701S. For the USIC all three are high to define mode seven, setting U1 for its most basic single-chip mode. Once the reset period is over, the pin 10 P22 input takes on a second function as the external clock input to the MC68701S for determining the baud rate of the serial transmission.

Port 2 bit 3, pin 11, is U1's serial data input receive line, while Port 2 bit 4, pin 12, is the serial data output transmit line. These connect U1 to the USIC's interface-unique circuitry for either 20-mA current loop, RS-232, or RS-422 I/O.

Before getting into that, let's look at baud-rate generation. The 2.4576-MHz crystal XL1 generates the clock frequency for U1's internal processing. U1 also divides the clock frequency by a factor of 4 and outputs the result on pin 40 as signal E to the pin 10 clock input of U8, a CD4040 12-stage binary ripple counter.

U1 requires that the frequency into its P22 be eight times the desired baud rate, which is taken care of in the U8 circuitry. Only one segment of SW2 should be ON at any time, and this setting must match the baud rate of the computer port connected to the USIC.

Each of the eight output lines used on U8 divides the clock input by a different factor of two, and passes through one of the eight segments of SW2. Turning on the segment for U8's pin 12 connects the input frequency to U1 pin 10, which provides a baud rate of 150. Likewise, turning on pin 13's segment gives 300 baud, and so on up to pin 7 for 19,200 baud.

The R9/C18 circuit ensures that the U1 RESET*/VPP line is held low during the power-up sequence. Likewise, the U7 inverter—along with R3, R4, and C19—holds the pin 11 reset line to U8 high for a period longer than U1's reset period. This initialization forces all the counter's outputs low, and these are inverted by U7 to set the P22 input high initially for mode 7 initialization of U1.

Returning to the interface-unique circuitry, install parts for only one of the three possible serial interface systems. For example, U11 is an RS-232 receiver IC that converts the incoming +/−12-Vdc signals to +5-Vdc and 0-V logic levels for U1. U12 is an RS-232 driver IC that converts the +5-Vdc and 0-V logic levels from U1 line TX to +/−12-Vdc outgoing signals. U12 needs an external +/−12-Vdc power supply, with + and − connected to pins 9 and 8, respectively, and the common ground to pin 7 of the right-angle header S15.

Receiver-driver pairs U9 and U10 are for RS-422. Resistor R9 keeps the U9 receiver chip always enabled, while the pin 4 enable line for the U10 driver chip is software controlled by the P43 line out of U1. That way, with multiple USICs the only one with an active transmitter driver on the line is the one from which input data are requested.

Optoisolators U13 and U14 are for the 20-mA current loop, and with them the inverters in U6 provide the proper signal polarity and prevent the optoisolator's circuitry from overloading U1. Because U6 is always installed, omit jumper P1 when using either the RS-232 or RS-422, to keep the pin 12 output of U6 from interfering with the U9 or U11 outputs.

The three LEDs, L1-L3, indicate the operational status of the USIC. L1 (green) blinks to show that U1's internal program is operating correctly, and also provides error codes when it is not operating correctly. L2 (amber) blinks when the USIC is receiving data, and L3 (red) blinks when USIC is sending data.

USIC Parts

Because for the serial case this is our first UCIS circuit card assembly, I'll go into more detail on it than I will for the rest of the UCIS cards. Ready-to-assemble USIC PC boards are available from JLC Enterprises, and fully assembled and tested boards from ECO Works, EASEE Interfaces, and AIA. See Appendix A for details, including USIC artwork for those wishing to fabricate their own card from scratch. The USIC is a double-sided PC card with circuit traces on each side interlaced via plated-through holes, and therefore would be very difficult to etch on your own.

Figure 4–5 shows the parts layout on the top side of USIC, and Table 4–1 is the parts list. You should refer to the parts layout as you select and mount parts. Keep the card oriented the same way as the drawing while you work to help you place each part in its proper holes with correct orientation. Take care to assure that you are installing the right components. Sometimes the markings are so small you need a magnifying glass to read them, but extra effort at this stage is worthwhile. Check resistors against the color codes included in the parts list.

Insert all components from the component, or A, side, and do all soldering on the back, or B, side. For a few parts the correct orientation on the card is very important. These are capacitors C14 through C19, three LEDs L1-L3, DIP switches SW1 and SW2, all IC sockets, and the ICs themselves. To remind you to check this as you build your USIC, I've marked the orientation-critical parts with a plus sign in brackets [+] in the instructions below.

Double check your selection and location of each component before soldering it in place. It is a lot easier to remove an incorrect component before you've soldered it to the card. This is especially true with plated-through-holes because, as you solder the part in place, the solder flows all the way through the hole.

The basic skill for building the USIC is PC card soldering. If that's a new one for you, make doubly sure that you have thoroughly digested the information on PC card soldering in Chap. 1.

Note that U1, the MC68701S, is static-sensitive. Before handling it you should ground your hands by touching them to a large metal object. This helps discharge any static charge on your body that could damage the chip.

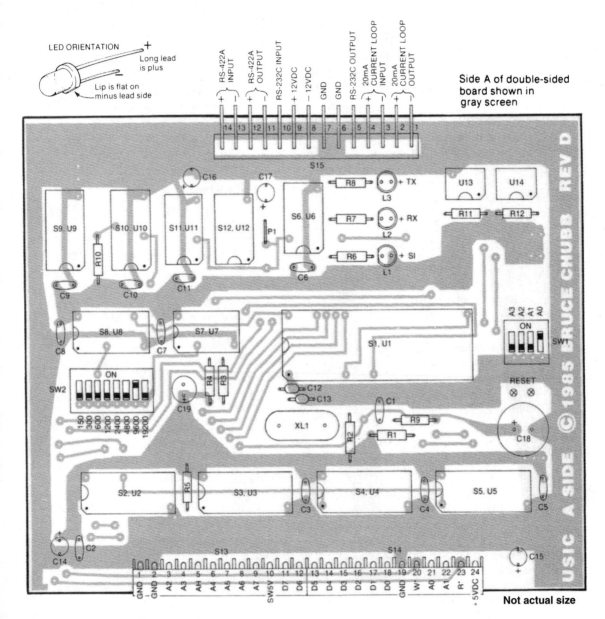

4-5 USIC parts layout.

USIC Assembly

The order of assembly is not critical, but for the sake of having a plan, follow the steps in order and check off the boxes as you complete each one. I'll list the common steps first, then the unique steps to configure your USIC for one of the three interface standards.

Table 4-1. USIC Parts List

Qnty.	Symbol	Description
2	R1,R2	4.7KΩ resistors [yellow-violet-yellow]
1	R3	470Ω resistor [yellow-violet-brown]
1	R4	2.2KΩ resistor [red-red-red]
1	R5	1.0KΩ resistor [brown-black-red]
3	R6-R8	330Ω resistors [orange-orange-brown]
1	R9	51Ω resistor [green-brown-black]
1	S1	40-pin DIP socket (Digi-Key A9340)
4	S2-S5	20-pin DIP sockets (Digi-Key A9320)
2	S6,S7	14-pin DIP sockets (Digi-Key A9314)
1	S8	16-pin DIP socket (Digi-Key A9316)
2	S13,S14	12-pin Waldom side entry connectors (Digi-Key WM3309)
1	S15	4- or 6-pin right angle header (see text to cut from Digi-Key WM4510)
1	SW1	4-segment DIP switch (Digi-Key CT2064)
1	SW2	8-segment DIP switch (Digi-Key CT2068)
8	C1-C8	100000pF, 50V ceramic disk capacitors (Digi-Key P4164)
2	C12,C13	22pF, 50V ceramic disk capacitors (Digi-Key P4016)
2	C14,C15	2.2µF, 35V tantalum capacitors (Digi-Key P2061)
1	C18	1000µF, 10V radial lead electrolytic capacitor (Digi-Key P6218 or equivalent with .197" lead spacing)
1	C19	100µF, 10V radial lead electrolytic capacitor (Digi-Key P6214 or equivalent with .098" lead spacing)
1	XL1	2.4576mHz crystal (Digi-Key X047 or equivalent with .486" lead spacing)
1	L1	Diffused green T1¾ size LED (Digi-Key P303)
1	L2	Diffused amber T1¾ size LED (Digi-Key P306)
1	L3	Diffused red T1¾ size LED (Digi-Key P300)
1	U1	MC68701S microcomputer unit with USIC programmed EPROM (The Chesapeake Computer Group Inc., see Appendix B)
4	U2-U5	74LS244 octal buffer/drivers, non-inverting (Jameco 47183)
2	U6,U7	74LS04 hex inverters (Jameco 46316)
1	U8	CD4040 12-state binary/ripple counter (Jameco 12950)

Used only for RS-232

2	S11,S12	14-pin DIP sockets (Digi-Key A9314)
1	C11	100000pF, 50V ceramic disk capacitor (Digi-Key P4164)
2	C16,C17	2.2µF, 35V tantalum capacitors (Digi-Key P2061)
1	U11	LM1489N quad line receiver, RS232 (Jameco 23181)
1	U12	LM1488N quad line driver, RS232 (Jameco 23157)

Used only for RS-422

1	R10	1.0KΩ resistor [brown-black-red]
2	S9,S10	16-pin DIP sockets (Digi-Key A9316)
2	C9,C10	100000pF ceramic disk capacitors (Digi-Key P4164)
1	U9	MC3486 quad differential line receiver, RS-422 (Jameco 25232) or may substitute LM3486, 26LS32 or SN75175
1	U10	MC3487 quad differential line driver, RS-422 (Jameco 25259) or may substitute LM3487, 26LS31 or SN75174

Used only for 20-mA current loop

1	P1	Program jumper, use no. 24 insulated wire
2	R11,R12	330Ω resistors [orange-orange-brown)
2	U13,U14	4N33 optoisolators (Digi-Key 4N33QT)

Mating connectors for USIC cable to computer

1	--	4- or 6-pin terminal housing (see text to cut from Digi-Key WM2110)
4 or 6	--	Crimp terminals (Digi-Key WM2301 for wire sizes 18-24 or WM2301 for wire sizes 22-26)

Author's recommendations for suppliers given in parentheses above with part numbers where applicable. Equivalent parts may be substituted. Resistors are ¼W, 5 percent and color codes are given in brackets.

☐ *Card test.* Use your VOM (volt-ohm meter; review Chap. 1 if you need assistance using a VOM) to test trace continuity on the circuit card. Make sure there are no open-circuit traces, such as a conductor with a hairline crack, and no shorts between adjacent traces or pads. You'll seldom find a problem in a bare card, but if you find one and correct it now you'll save time in later debugging. If you find an open-circuit trace, solder a small piece of wire across the crack. If you find a short circuit, use a knife to scrape away the offending material.

☐ *R1-R9.* Make 90-degree bends in the leads of each resistor so it is centered between its two holes and the leads just fit. Insert the part and solder it while holding the part flat against the card, then trim its leads flush with the tops of the solder tents on the back side of the card.

☐ *S1-S8[+].* To avoid mixing them up, install 40-pin IC socket S1 first, then 20-pin sockets S2-S5, then 16-pin socket S8, followed by 14-pin sockets S6 and S7. Make certain that pin 1 orientation is correct for each. As for any multipin part, solder only a couple pins first, those on opposite corners of each socket. Reheat as necessary to make certain that the socket is firmly against the board, then solder the remaining pins. If in doubt as to the correct orientation, see Fig. 1–8 for help.

☐ *S13, S14.* Install these 12-contact side-entry connectors by first hooking their nylon retaining fingers over the card edge, then feeding the metal contact pins through the card holes. Make sure all 12 pins of each connector pass through the holes. Hold the connector shell tightly against the card as you solder. We'll hold off installing S15, because which pins are required is a function of the selected interface standard.

☐ *SW1, SW2[+].* Mount the two DIP switches using your VOM to be sure they are oriented so their contacts are closed when the switches are thrown toward the ON label on the USIC. Hold the switches securely against the card as you solder.

☐ *C1-C8.* Insert this component with capacitor disk standing perpendicular to the card, solder, and trim leads.

☐ *C12, C13.* Insert this component with capacitor disk standing perpendicular to the card, solder, and trim leads.

☐ *C14, C15[+].* Make sure that the + leads go into the + holes as shown in Fig. 4–5. Incorrect polarity will damage these capacitors. Solder and trim leads.

☐ *C18, C19[+].* Make sure that the + leads go into the + holes as shown in Fig. 4–5. Incorrect polarity will damage these capacitors. Solder and trim leads.

☐ *XL1.* Hold the case firmly against the card while soldering the leads, then trim the leads.

☐ *L1-L3[+].* Install the LEDs with the + leads in the holes closest to U13. The insert drawing in Fig. 4–5 shows how to identify LED leads. Solder and trim the leads.

☐ *U2-U8[+].* See Fig. 1–7 for IC insertion and extraction procedures. Be sure you have the ICs specified, that you put them in the right sockets with correct pin 1 orientation, and that all pins go into the socket.

That completes the general USIC assembly. Next install only the parts for the one interface standard you've chosen: 20-mA current loop, RS-232, or RS-422.

20-mA Current Loop

☐ *P1.* This is the program jumper. Make it from no. 24 uninsulated bus wire (Belden no. 8022). Form the wire with 90-degree bends at both ends, so that the end leads just fit in the holes. Insert it so the jumper is straight and taut, then solder and trim the ends.

☐ *R11, R12.* Insert, solder, and trim leads.

☐ *U13, U14[+].* Insert these six-pin ICs directly into the PC board. Double check for correct pin 1 orientation, then hold each one firmly against the card while soldering each lead.

☐ *S15.* Cut off a four-prong section of the right-angle header (Digi-Key WM4510) and install in S15 holes 1 through 4. Make sure the header base is held tightly against the card while soldering each pin. Skip down to U1.

RS-232

☐ *S11, S12[+].* Install these 14-pin IC sockets making certain that pin 1 orientation is correct for each. Push each socket tight against the card and hold it that way as you solder.

☐ *C11.* Insert with capacitor disk standing perpendicular to the card; solder and trim leads.

☐ *C16, C17[+].* Make sure that the + leads go into the + holes as shown in Fig. 4–5. Incorrect polarity will damage these capacitors. Solder and trim the leads.

☐ *U11, U12[+].* Be sure you have the ICs specified, that you put them in the right sockets with correct pin 1 orientation, and that all pins go into the socket.

☐ *S15.* Cut off a six-prong section of the right-angle header (Digi-Key WM4510) and install it in S15 holes 5 through 10. Make sure the header base is held tightly against the card while soldering each pin. Skip down to U1.

RS-422

☐ *R10.* Insert, solder, and trim leads.

☐ *S9, S10[+].* Install these 16-pin IC sockets making certain that pin 1 orientation is correct for each. Push each socket tight against the card and hold it that way as you solder.

☐ *C9, C10.* Insert with capacitor disk standing perpendicular to the card, solder, and trim.

☐ *U9, U10[+].* Be sure you have the ICs specified, that you put them in the right sockets with correct pin 1 orientation, and that all pins go into the socket.

☐ *S15.* Cut off a four-prong section of the right-angle header (Digi-Key WM4510) and install them in S15 holes 11 through 14. Make sure the connector base is held tightly against the card while soldering each pin.

☐ *U1[+].* The MC68701S is a static-sensitive chip that must be handled with care. Avoid unnecessary handling, and keep it protected in its conductive wrapper until you are ready to insert it. Before touching it, ground your hands by touching a large metal object to help discharge any static charge on your body that could damage the chip. Make sure you have the correct pin 1 orientation and that all pins go into the socket.

☐ *Cleanup and inspection.* For a professional-looking job and to help ensure that your card functions properly, follow the steps in Chap. 1 regarding cleanup and inspection. This is vital, so don't cut it short!

We'll hold off testing the USIC until Chap. 11. If you're using the 20-mA current loop or RS-232 I/O, you can skip the rest of this chapter and go directly to Chap. 8.

RS-232/422 Conversion Card

If you want to use RS-422 I/O with a computer not equipped for it, assemble an RS-232/422 conversion card as shown in Fig. 4–6. Ready-to-assemble RS-232/422 cards are available from JLC Enterprises and fully assembled-and-tested cards are available from ECO Works, EASEE, and AIA. See Appendix A for details.

The parts layout is shown in Fig. 4–7, and the parts list is given in Table 4–2. The assembly steps are as follows:

☐ *Card inspection.*
☐ *Terminals.* Solder wires directly to terminal pads, or enlarge holes with no. 33 drill for 4-40 terminal screws with brass nuts for soldering to pads.

Table 4-2. RS-232/422 Conversion Card Parts List

Qnty.	Symbol	Description
11	-	4-40 x ¼" pan-head machine screws (Digi-Key H142)
11	-	4-40 brass nuts
2	R1, R2	330Ω resistors [orange-orange-brown]
4	R3-R6	2.2KΩ resistors [red-red-red]
2	S1,S2	16-pin DIP sockets (Digi-Key A9316)
3	S3-S5	14-pin DIP sockets (Digi-Key A9314)
4	C1-C4	100000pF, 50V ceramic disk capacitors (Digi-Key P4164)
3	C5-C7	2.2μF tantalum capacitors (Digi-Key P2061)
1	L1	Diffused amber LED (Digi-Key P306)
1	L2	Diffused red LED (Digi-Key P300)
1	U1	MC3487 quad differential line driver, RS422 (Jameco 25259) or may substitute LM3487, 26LS31 or SN75174
1	U2	MC3486 quad differential line receiver, RS-422 (Jameco 25232) or may substitute LM3486, 26LS32 or SN75175
1	U3	7407 hex buffer/driver, noninverting (Jameco 49120)
1	U4	LM1489N quad line receiver, RS-232 (Jameco 23181)
1	U5	LM1488N quad line driver, RS-232 (Jameco 23157)

Author's recommendations for suppliers given in parentheses above with part numbers where applicable. Equivalent parts may be substituted. Resistors are ¼W, 5 percent and color codes are given in brackets.

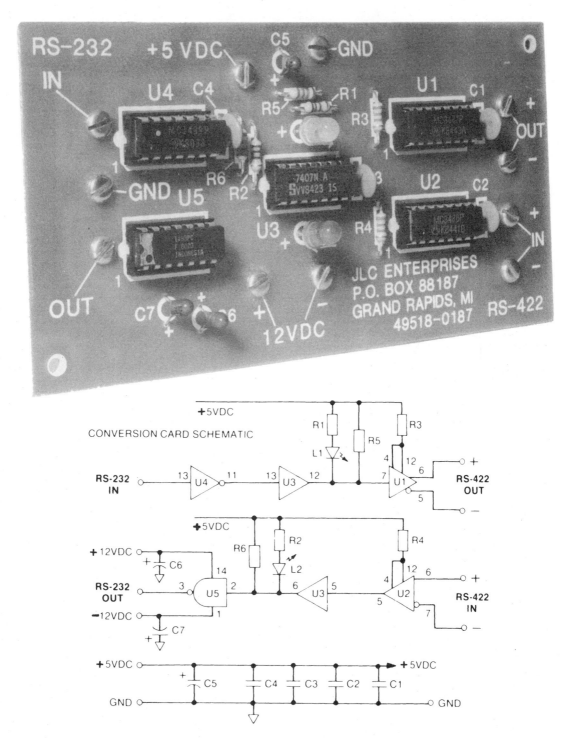

4-6 RS-232/422 conversion card and schematic.

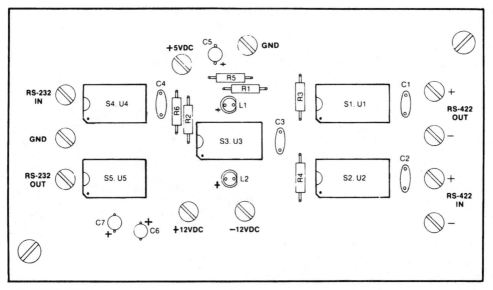

4-7 RS-232/422 conversion card parts layout.

**Table 4-3. RS-232/422
Conversion Card IC Power Tests**

✔	IC	+ METER LEAD ON PIN NO.	− METER LEAD ON PIN NO.	VOLTAGE READING
	U1	16	8	+ 5VDC
	U2	16	8	+ 5VDC
	U3	14	7	+ 5VDC
	U4	14	7	+ 5VDC
	U5	14	7	+ 12VDC
	U5	7	1	+ 12VDC

□ *R1-R6.*
□ *S1-S5[+].*
□ *C1-C4.*
□ *C5-C7[+].*
□ *L1, L2 [+].*
□ *U1-U5[+].*
□ *Cleanup and inspection.*

5

Configuring and Installing the IBEC

In this chapter we'll do the initialization work required to install an IBEC in an $x86$ class computer. The same procedures also work for the IBM PC, XT, and their abundance of compatibles, plus the IBM System/2 Model 30. Figure 5–1 shows a typical computer end of the UCIS, an IBEC plugged into a Pentium computer with a tower case. All the IBEC's DIP switches, the potentiometer, and the two test LEDs are placed along the top edge of the card, making it easy to set the I/O mode select and UCIS starting address DIP switches and to observe the test LEDs while keeping the card fully plugged into its ISA bus slot.

Being able to keep IBEC in its slot is an important feature. First, every time you insert or remove any adapter or expansion card, it is vital that you turn off your computer. Keeping power on can easily damage your computer as well as the IBEC. Thus, having all the controls and LED test indicators easily accessible is a big time saver. Second, every time you place your hands and tools inside the computer case you run some risk of goofing something up, and this is true no matter how great an expert you are on computers and electronics. Fundamentally, it is best to insert and remove the IBEC, and any other card plugged into the bus, as few times as possible.

Before plugging IBEC into your computer, it's best to perform some stand-alone card testing. This will help ensure interface success, plus possibly avoid damage to your computer once power is turned on with the inserted IBEC.

Stand-alone IBEC Testing

The IBEC is a rather complex card, performing many functions as presented in Chap. 3. On the computer side it makes 35 connections to the computer's bus, and on the UCIS side 21 connections leading to the external hardware via the I/O motherboard. To test it thoroughly on a stand-alone basis would require exercising every combination of input and making certain each output responded correctly. To do so

5-1 IBEC installed in a Pentium.

efficiently would require a special test fixture to generate the inputs and provide a method of reading and comparing the outputs to check that they are correct.

College students at WITC in New Richmond, Wisconsin built up a fixture for testing UBEC/CBAC combinations, and I expect they will do the same for IBEC. Using it they set switches to simulate each input and read LEDs to check each output. Each student, after building up a set of UCIS cards, is required to hook them up to the test fixture and run through a predefined set of input/output tests. This process helps ensure that the cards are working correctly before plugging them into the computer. They nicknamed the test box a "Chubb tester."

Such a test process is great, but because it's not very practical for you to build one for testing a single IBEC, I've taken special care in helping you minimize assembly errors. This is the reason for including the printed-on parts legend. Hopefully it should help ensure that parts are installed in their correct position with the correct orientation. Also, solder-masking is incorporated on both sides of the IBEC board. The application of this solder-resistant green film to all board areas except the pads where solder needs to be applied reduces the possibility of solder bridges.

Having the IBEC assembled correctly is essential to proper UCIS operation and helps ensure that no damage occurs to your computer. It may not be practical for you to build up a stand-alone Chubb tester, but we will be doing this sort of thing at the system level on a pretty much automated basis when we arrive at the point of running the interface wraparound test in Chap. 9. At this point our testing is limited to seeing that power and ground are correctly reaching each IC, and that the two test indicator LEDs function.

To perform this initial IBEC testing, you need access to a +5-Vdc regulated power supply. If you do not have one handy, I suggest you jump ahead to Chap. 8 and read up on how to build one. Once you have your supply, come back to this point for continuing with IBEC. To do so, follow these steps:

☐ *Component side checking.* With boards sent in for repair, it is amazing the number of errors I find by making a detailed visual inspection. You can do the same. Hopefully it has been a while since you completed the IBEC assembly. When you leave a card for a few days, or longer, and then go back and recheck it, you are more likely to find errors. It's almost like you are checking someone else's work and not your own. If you can establish the mind-set that you are checking someone else's work and receiving a big bonus for each error you find, it's surprising how many errors you will find. The bonus is real in that you will more quickly have a fully working system.

Always use a bright light *and* a magnifying glass when performing visual testing. Starting with the component side, and using the parts list and parts layout diagrams from Chap. 3, make sure that every part is correct and correctly installed. Does every IC have the correct part number, is it installed with the correct pin 1 orientation, and are all pins correctly inserted into the socket? Study every pin of every part before moving on to the next.

Does C13 have the correct part value? Do parts C14-C17 have the correct part value and correct polarity orientation? Are the two LEDs installed with the flat portion of their lip facing away from the + sign marking? Are the DIP switches all facing in the same and correct closed-circuit direction? Do each of the resistor networks have their pin 1 + markings oriented as in the parts layout diagram? Does each resistor have the correct value?

☐ *Solder side checking.* Now turn the card over and perform a detailed visual inspection of the card's solder side. Use your bright light and magnifying glass to carefully re-examine every solder joint. Is the total board free of all traces of solder flux? If you find some flux, repeat the cleaning process.

Once the board is perfectly clean, re-examine every solder joint. Look at each joint from different angles. Is there a perfect solder tent around each lead? Does the lead protrude out the top of each tent? Does the solder totally fill the hole all the way around the lead? Are there any signs of flux residue between the lead and the solder or between the pad and the solder? For any questionable joint, reheat it and, if needed, apply a bit more solder to ensure a perfect joint.

With the solder masking there shouldn't be any solder bridges, but make sure that any areas of the mask that might have become damaged during your assembly do not contain any solder bridges between pads. Be especially diligent in your examination of the S13 header pads, because this area contains 40 pins that are closely spaced.

☐ *Hooking up IBEC power.* The IBEC normally receives 5 Vdc from your computer; however, because we want to perform this test before plugging the card into the computer, you will need to temporarily apply power using a separate +5-Vdc regulated supply. Before putting it to use, re-check to make sure it is putting out +5 Vdc (acceptable range 4.9–5.1 Vdc) and that you know which terminal is the ground and which is the +5 Vdc. Then attach a clip lead between the ground terminal of your supply and the IBEC mounting bracket.

Temporally solder an insulated wire, like size 22 or thereabouts, to the (+) pad of C16. Connect the other end of this wire to the positive terminal of your 5-Vdc supply. Make sure your card is lying on an insulated surface, clear of any conductive items including bits of leftover solder, clipped lead remnants, and so forth.

☐ *IBEC IC power test.* Once everything is checked and your card is resting component-side-up, turn on your 5-Vdc supply. With the supply on, follow Table 5–1 by placing the positive meter lead on its indicated pin while the negative meter lead is placed on its pin. For each IC you should measure +5 Vdc. If you find a reading outside the acceptable range of 4.9–5.1 V, then you have a problem. For example, if everything reads near zero including the voltage across your supply, your card has a short. Look for an IC or a capacitor C14-C17 in backward, or a soldering problem. For any particular IC that doesn't pass the test, follow along the traces by touching your meter to exposed pads and feed-through-holes to locate where you have a poor connection, then correct it.

Table 5-1. IBEC IC Power Tests

✓	IC	+ METER LEAD ON PIN NO.	- METER LEAD ON PIN NO.
	U1	20	10
	U2	20	10
	U3	20	10
	U4	20	10
	U5	20	10
	U6	20	10
	U7	14	7
	U8	14	7
	U9	14	7
	U10	14	7
	U11	14	7
	U12	8	1

Each line should read +5Vdc

☐ *RRON sensor test.* Use another clip lead to connect the +5-Vdc side of your supply to the resistor lead from R2 that is next to R1. The L2 LED should light. If not, check for Q1 inserted wrong, L2 in backwards, or soldering problems. Correct as required so that the LED lights with the clip lead attached and goes out when the lead is removed.

☐ *LED 1 test.* Temporarily remove U10 and use a clip lead to temporarily ground pin 2 of its socket. This should light L1. If it doesn't, look for the LED in backward, a faulty LED, or a soldering problem. Replace U10 in its correct orientation.

☐ *IBEC DIP switch test.* Connect your (−) meter lead to ground and move the (+) lead to touch the IC socket pinouts listed in Table 5–2. At each position moving the DIP switch segment to Logic 1 you should measure +5V dc on the IC pin and with switch segment set to Logic 0 you should measure 0 V. If all readings are reversed, you have incorrect DIP switches installed. Replace them with those recommended in the parts list. If only some segments are incorrect, check around for poor soldering.

That completes the stand-alone testing of the IBEC. Disconnect your supply and set the IBEC aside for a bit, while we look into configuring the various DIP switches.

Table 5-2. IBEC DIP Switch Tests

✔	IC	PIN NO.	DIP SWITCH SEGMENT
	U5	3	A8
	U5	5	A9
	U5	7	A10
	U5	9	A11
	U5	12	A12
	U5	14	A13
	U5	16	A14
	U5	18	A15
	U6	3	A16
	U6	5	A17
	U6	7	A18
	U6	9	A19

With switch segment set to "Logic 1" IC pin should read +5Vdc.
With switch segment set to "Logic 0" IC pin should read 0Vdc.
Where:

Logic 1 position is with levers moved up away from circuit board
Logic 0 position is with levers moved down toward circuit board

Selecting the Desired I/O Mode

When a computer uses any of the memory control lines to control peripheral input and output devices, the peripherals are treated as extensions of the memory address space. This is memory-mapped I/O. The computer "thinks" it's communicating with memory, but we find a small range of addresses where memory isn't installed and we tuck in our interface. We then set up our software to talk to these addresses, just as if memory is present, and we are able to directly communicate with our interface. The result provides a very elegant and fast method of communicating with external hardware.

When a computer uses I/O read and write control lines to control peripheral input and output devices, the peripherals are treated as I/O ports. The same address bus is used so that all devices, whether considered memory or ports, decode the same addresses. With port I/O, only those devices set up to use the I/O read and write control lines respond. Because the same address bus is used to map input and output to I/O ports via address bus decode, we call the procedure port-mapped I/O. To use it, we find a small range of unused addresses in the I/O port space and tuck in our interface, and then set up our software to talk to these addresses.

Here are some pros and cons associated with each mode.

Memory-Mapped I/O

Because all machines have conventional memory (memory below 1 Mbyte), and most have a sufficient addressing hole available, this mode is the most popular with UCIS users. Also, with some computers memory bus cycles employ fewer built-in wait states (as compared to I/O bus cycles); therefore selecting this mode can result in a slightly faster system. On the negative side though, it takes a bit of effort to find an available address hole. Then, even when you do so and everything is working perfectly, at some point in the future when you install new software and/or hardware, you might need to reconfigure your IBEC address DIP switches and change your interface software's UCIS starting address.

Memory management software packages, such as EMM386.EXE, 386Max, and QEMM, can cause conflict problems with this I/O mode. These packages map expanded memory (memory located on adapter cards) and/or extended memory into the vacant memory holes between 640K and the 1-Mbyte boundary. This is great from the perspective that it gives more effective conventional memory space for application programs, but it can fill up the holes needed by UCIS. Therefore, if your computer setup uses a memory manager package, be sure to disable it or add the appropriate EXCLUDE statements in your DOS CONFIG.SYS file. The latter will manually exclude the specific address locations you set aside for UCIS from being mapped into added RAM. Alternatively, you can go ahead and convert all the open spaces to RAM, and simply use the port-mapped I/O mode for UCIS.

In BASIC programming, you use the PEEK and POKE instructions with memory-mapped I/O. Equivalent instructions exist in Pascal and C.

Port-Mapped I/O

With the original PCs and XTs, this mode isn't very attractive because of limited port address availability. This situation changes with newer computers starting with the AT/286, where the I/O port address space is expanded to 64K. Assuming you have one of the newer computers, it becomes relatively easy to find an open hole for UCIS. However, such holes are sometimes smaller than the 256 bytes we would like.

While holes in the case of memory-mapped I/O are typically many, many kilobytes in size, those in port-mapped I/O tend to be more frequent but much smaller. Their effective size is further reduced because we want our UCIS starting address, when expressed in hex, to have the two least significant digits be zeros. This enables the DIP switch settings on all our I/O cards to be independent of the UCIS starting address. This makes the initial system setup easier, and if you ever have to change starting address it makes it a *whole lot* easier.

The net result: on some computers the largest available port-mapped I/O hole can be less than the desired 256 bytes. For example, on my 286 the largest port-mapped hole I can find is 241 bytes. Even so, this can handle 60 I/O cards and that should be adequate for most systems. In contrast, my Pentium has 64 holes available with sizes of over 800 bytes each.

Another factor with port-mapped I/O is that port addresses adjacent to the hole tend to contain very critical data, so if you accidentally overlap into them you may very well "lock up" your computer, requiring a reboot. Also, some computers build-in added wait states to I/O bus cycles; therefore communication can be a bit slower using port-mapped I/O. The one wait state difference that sometimes exists adds maybe 100 ns to each UCIS transaction. You'll probably never see the difference; therefore port-mapped I/O should be plenty fast for most users.

In BASIC programming, you use the IN (or INP) and OUT instructions with port-mapped I/O. Equivalent instructions exist in Pascal and C.

Determining UCIS Starting Address

No matter which mode you select, it is essential to find an appropriate hole in either the computer's memory if using memory-mapped I/O, or in the I/O port address space if using port-mapped I/O. By a hole, we mean a block of addresses where *nothing* else is installed to decode this block of addresses for your selected I/O mode. The first address in the hole you use for UCIS is your UCIS starting address.

The minimum acceptable size of the hole is four times the number of I/O cards you plan to use, to a maximum of 256 bytes. This is the maximum because address lines A0–A7 (eight bits) are decoded by the particular I/O card being addressed. With eight bits you require 2^8 or 256 bytes. With computers using an IBEC you'll ordinarily have no problem locating a hole of 256 contiguous bytes.

Figure 5–2 shows a graphical representation of the memory and I/O address space. Even though this map is based on a Pentium, it still reflects the strong influence of the original IBM PC, using 20 address lines with the corresponding 1-Mbyte address space. Newer computers and associated operating systems desiring compatibility with older models still pay attention to this same 1-Mbyte boundary. In fact, such computers operate quite differently when addressing conventional memory below and extended memory above this 1-Mbyte boundary. The conventional memory space is further subdivided into two regions, typically called upper memory blocks when above 640K and lower memory blocks when below 640K.

Early PCs with 16-bit internal registers could most easily only address a block of 2^{16} or 65,536 memory locations (64K). To maintain software compatibility, this limit also is a factor in how newer hardware and software look at memory addressing. For example, it takes 16 of the 64K blocks to fill the 1-Mbyte conventional memory space; $16 \times 64K$ equals 1024K, which is 1 Mbyte. It's handy for discussion to use hexadecimal notation when referring to these 16 memory blocks. The lower ten-block group from hex 0 through 9, which is referred to as lower memory, is filled with 640K of RAM. The remaining six blocks, hex A through F, are reserved for other functions that vary based upon particular computer configuration requirements.

The lower memory region is used by application programs and their associated data and device drivers. It also typically contains portions of DOS, interrupt vectors, and the like. The hex address range is 00000h through 9FFFFh, and because this region is full of RAM it is off-limits for UCIS interfacing.

The first two upper memory blocks, A and B, are reserved for use by video cards to handle graphics and text data. Depending on the type of monitor and the format your video card uses (EGA, VGA, SVGA, and so on), different portions of this space are used. The A block address range is A0000h–AFFFFh. The B block range is B0000h–BFFFFh. Within the B block, monochrome monitors typically start at B0000h and color monitors at B8000h.

The next two blocks, C and D, are reserved for use by installable ROM that might be on adapter cards plugged into the ISA bus. Frequently these device ROMs are in small (say 2K or so) sizes, leaving plenty of free space. The C block address range is C0000h–CFFFFh and the D block is D0000h–DFFFFh. Any unused group of 256 contiguous bytes in A, B, C, or D blocks is usable by the UCIS.

The last two blocks, called E and F, are reserved for ROM on the system board. The F block, addresses F0000h–FFFFFh, is fully occupied by two 32K ROMs containing the BIOS (*basic input/output system*). Every time you turn on your computer, the BIOS firmware program boots-up the system to get everything working properly, including performing a POST (*power on self test*) to check for proper operation of all major components. It also includes the core code used to facilitate the transfer of data between peripherals. Because this area is totally occupied by ROM, the F block is off-limits.

On older computers using an IBEC, like the IBM AT and its clones, the E block (addresses E0000h–EFFFFh) is reserved for optional system ROM and typically left blank. However, newer computers are built with system ROM totally occupying the E and F blocks. Even if your particular machine may have the E block open, it's better to select a hole in one of the other blocks, because future updates will likely fill the E block.

Many computer manufacturers provide detailed tabular versions of the memory and I/O port maps as part of their technical reference manuals. Sometimes these are

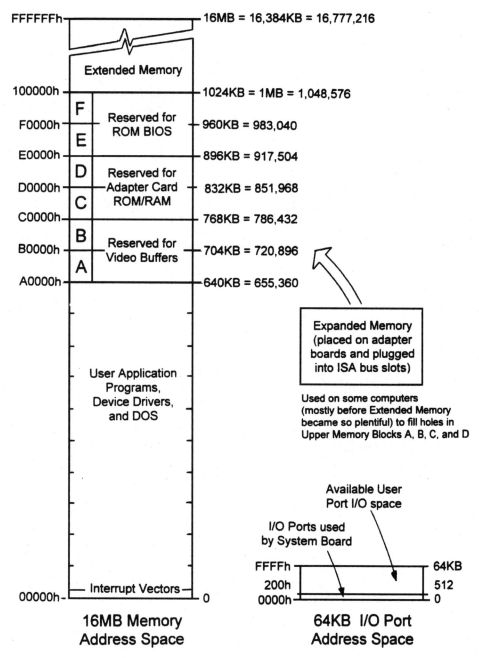

5-2 Memory and port I/O maps.

obtainable from your computer dealer, through a computer service center, or from the customer technical support group provided by your computer's manufacturer. However, with the tight competitive pricing that exists in today's market, many manufacturers cut back technical documentation provided with their computers. If you can obtain a set of maps specific to your particular computer, they can provide a great starting point for selecting your UCIS starting address.

Table 5–3 shows an I/O port address map for an IBM 286 covering the range up to decimal address 1024, or 400h. Newer machines, with the strong desire for backward compatibility, look much the same in this address range. Port addresses you might recognize are 3F8h (decimal 1016 for the serial COM1 port), and 2F8h (decimal 760 for the serial COM2 port). These are used with the serial version of UCIS.

Table 5-3. AT-286 Port Map

Hex Range	Decimal Range	Device
000-01F	0000-0031	DMA controller 1, 8237A-5
020-03F	0032-0063	Interrupt controller 1, 8259A, Master
040-05F	0064-0095	Timer, 8254.2
060-06F	0096-0111	8042 (Keyboard)
070-07F	0112-0127	Real-time clock, NMI(non-maskable interrupt) mask
080-09F	0128-0159	DMA page register, 74LS612
0A0-0BF	0160-0191	Interrupt controller 2, 8259A
0C0-0DF	0192-0223	DMA controller 2, 8237A-5
0F0	0240	Clear Math Coprocessor Busy
0F1	0241	Reset Math Coprocessor
0F8-0FF	0248-0255	Math Coprocessor
1F0-1F8	0496-0504	Fixed Disk
200-207	0512-0519	Game I/O
278-27F	0632-0639	Parallel printer port 2
2F8-2FF	0640-0767	Serial port 2
300-31F	0768-0799	Prototype card
360-36F	0864-0868	Reserved
378-37F	0888-0895	Parallel printer port 1
380-38F	0896-0911	SDLC, bisynchronous 2
3A0-3AF	0928-0943	Bisynchronous 1
3B0-3BF	0944-0959	Monochrome Display and Printer Adapter
3C0-3CF	0960-0975	Reserved
3D0-3DF	0976-0991	Color/Graphics Monitor Adapter
3F0-3F7	1008-1015	Diskette controller
3F8-3FF	1016-1023	Serial port 1

Most new operating systems and supporting utility packages provide the capability to display your I/O port and memory maps. For example, with DOS you can obtain a top-level memory map by simply entering MEM after the C:> prompt. With Windows 95 you can obtain more detailed maps by following this procedure:

☐ Click on the Start button.
☐ Click on the Settings control panel.
☐ Double click on System.
☐ Click on Device Manager.

☐ Double click on Computer.
☐ Under the View Resources tab, select either "Input/Output (I/O)" to see the I/O Port Map or "Memory" to see the Memory Map.

Both displays provide a list of hex address range settings and a description of the device using each address range.

Although convenient to acquire, I find the resulting maps lack detail and include numerous ambiguous messages such as "address used by unknown device" or "address range not usable." And, when I try using portions of these address ranges with the UCIS, they often work just fine. Likewise, they tend to define the address range that might be reserved for a given device, rather than the address range that is actually used by the device. This can be a big difference. For example, on my Pentium the memory map lists all the A and B block and about half of the C block as occupied by my Matrox MGA Millenium video card. However, when you dig into it deeper (as we'll do shortly), you'll find that only 40 percent of the space is actually used, leaving 60 percent free for UCIS.

For Windows 95 users, it is very helpful to exercise this easy access to Memory and Port I/O maps. For example, even if we took its defined space allocation requirements for the video card literally, it still shows we have half the C block and the whole D block open. This gives us a 96K hole, and that's 98,304 bytes (96 × 1024) available for UCIS when we need only 256 bytes!

An entirely different approach to find our hole is to directly read the contents of address locations to see what's there. For example, CORELOOK, a public domain program, lets you type in an address and it displays the contents of the address in hexadecimal. DOS's DEBUG utility can do the same thing. A work of caution, though, with DEBUG: With it you have access to the innermost system functions. A carelessly typed command can cause havoc with your system. I suggest you read about DEBUG, either in your computer's DOS manual or in one of the many books on DOS, before you try your hand at using it.

Figure 5–3 shows an example screen display using DEBUG to explore what's in the first parts of the A block and the E block on my Pentium. Every memory location in the interrogated portion of the A block contains FFh, meaning that each of the memory location's eight bits is a logic 1. Because the ISA bus data lines float high to logic 1 when no device is driving it, its quite likely that you locate a hole whenever you find a large block of contiguous addresses displaying FFh. Such groupings, if at least 256 bytes in length, are good candidates for using with UCIS.

In stark contrast, the memory locations in the interrogated portion of the E block contain specific data other than FFh, and hence most likely contains RAM or ROM. It therefore is unacceptable for use with UCIS. In this case the printout simply confirms what we said earlier, that with newer computers the E Block is occupied by ROM BIOS. Performing the same printout on my Compaq 286 shows the E Block all FFh, indicating that its "Reserved for Optional System ROM" space is still vacant. Pretty neat finding, isn't it?

Rather than poke around with DEBUG, you can use the BASIC program shown in Fig. 5–4 to help find a memory hole. I wrote this one in QuickBASIC. Although statement numbers are not required, I included them as an aid in describing the program.

```
C:\WINDOWS>DEBUG
-D A000:0000
A000:0000  FF FF FF FF FF FF FF FF-FF FF FF FF FF FF FF FF
A000:0010  FF FF FF FF FF FF FF FF-FF FF FF FF FF FF FF FF
A000:0020  FF FF FF FF FF FF FF FF-FF FF FF FF FF FF FF FF
A000:0030  FF FF FF FF FF FF FF FF-FF FF FF FF FF FF FF FF
A000:0040  FF FF FF FF FF FF FF FF-FF FF FF FF FF FF FF FF
A000:0050  FF FF FF FF FF FF FF FF-FF FF FF FF FF FF FF FF
A000:0060  FF FF FF FF FF FF FF FF-FF FF FF FF FF FF FF FF
A000:0070  FF FF FF FF FF FF FF FF-FF FF FF FF FF FF FF FF

-D E000:0000
E000:0000  2C 48 8E E1 F6 91 26 EE-B3 5F 73 33 37 37 77 74
E000:0010  37 75 51 15 55 D0 54 75-8A B3 62 84 8C 14 11 14
E000:0020  08 29 16 3A 1B 10 00 88-05 24 C2 56 DB 6C BB B9
E000:0030  8B 27 23 2C C0 D8 C8 1B-8C 6D FB 96 DB 2D 96 4B
E000:0040  56 17 92 CE 32 DB F1 58-46 02 48 06 B8 20 B1 A5
E000:0050  8A B2 58 48 C9 22 89 19-EB 31 8D 22 C4 37 55 4D
E000:0060  FF EE 6E 81 19 6F DC BF-73 CE F7 CE 79 DE 79 CF
E000:0070  3B CE FE 1E FE CF BB BB-59 1B 92 36 CF A9 6F 66
```

5-3 DEBUG output of memory blocks A and F.

Line 10 declares a user function to convert a hex input character to binary. Lines 20 through 40 clear the screen and print column headers. Line 50 starts a loop going from memory location 640K to 892K, incrementing in 4K steps. Line 60 calculates the decimal address being checked, that is the K address times 1024. Line 70 calculates the corresponding hexadecimal address using BASIC's built-in function HEX$.

Line 80 separates out the left two hex characters because these are the only two that affect our DIP switch settings; all others are zero. Lines 90 and 100 separate the two remaining into HEXHI$ (used to set SW3) and HEXLO$ (used to set SW2). Lines 110 and 120 convert the hex characters to the required binary switch segment settings.

Line 130 calculates the segment value corresponding to the decimal address being checked by performing an integer divide by 16 and then line 140 places this segment address into the required DEF SEG format. Line 150 then reads the contents of the memory location using PEEK, and stores the result in variable location MEM. Because in this case DEF SEG completely defines the DECADD value (DESEG × 16 equals DECADD), the offset entry in the PEEK instruction is zero.

Line 160 checks to see if the memory location contains a FFh (decimal 255), which is eight binary 1s. If it does, then we execute the FOR loop defined by lines 170 through 210 to determine how many adjacent higher-addresses also contain FFh, that being the size of the potential hole. If a non-FFh value is found before we get to the end of the loop, we jump out to EHOLE with the hole size defined as variable *HOLESZ*.

If line 160 finds the initial MEM value to be other than FFh, it executes line 230, in which hole size is set to zero. At line 250 hole size is properly defined, and line 260 defines the format for our printout. Lines 270 and 280 perform the printout on the screen and printer. Simply REM out the LPRINT lines if you don't desire the printer

```
10    DECLARE FUNCTION BINARY$ (HEXIN$)
20    CLS              'Clear screen and print out column headers
30        PRINT "   KB DECADD MEM HEXADD  A19-A16  A15-A12 SIZEHOLE DEF SEG"
40        LPRINT "   KB DECADD MEM HEXADD  A19-A16  A15-A12 SIZEHOLE DEF SEG"
50    FOR KB = 640 TO 892 STEP 4        'Step thru A-D blocks at 4KB INCREMENTS
60        DECADD = KB * 1024            'Calculate decimal address from KBytes
70        HEXADD$ = HEX$(DECADD)        'Calculate hexadecimal address
80        HEXSW$ = LEFT$(HEXADD$, 2)  'Strip off hex digits set by DIP switches
90        HEXHI$ = LEFT$(HEXSW$, 1)   'Separate off nibble to be set by SW3
100       HEXLO$ = RIGHT$(HEXSW$, 1)  'Separate off nibble to be set by SW2
110       BHI$ = BINARY$(HEXHI$)      'Convert to binary for setting switches
120       BLO$ = BINARY$(HEXLO$)      'Convert to binary for setting switches
130       DESEG = DECADD \ 16          'Calculate segment offset to desired address
140       DEF SEG = DESEG              'Set up DEF SEG statement format
150       MEM = PEEK(0)                'Read in memory value at DECADD address
160       IF MEM = 255 THEN            'If value is 255 then execute a loop to
170           FOR HOLECK = 1 TO 4095   '....see how far the hole extends
180               MEMCK = PEEK(HOLECK)
190               IF MEMCK <> 255 THEN HOLESZ = HOLECK + 1: GOTO EHOLE
200               HOLESZ = HOLECK + 1
210           NEXT HOLECK       '...Still reading 255 so increment to next
220       ELSE                          'address so can check it
230           HOLESZ = 0       'MEM(at 4KB increment)<>255 so...
240       END IF                        '...set hole size = 0
250    EHOLE:                    'Printout data for the 4KB increment
260       FORMAT$ = "   ### ###### ### \    \   \    \   \    \   ####    #####"
270    PRINT USING FORMAT$; KB; DECADD; MEM; HEXADD$; BHI$; BLO$; HOLESZ; DESEG
280    LPRINT USING FORMAT$; KB; DECADD; MEM; HEXADD$; BHI$; BLO$; HOLESZ; DESEG
290       FOR IDELAY = 1 TO 50000: NEXT IDELAY    'Slow down to read screen
300    NEXT KB          'Increment to next 4KB step


FUNCTION BINARY$ (HEXIN$)
IF HEXIN$ = "0" THEN BINARY$ = "0000"
IF HEXIN$ = "1" THEN BINARY$ = "0001"
IF HEXIN$ = "2" THEN BINARY$ = "0010"
IF HEXIN$ = "3" THEN BINARY$ = "0011"
IF HEXIN$ = "4" THEN BINARY$ = "0100"
IF HEXIN$ = "5" THEN BINARY$ = "0101"
IF HEXIN$ = "6" THEN BINARY$ = "0110"
IF HEXIN$ = "7" THEN BINARY$ = "0111"
IF HEXIN$ = "8" THEN BINARY$ = "1000"
IF HEXIN$ = "9" THEN BINARY$ = "1001"
IF HEXIN$ = "A" THEN BINARY$ = "1010"
IF HEXIN$ = "B" THEN BINARY$ = "1011"
IF HEXIN$ = "C" THEN BINARY$ = "1100"
IF HEXIN$ = "D" THEN BINARY$ = "1101"
IF HEXIN$ = "E" THEN BINARY$ = "1110"
IF HEXIN$ = "F" THEN BINARY$ = "1111"
END FUNCTION
```

5-4 Memory hole-finding program listing (MEMFIND).

output. Line 290 inserts a time delay to read the screen, and line 300 loops back to examine the next 4K memory block. You might want to adjust the delay size based on the speed of your computer.

The function BINARY$ takes any hex input character and converts it into the equivalent four-bit binary pattern. Both these program modules are included on the enclosed disk. Run them to check out the configuration of your A through D memory blocks.

Figure 5–5 shows sample output for my Pentium operating under DOS. It shows every location to be FFh except for addresses B8000h through C7FFFh, the actual addresses used by my Matrox video card. The printout from my 286 is identical, except that the video card's usage ends at C3FFFh. Repeating the run on my Pentium when operating under the DOS prompt under Windows 95, rather than executing a Windows 95 shut down and reboot under DOS, shows the C7FFFh–DFFFFh occupied, meaning that memory management has mapped extended memory into this area. This still left the A0000h–B8000h range open for UCIS, or you could include an EXCLUDE statement and gain back part of the C or D block for the UCIS.

Typically, this output provides all you need to successfully pick and set up your UCIS starting address. For example, using my Pentium, I picked the first opening (which was at 640K), set the DIP switch segments A19 and A17 to ON and the rest to OFF, and used the corresponding DEF SEG = 40960 to define my software, all entries found in the first.row of Fig. 5–5. If your configuration shows this 4K region occupied, then simply pick another row that shows a hole size of 256 bytes or larger. If for some reason a hole you try in one area doesn't work, try picking another in an entirely different address block.

Figure 5–6 lists a similar program that explores the I/O port-mapped address space from 1024 up to 65280. Here I use a step size of 256 bytes—and for a good reason. This keeps the address lines A0–A7 at zero for each starting address being explored. This makes it very handy to set the incremental I/O card address, and also makes them independent of any future changes in your UCIS starting address.

The two programs are very similar, so I'll only cover significant differences. Because the decimal address range 1024–65280 corresponds to 400h–FF00h, the operations to separate out the hex characters needed for switch settings are a bit more complicated. Lines 70 through 140 take care of this as a function of whether the hex address length is three or four characters.

Also, because this program uses port-mapped I/O, the INP instruction is used in place of the PEEK. Note that some versions of BASIC use IN rather than INP, so if yours does you'll have to change lines 170 and 200. With INP and OUT instructions the address range is limited to 65535, so the DEF SEG isn't used and the decimal address is placed directly in the INP statement.

Figure 5–7 shows example output for the first portions of each printout for my Pentium and 286. The Pentium shows a candidate hole at address 1024, but the FFh continues for only 113 bytes. This would support an interface size up to 28 cards. However, using address 1280, there are FFh values for the next consecutive 768 bytes (3 × 256). Therefore, picking 1280 as the candidate to try first is a better choice. Also, the I/O port map provided by Windows 95 listed nothing at these address ranges except for the ambiguous statement "occupied by unknown device." Well, we just made one of those unknown devices on my Pentium be the UCIS, and everything works just fine.

The Compaq 286 printout shows the 1024 address to be occupied, so it is off-limits to use as the UCIS starting address. However, the 1280, the 2304, and others provide 241 consecutive FFh readings, so are good candidates. Both these programs are included on the 3.5-in disk, as are all the other programs presented in this second edition. Many are included in different languages. Be sure to read the README file for a description of the disk's content.

With the above information and programs you should be all set to pick an I/O mode and to set your UCIS starting address. Even if you aren't 100 percent confident you have everything optimized, simply pick one of the possibilities and try it for awhile. It's easy to change later as you gain more system insight.

KB	DECADD	MEM	HEXADD	A19–A16	A15–A12	SIZEHOLE	DEF SEG
640	655360	255	A0000	1010	0000	4096	40960
644	659456	255	A1000	1010	0001	4096	41216
648	663552	255	A2000	1010	0010	4096	41472
652	667648	255	A3000	1010	0011	4096	41728
656	671744	255	A4000	1010	0100	4096	41984
660	675840	255	A5000	1010	0101	4096	42240
664	679936	255	A6000	1010	0110	4096	42496
668	684032	255	A7000	1010	0111	4096	42752
672	688128	255	A8000	1010	1000	4096	43008
676	692224	255	A9000	1010	1001	4096	43264
680	696320	255	AA000	1010	1010	4096	43520
684	700416	255	AB000	1010	1011	4096	43776
688	704512	255	AC000	1010	1100	4096	44032
692	708608	255	AD000	1010	1101	4096	44288
696	712704	255	AE000	1010	1110	4096	44544
700	716800	255	AF000	1010	1111	4096	44800
704	720896	255	B0000	1011	0000	4096	45056
708	724992	255	B1000	1011	0001	4096	45312
712	729088	255	B2000	1011	0010	4096	45568
716	733184	255	B3000	1011	0011	4096	45824
720	737280	255	B4000	1011	0100	4096	46080
724	741376	255	B5000	1011	0101	4096	46336
728	745472	255	B6000	1011	0110	4096	46592
732	749568	255	B7000	1011	0111	4096	46848
736	753664	32	B8000	1011	1000	0	47104
740	757760	32	B9000	1011	1001	0	47360
744	761856	32	BA000	1011	1010	0	47616
748	765952	32	BB000	1011	1011	0	47872
752	770048	32	BC000	1011	1100	0	48128
756	774144	32	BD000	1011	1101	0	48384
760	778240	32	BE000	1011	1110	0	48640
764	782336	32	BF000	1011	1111	0	48896
768	786432	85	C0000	1100	0000	0	49152
772	790528	191	C1000	1100	0001	0	49408
776	794624	206	C2000	1100	0010	0	49664
780	798720	141	C3000	1100	0011	0	49920
784	802816	24	C4000	1100	0100	0	50176
788	806912	0	C5000	1100	0101	0	50432
792	811008	0	C6000	1100	0110	0	50688
796	815104	57	C7000	1100	0111	0	50944
800	819200	255	C8000	1100	1000	4096	51200
804	823296	255	C9000	1100	1001	4096	51456
808	827392	255	CA000	1100	1010	4096	51712
812	831488	255	CB000	1100	1011	4096	51968
816	835584	255	CC000	1100	1100	4096	52224
820	839680	255	CD000	1100	1101	4096	52480
824	843776	255	CE000	1100	1110	4096	52736
828	847872	255	CF000	1100	1111	4096	52992
832	851968	255	D0000	1101	0000	4096	53248
836	856064	255	D1000	1101	0001	4096	53504
840	860160	255	D2000	1101	0010	4096	53760
844	864256	255	D3000	1101	0011	4096	54016
848	868352	255	D4000	1101	0100	4096	54272
852	872448	255	D5000	1101	0101	4096	54528
856	876544	255	D6000	1101	0110	4096	54784
860	880640	255	D7000	1101	0111	4096	55040
864	884736	255	D8000	1101	1000	4096	55296
868	888832	255	D9000	1101	1001	4096	55552
872	892928	255	DA000	1101	1010	4096	55808
876	897024	255	DB000	1101	1011	4096	56064
880	901120	255	DC000	1101	1100	4096	56320
884	905216	255	DD000	1101	1101	4096	56576
888	909312	255	DE000	1101	1110	4096	56832
892	913408	255	DF000	1101	1111	4096	57088

5-5 Memory hole-finding program sample output from a Pentium.

```
10    DECLARE FUNCTION BINARY$ (HEXIN$)
20    CLS           'Clear screen and print out column headers
30      PRINT "    DECADD MEM HEXADD   A15-A12   A11-A8   SIZEHOLE"
40      LPRINT "   DECADD MEM HEXADD   A15-A12   A11-A8   SIZEHOLE"
50    FOR DECADD = 1024 TO 65280 STEP 256 'Step ports at 256bytes
60        HEXADD$ = HEX$(DECADD)   'Calculate hexadecimal address
70        IF LEN(HEXADD$) = 3 THEN
80            HEXHI$ = "0"
90            HEXLO$ = LEFT$(HEXADD$, 1)
100       ELSE
110           HEXTEM$ = LEFT$(HEXADD$, 2)
120           HEXHI$ = LEFT$(HEXTEM$, 1)
130           HEXLO$ = RIGHT$(HEXTEM$, 1)
140       END IF
150       BHI$ = BINARY$(HEXHI$) 'Convert to binary to set switches
160       BLO$ = BINARY$(HEXLO$) 'Convert to binary to set switches
170       MEM = INP(DECADD)    'Read memory value at DECADD address
180       IF MEM = 255 THEN    'If value is 255 then execute a loop
190           FOR HOLECK = 1 TO 255  '..to see how far the hole extends
200               MEMCK = INP(DECADD + HOLECK)
210               IF MEMCK <> 255 THEN HOLESZ = HOLECK + 1: GOTO EHOLE
220               HOLESZ = HOLECK + 1
230           NEXT HOLECK    '...Still reading 255 so increment to next
240       ELSE                     'address so can check it
250           HOLESZ = 0     'MEM(at 256 byte increment)<>255 so...
260       END IF                   '...set hole size = 0
270   EHOLE:                   'Printout data for the 4KB increment
280       FORMAT$ = "   #####  ###   \   \   \ \   \ \     ####"
290       PRINT USING FORMAT$; DECADD; MEM; HEXADD$; BHI$; BLO$; HOLESZ
300       LPRINT USING FORMAT$; DECADD; MEM; HEXADD$; BHI$; BLO$; HOLESZ
310       FOR IDELAY = 1 TO 50000: NEXT IDELAY 'Slow down to read screen
320   NEXT DECADD          'Increment to next 256 byte step
```

5-6 Port hole-finding program listing (PORTFIND).

Setting IBEC DIP Switches

The IBEC uses three DIP switches, one for selecting the I/O mode and two to define the UCIS starting address. The recommended side-actuated switches make for easy switch setting, plus easy reading, all while the card is installed. Because this installed-view is the view you'll have most all the time, I recommend that, even with the card out of the computer, you keep the top edge of the card facing you anytime you set the IBEC DIP switches.

It minimizes confusion if we treat the IBEC switches a bit differently than we treat all others as spelled out in Chap. 1. When working on the IBEC only, forget the terminology ON and OFF. Forget it even if that is the nomenclature placed on the DIP switch by the manufacturer! For the IBEC, use the terminology *open* and *closed* for I/O mode select switch SW1, and for the address switches (SW2, SW3) use the terminology Logic 1 and Logic 0.

Logic 1 and open are with the switch segment levers moved up away from the circuit board. Logic 0 and closed are with the levers moved down toward the circuit board. If you substituted switches different from those recommended and they don't function with closed contacts when the levers are moved toward the circuit card, then I really recommend they be changed to those specifically recommended in the parts list. Doing so makes everything easier, and there is much less chance of error. We'll look at setting each switch.

```
            Output  sample  from  Pentium

DECADD  MEM  HEXADD   A15-A12   A11-A8   SIZEHOLE
 1024   255    400     0000      0100      113
 1280   255    500     0000      0101      256
 1536   255    600     0000      0110      256
 1792   255    700     0000      0111      256
 2048   255    800     0000      1000      113
 2304   255    900     0000      1001      256
   :     :      :        :         :         :
   :     :      :        :         :         :

            Output  sample  from  286

DECADD  MEM  HEXADD   A15-A12   A11-A8   SIZEHOLE
 1024    0     400     0000      0100        0
 1280   255    500     0000      0101      241
 1536   255    600     0000      0110       61
 1792   255    700     0000      0111      189
 2048    0     800     0000      1000        0
 2304   255    900     0000      1001      241
 2560   255    A00     0000      1010       61
   :     :      :        :         :         :
   :     :      :        :         :         :
```

5-7 Sample outputs from the PORTFIND program.

Setting the Mode Select DIP Switch SW1

Assuming you have picked your I/O mode, the next step is to set the I/O mode select DIP switch. Even if you haven't yet finalized on which mode you like best, go ahead and start using one of the modes. It doesn't take much more than a flip of a switch to change modes if you decide to do so later on.

Figure 5–8 illustrates which two segments of DIP switch SW1 need to be closed to select the I/O mode. Only the designated pair should be closed, and the remaining pair need to be in the open position. If not, you will be shorting together the computer's control lines for different modes and incorrect operation will result, along with possible damage to U11. Any time you set SW1, the correct procedure is to first set all four segments open, and then set the two required segments closed. That's all there is to setting the mode select DIP switch.

Setting the Address Decode Dip Switches SW2 and SW3

If you pick your UCIS starting address from Fig. 5–5 or 5–7, or the equivalent program outputs from your computer, then setting the address DIP switches is very straightforward. The address line label corresponding to each switch segment is printed right on IBEC: A8, A9, 10, 11, 12, and so on up to 19. Note for the two-digit

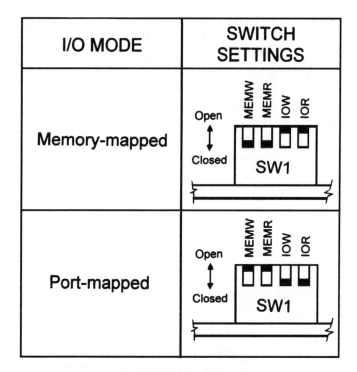

5-8 I/O mode selector DIP switch settings—SW1.

numbers the A notation, for address, is dropped to conserve board space. The address decode DIP switch settings required for each I/O mode are summarized as:

Memory-mapped I/O
☐ Set segments A11–A8 to Logic 0.
☐ Set segments A19–A12 per Fig. 5–5.
☐ For software in Basic, use the DEF SEG value defined in Fig. 5–5 along with your PEEK and POKE statements, to be explained later.

Port-mapped I/O
☐ Set segments A15–A8 per Fig. 5–7.
☐ Set segments A19–A16 (not used) to Logic 0.
☐ For software in Basic, use the IN (or INP) and OUT statements with the directly inserted UCIS starting address defined in Fig. 5–7 as DECADD. DEF SEG is not used.

Figure 5–9 illustrates SW2 and SW3 settings for three example UCIS starting addresses for both memory-mapped and port-mapped I/O.

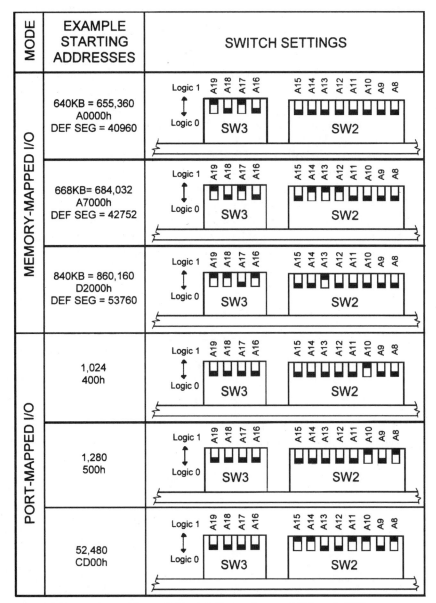

5-9 UCIS starting address DIP switch settings—SW2 and SW3.

Inserting the IBEC into Your Computer

With your stand-alone testing complete and your DIP switches all configured, its time to insert the IBEC into your computer. Follow these steps.

Warning: Always make sure your computer is turned off whenever you insert or remove your IBEC, and whenever you attach or remove the ribbon cable between IBEC and IOMB. You do not have to turn off your computer when setting DIP switches. However, as you change settings you might occasionally lock up the computer, thereby requiring a reboot.

☐ *IBEC insertion.* With your computer turned off, open the case and insert IBEC into an available 8- or 16-bit ISA card slot. Make sure everything fits properly, that the card is properly in the 62-contact bus socket, nothing is touching adjacent card parts, that the bottom tab of the IBEC mounting bracket is correctly in its notch, and the notch on the side of the top tab lines up with the mounting screw hole.

☐ *Computer operation test.* With IBEC inserted correctly, turn your computer power on. It is normal for the red address selected test LED to blink a bit during the power-up sequence; this results from the computer's self test program checking memory and port locations. Once the computer is up and running, the red LED should remain off. Use your computer for some of its normal operating functions. Exercise the different devices on the bus such as sound cards, fax modems, keyboard, printers, and so on. Assuming everything works fine, then your IBEC installation is correct, and you can skip to the last paragraph of this chapter.

A problem at this stage is rare, but if having IBEC inserted causes problems, then it's most likely a solder bridge touching one of the address lines, data lines, or control lines. Another possibility is that one of the ICs that connect directly to the computer bus is inserted incorrectly or is faulty.

The first solution step is to repeat the visual examination of both sides of the IBEC as presented in the first part of this chapter. If that doesn't locate the solution, remove the ICs from their sockets one by one—turning the computer off and on again each time—to see if your problem is in the part or in the circuit itself. Refer to the IBEC schematic in Fig. 3–4, and start with the ICs that have a direct connection to the computer bus, like U1, U3, U4, U5, U6, and U11. If you do trace a problem to an IC, replace it with a new part. It's handy to keep at least one spare of each IC type in your UCIS for debugging and replacement.

If you still have a problem when all the ICs are out, then use your VOM to check that you have isolation between each of the data, address, and control lines as listed down the left side of Fig. 3–4. That is reading between D0 and each of the other 34 lines, including 5 Vdc and GND, and they should each be an open circuit. Likewise, check with D1, then D2, and so on until you have checked each of the 35 lines with all the others. Once you find the offending lines, correct the problem. Following all these procedures you should be able to find what is bad and then correct it.

Keep the IBEC installed in your computer, and (assuming you are not also building a UBEC) skip to Chap. 8, where we'll start building the general-purpose I/O cards.

6
CHAPTER

The Universal Bus Extender Card

In this chapter I'll discuss building the universal bus extender card (UBEC) as shown in Fig. 6–1. I assembled this UBEC on a ready-to-assemble PC card made by JLC Enterprises, which comes with the parts layout printed on its component side. This one I configured for any of the Apple II family of computers, but I'll also explain how to tailor its construction for a much larger variety of computers so that we'll be ready to hook it up to your computer in Chap. 7.

You don't need to understand electronic circuits to build the UCIS. The circuit cards are readily available, and all you have to do is follow my instructions. To explain how the UBEC works and to help in debugging in case you have any trouble, I'll explain what this card does and give you a tour of its schematic. Many parts of UBEC are similar to IBEC, so if you find you desire more detail than provided in this chapter, I suggest you review the material in Chaps. 3 and 5. The IBEC material is especially helpful if you are using a UBEC/CBAC for interfacing to an IBM PC, x86, or compatible. If you really just can't wait to start building your UBEC, then skip ahead.

UBEC Functions

Figure 6–2 is a functional look at what the UBEC does. Its three functions are signal buffering, address decoding, and line enabling/direction control. The arrows indicate the direction of signal flow; double lines with a slash and number indicate that multiple parallel wires are involved—eight in this example case. The low-order address lines, A0 through A7, pass directly through the buffers from the computer side to the interface I/O side.

High-order address lines A8 through A15 are decoded by the UBEC. Decoding means that when input lines A8–A15 match the UCIS starting address set on the UBEC's DIP switch, decoding output line AH (address lines high-order bits) goes high. I'll explain how to define and set the UCIS starting address and how to handle computers that have more than 16 address lines in Chap. 7. When AH is high, the

6-1 Universal bus extender card (UBEC).

computer is set to communicate with the external hardware; when low, the computer is not communicating with the external hardware.

Unlike with the IBEC, the choice between using either port-mapped I/O or memory-mapped I/O is pretty much fixed with the UBEC as a function of which computer you are using. In memory-mapped I/O, the external hardware is treated as an expansion of the computer memory, while in port-mapped I/O the same address bus is used without taking up any space allocated to memory.

For memory-mapped I/O, UBEC uses the memory read and write control lines MEMR* and MEMW*, while for port-mapped I/O it uses the I/O read and write control lines IOR* and IOW*. The asterisk indicates that the lines are active low, which is the case for most of the computers we'll be looking at in Chap. 7. Whichever set is used, they pass through buffers before being sent out to the external hardware as R* and W*. The result is that when R* is low and AH is high simultaneously, the computer reads data *from* the external hardware. When W* is low and AH is high, the computer sends data *to* the external hardware.

The data are sent back and forth on lines D0 through D7, the data bus. Because data flow must be bidirectional, the data lines are double-buffered. One group of eight buffers, labeled IN in Fig. 6–2, handles data flow into the computer, and the group labeled OUT handles data out to the external hardware.

Control gates 1 and 2 logically determine which set of buffers is active by combining the AH address and control signals R and W with the signal remote-ready-on, RRON, which reads the external hardware's +5-Vdc power supply. The control gate logic is summarized in Table 3–1 in Chap. 3.

The UBEC Schematic

The schematic in Fig. 6–3 is simply an expansion of the function diagram in Fig. 6–2. Digital circuits can appear complicated because there are a lot of parallel lines doing basically the same thing. Once you understand the function of a few of the separate lines, however, you quickly pick up the understanding of the companion lines, and soon the overall diagram becomes understood.

The 16 lines of the address bus, A0 through A15, allow the computer to access 65,536 different locations or 64K. Using the CBAC for the IBM PC, XT, 286/AT, 386, 486, and Pentium, we'll extend this address capability to 1 Mbyte by decoding an additional four address lines. The majority of these addresses are taken up directly by the computer memory, but you need only a small subset of locations, 256 bytes, for your external hardware I/O.

Lines A2-A7 carry six-bit addresses for the individual I/O cards. Six bits can address 64 I/O cards, which gives you direct access to 1536 lines (64 × 24) connected to your external hardware. Address lines A0 and A1 determine which of three ports is selected on the addressed card.

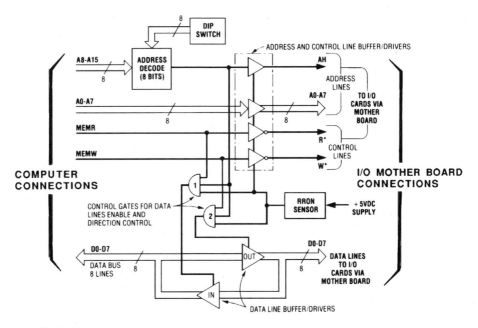

6-2 UBEC functions.

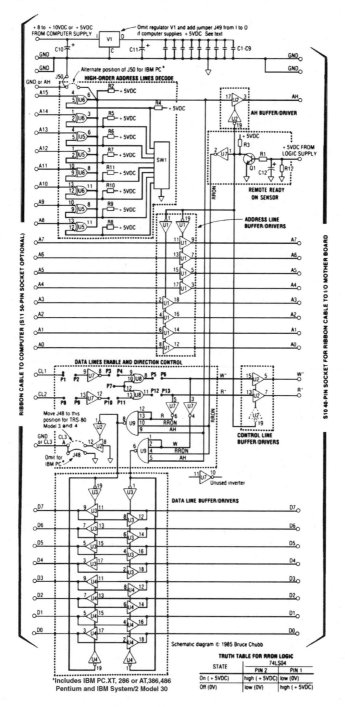

6-3 UBEC schematic.

Lines A8–A15 carry the high-order bits defining the starting address for the interface I/O cards. Each of these lines is fed into the address decode circuit, ICs U5 and U6. Pin 17 of the U2 IC goes high only if every input A8–A15 matches the corresponding settings on the eight-segment DIP switch. This input is then further buffered by U2 and sent out to the I/O cards as AH.

The important point is that AH is high if the computer is addressing the external hardware. With AH high, the specific I/O card that is being addressed is defined by A2 through A7, and the specific I/O port on the card is defined by A0 and A1. I'll provide more on this later when I talk about the specifics of the general-purpose I/O cards.

For now, look at the UBEC's second function, buffering. The four ICs U1, U2, U3, and U4 provide the desired buffering of the address, control, and the data bus lines. Each of these ICs is an octal buffer/line driver; octal means each IC can handle eight lines. These ICs provide tri-state logic, as explained in Fig. 3–5 and the associated text in Chap. 3.

The IBEC's address bus and control lines are unidirectional, as are their buffers. The data bus must be bidirectional, so two buffers are hooked up back-to-back for each of the eight data lines. If the rightmost vertical column of data buffers in Fig. 6–3 is enabled (pin 1s are low), data can be transmitted from the computer to the external hardware. If pin 19s are low, data can be transmitted from the external hardware to the computer. If all of the pins 1 and 19 are high simultaneously, then all the buffers are open circuits, and there is no data connection between the computer and external hardware. The U9 IC drives the pin 1 and 19 inputs following the logic in Table 3–1.

Note that in Fig. 6–3 all of the UBEC outputs to the interface I/O cards are noninverted (meaning that they are active high) except for R* and W*, which are active low. Thus, the interface I/O cards work with noninverted signals for the most part.

In this example, I'm assuming noninverted address and data signals from the computer, which is true for most. The schematic thus shows U1–U4 as 74LS244s, because that IC doesn't have built-in logic inversion. However if your computer happens to be like my Heathkit H-8, and has the address and data lines inverted, simply substitute 74LS240s for the affected 74LS244s, providing the same functions but with inversion of the signals, and reverse the settings of DIP switch SW1. Typically, every IC function is available with either inverted or noninverted outputs.

Because of the differences in control lines between machines, the UBEC has terminals CL1, CL2, and CL3, as well as 13 program jumper pads P1 through P13. By connecting CL1 through CL3 to the appropriate computer control lines and installing appropriate jumpers, we can handle the peculiarities of various computer control lines.

UBEC Parts

A key advantage to the UCIS is that ready-to-assemble printed circuit boards are available from JLC Enterprises. If you elect to buy them assembled and tested, they are available from ECO Works, EASEE Interfaces, or AIA. See Appendix A for details, including the UBEC artwork for those wishing to fabricate their own card from scratch.

The parts layout for the UBEC is shown in Fig. 6–4, and the parts are listed in Table 6–1. The 50-pin header S11 is for a ribbon-cable connection to the computer bus adapter card (CBAC), but it is optional. You might choose instead to hardwire the CBAC cable directly to the UBEC by soldering the cable's conductors into holes provided in the card. It is easier, and there will be less chance of error, if you use S11 and let the CBAC make the connections to the computer for you. However, cable connectors are always less reliable than good hardwiring. If you like, you can wait until you see what's involved in Chap. 7 and decide then whether or not to install S11.

Figure 6–4 shows the parts locations on the top or component side of the UBEC; you should refer to this diagram as you select and mount parts. Keep the card oriented the same way as the drawing while you work, to help you place each part in its proper hole with correct orientation. Also, take special care to see that you are installing the right components. Sometimes their markings are so small that you need a magnifying glass to read them, but extra effort at this stage is worthwhile. Check

Table 6-1. UBEC Parts List

Qnty.	Symbol	Description
1	J1-J51***	Jumpers, make from no. 24 uninsulated bus wire (Belden no. 8022)
2	R1,R2	1.0KΩ resistors [brown-black-red]
9	R3,R11*	2.2KΩ resistors [red-red-red]
1	R12	110Ω ½W resistor [brown-brown-brown]
4	S1-S4	20-pin DIP sockets (Digi-Key A9320)
5	S5-S9*	14-pin DIP sockets (Digi-Key A9314)
1	S10	40-pin double row header (Jameco 53532)
1	S11**	50-pin double row header (Jameco 53559)
1	SW1*	8-segment DIP switch (Digi-Key CT2068)
1	V1*	5V, 1A positive voltage regulator (Jameco 51262)
1	H1*	Heat sink bracket for H1, make as shown in Fig. 6-5
9	C1-C9*	100000pF, 50V ceramic disk capacitors (Digi-Key P4164)
2	C10,C11	2.2µF, 35V tantalum capacitors (Digi-Key P2061)
1	C12	100µF, 10V radial-lead electrolytic capacitor (Digi-Key P6214 or equivalent with .098" lead spacing)
1	Q1	2N3904 NPN small signal transistor (Jameco 38359)
—	P1-P13*	Program jumpers, make from no. 24 insulated wire per Fig. 6-6
4	U1-U4*	74LS244 octal buffer/drivers (Jameco 47183)
2	U5,U6	74LS136 quad EXCLUSIVE-OR gates with open collector (Jameco 46586) or (JDR 74LS136)
1	U7	74LS04 hex inverter (Jameco 46316)
1	U8*	74LS32 quad 2-input OR gate (Jameco 47466)
1	U9	74LS20 dual 4-input NAND gate (Jameco 47095)

Author's recommendations for suppliers given in parentheses above with part numbers where applicable. Equivalent parts may be substituted. Resistors are ¼W unless otherwise specified, color codes are given in brackets.
*Quantity and/or specifications vary for different computers. See text.
**Not required if ribbon cable from computer is hard-wired to UBEC.
***Notes on jumpers:
- J44-48 not required if computer cable is hard-wired to UBEC.
- Alternate location of J48 is for TRS-80 Models 3 and 4.
- For IBM System/2 Model 30, IBM PC, XT, AT/286, 386, 486 and Pentium omit J48 altogether, use alternate location for J50, and add jumper J51.

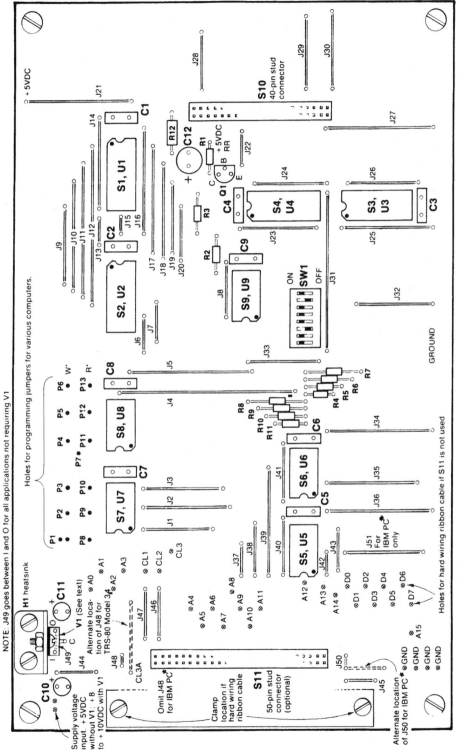

6-4 UBEC parts layout.

resistors against the color code included in the parts list. In addition to the three color bands there should also be a gold band, indicating a tolerance of 5 percent.

In most every case UBEC receives its 5-Vdc power from the computer via the 50-conductor ribbon cable, and for these cases you leave off V1 and simply install J49. If your computer doesn't happen to supply 5-Vdc, as my H-8 does not, then install V1 with its heat sink bracket H1 as illustrated in Fig. 6–5.

For the following parts, the correct orientation on the card is very important: capacitors C10, C11, and C12; transistor Q1; DIP switch SW1; and both IC sockets S1–S9 and the ICs themselves, U1–U9. To remind you to check this as you build your UBEC, I've marked the orientation-critical parts with a plus sign in brackets [+] in the following instructions. If you need additional help, Fig. 1–8 shows how to determine the correct orientation for different parts. Double- check your selection and location of each component before soldering it in place. It is a lot easier to remove an incorrect component before you've soldered it to the card.

UBEC Assembly

The basic skill for building the UCIS is PC card soldering. If that's a new one for you, make doubly sure that you have thoroughly digested the information on PC card soldering in Chap. 1, so you know the tools and techniques of good soldering. Do all soldering on the back or bottom side of the card—the surface with the metal circuit traces.

The order of parts assembly on the UBEC is not critical, but for the sake of having a plan, follow the steps in order and check off the boxes as you complete each one.

☐ *Card test.* Use your VOM (volt-ohm meter) to test trace continuity on the circuit card. Make sure there are no open circuit traces, such as a conductor with a hairline crack, and no shorts between adjacent traces or pads. You'll seldom find a problem in a bare card, but if you find one and correct it now you'll save time in later debugging. If you find an open-circuit trace, solder a small piece of wire across the crack. If you find a short circuit, use a knife to scrape away the offending material.

☐ *J1 through J51.* Be sure to follow the notes at bottom of the parts list defining changes required in jumper locations for adapting UBEC to different computers. Form each jumper wire with 90-degree bends at both ends, so that the end leads just fit in the holes. Each jumper should be straight and taut with no chance of touching an adjacent wire or component. After soldering each jumper in place, trim its leads flush with the tops of the solder tents on the back side of the card.

☐ *R1-R12.* Make 90-degree bends in the leads of each resistor so it is centered between its two holes and the leads just fit. Insert and solder it while holding the part flat against the card, then trim the leads.

☐ *S1-S9[+].* To avoid mixing them up, install 20-pin IC sockets S1–S4 first, then the 14-pin sockets, making certain that pin 1 orientation is correct for each. Push each socket tight against the card and hold it that way as you solder. Do not install the ICs at this time. You may need to do some card and

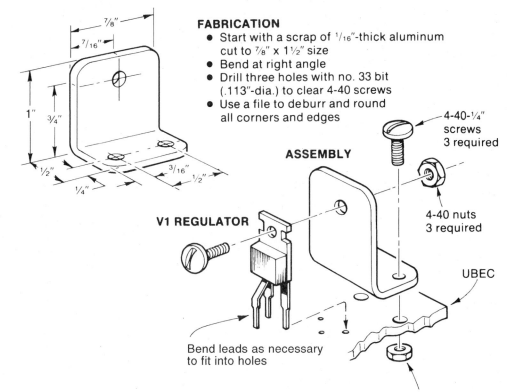

FABRICATION

- Start with a scrap of $1/16$"-thick aluminum cut to $7/8$" x $1\frac{1}{2}$" size
- Bend at right angle
- Drill three holes with no. 33 bit (.113"-dia.) to clear 4-40 screws
- Use a file to deburr and round all corners and edges

ASSEMBLY

4-40-$\frac{1}{4}$" screws 3 required

4-40 nuts 3 required

UBEC

V1 REGULATOR

Bend leads as necessary to fit into holes

NOTE: Solder and trim regulator leads AFTER heatsink (with V1 attached) is firmly mounted in place on card

Solder two mounting nuts to bottom of board after assembly (optional)

6-5 Constructing the V1 heat sink bracket (H1).

cable continuity checking in Chap. 7, and this is better accomplished without the ICs installed.

☐ *SW1[+]*. Mount the DIP switch using your VOM to be sure it is oriented so the contacts are closed when the switches are thrown toward U9. This direction will be on, and away from U9 will be off. Hold the switch securely against the card as you solder it.

☐ *S10.* Install the 40-pin header for the ribbon cable to the I/O motherboard, holding it securely against the card as you solder. Solder a couple pins first near each end, and once you have verified that it is tight against the card, solder the remaining pins.

☐ *S11.* Assuming you decide to use this header instead of hard-wiring, install it just as you did S10.

☐ *C1-C9.* Insert with capacitor disk standing perpendicular to the card, solder, and trim.

☐ *C10-C12[+].* Make sure that the positive leads go into the (+) holes as shown in Fig. 6–4. Incorrect polarity will damage these capacitors. Solder and trim.

☐ *Q1[+].* Spread the leads of this transistor slightly to fit the three holes, making sure the center (base) lead goes into the hole closest to S10, and

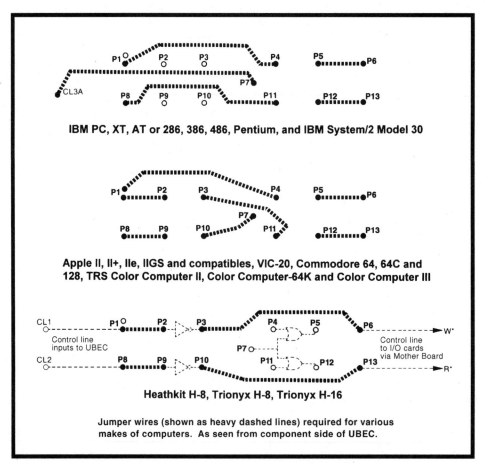

IBM PC, XT, AT or 286, 386, 486, Pentium, and IBM System/2 Model 30

Apple II, II+, IIe, IIGS and compatibles, VIC-20, Commodore 64, 64C and 128, TRS Color Computer II, Color Computer-64K and Color Computer III

Heathkit H-8, Trionyx H-8, Trionyx H-16

Jumper wires (shown as heavy dashed lines) required for various makes of computers. As seen from component side of UBEC.

6-6 Program jumper wiring.

that the flat side of Q1 also faces this same connector. Push in only far enough to fit snugly without stressing the leads. Solder and trim.

☐ *P1-P13.* Install the appropriate program jumpers to suit your computer, as shown in Fig. 6–6.

☐ *Cleanup and inspection.* For a professional-looking job and to help ensure that your card functions properly, follow the specific steps covered in Chap. 1 regarding cleanup and inspection. This is a most vital step, so don't cut it short! If you will be hardwiring your computer cable, postpone the finish spraying until you've completed all the UBEC soldering.

UBEC Enclosures

I suggest that you simply mount your UBEC on a small piece of plywood until you've completed testing and debugging. Having it secured on a block of wood helps avoid accidental shorts against the underside of the card. Drill out the four corner holes with a #25 bit or larger to clear four 6-32 screws, and mount the card with standoffs as shown in Fig. 6–7.

Once you get everything working, it's a good idea to mount your UBEC in some type of enclosure or box to keep things neat around the computer. If you choose a metal enclosure, use standoffs and make sure that none of the circuit traces or components touch the enclosure. The UBEC should be close enough to your computer so the ribbon cable is no more than about 18 in long. The motherboard with its I/O cards can be mounted up to 10 ft from your UBEC. The wiring from the I/O cards to your external hardware can be as long as you wish.

With your UBEC complete, you've taken the first big step in building your own UCIS. In the next chapter I'll show you how to connect the UBEC to your computer.

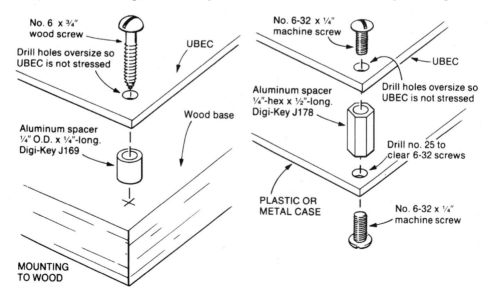

6-7 Mounting the UBEC.

7
CHAPTER

Connecting UBEC
to Your Computer

In this chapter we'll work on the computer end of the UCIS and connect the UBEC to your personal computer. I'll provide specific connections covering many popular computers including the IBM PC, x86 series, and clones, plus the TRS Color Computer family, the Apple II family, the Commodore family, and the IBM System/2 Model 30.

Figure 7–1 shows a typical computer end of the UCIS, a Universal Bus Extender Card connected to an Apple II+ by a special computer bus adapter card (CBAC) plugged into bus slot 4 in the Apple motherboard. We'll move on to the input/output-card-end of the UCIS in Chap. 8.

As in the previous chapters, you don't need to understand exactly what goes on inside the computer, or why the UCIS is connected as it is. With CBACs and ready-to-use ribbon cables you'll find it easy to plug the UCIS right into most computers. At the end of this chapter, I'll even show an example with a UBEC *hardwired* directly into a computer. This technique comes in handy for interfacing to computers that might not have an available JLC adapter card.

I'll explain how computers work with UBEC and CBAC. Such knowledge should be helpful if you run into any trouble. However, if you prefer to just get started hooking UBEC to your computer, then jump ahead in this chapter. Also, when you get to the instructions for specific computers you can ignore everything except your own machine; then jump on to the last paragraph of this chapter.

Memory-Mapped or Port-Mapped I/O

For the most part, the main buses of the computers attaching to UBEC are standardized with 16, 20, or 24 address lines and 8 or 16 data lines. I'll show you how to handle these differences later. In terms of bus signals, the main difference from computer to computer is in the control lines as shown by Fig. 7–2. For example, the IBM PC through the Pentium, the Heathkit/Trionyx H-8/H-16, and the IBM System/2

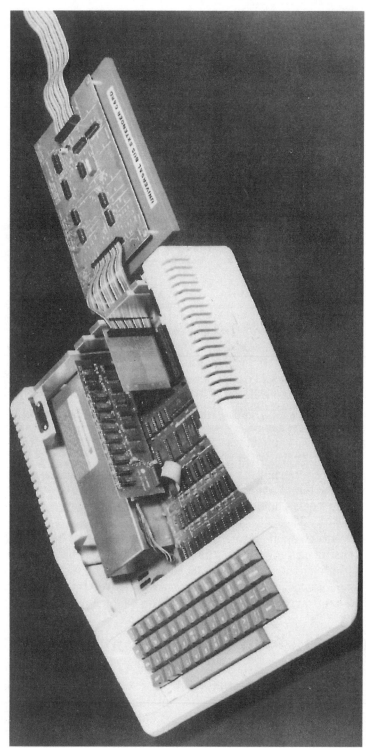

7-1 UCIS connected to Apple II+.

CONTROL LINES/SIGNALS	PENTIUM, 486, 386, 286 or AT	IBM PC, XT, and SYSTEM/2 MODEL 30	APPLE II, II+, IIe, IIGS	TRS COLOR COMPUTERS	COMMODORE 64, 65C and 128	COMMODORE VIC-20	HEATHKIT/ TRIONYX H-8s and TRIONYX H-16
Memory read	SMEMR* (B12)	MEMR* (B12)	—	—	—	—	MEMR (28)
Memory write	SMEMW* (B11)	MEMW* (B11)	—	—	—	—	MEMW (23)
I/O read	◸ IOR* (B14)	◤ IOR* (B14)	—	—	—	—	◤ I/OR (26)
I/O write	◸ IOW* (B13)	◤ IOW* (B13)	—	—	—	—	◤ I/OW (21)
Combined memory read/write*	—	—	R/W* (18)	R/W* (18)	CR/W* (18)	R/W* (5)	—
Clock signal	—	—	Φ0 (40)	E (6)	SΦ2 (V)	Φ2 (E)	—

Filled triangle upper left corner indicates signal available but not used by UBEC/CBAC.
Open triangle upper left corner indicates signal available and useable by IBEC but not UBEC, unless alter CBAC per text.
Asterisks indicate signals that are active low (inverted logic)
Dashes indicate signals not provided by the computer.
Computer connector pinouts for signals are given in parentheses.

7-2 Control line variations.

Model 30 each have four control lines of interest. These are referred to as MEMORY-WRITE, MEMORY-READ, I/O WRITE, and I/O READ. The CPU sets the appropriate control line high (or low) for a fraction of a second at the instant at which data is to be transferred. The two memory control lines determine whether the transfer is to be to or from memory. The I/O control lines determine whether the transfer is to be to or from one of the peripheral I/O device ports.

On the other hand, the Apple II family, the VIC-20, the Commodore family, plus all TRS Color Computers don't have separate MEMORY-READ and MEMORY-WRITE control lines. They use a single line, READ/WRITE*, which is set low when the CPU sends data to memory and high when reading data from memory. A separate line, called the clock signal, is set high at the instant the data are to be transferred.

When a computer uses any of the memory control lines to control peripheral input and output devices, the peripherals are treated as extensions of the memory address space. This is memory-mapped I/O. The opposite is port-mapped I/O, the kind of operation in which the I/O WRITE and I/O READ lines control the data transfer through addresses designated as external I/O connections, or ports.

To summarize, we generally have a choice between two types of I/O: memory-mapped I/O and port-mapped I/O. If the control-line terminals on the UBEC connect to the computer's memory control lines, we are using memory-mapped I/O, and if they connect to the computer's input-output control lines, we are using port-mapped I/O. Some computers limit you to one or the other, while some machines give you the choice.

In BASIC programming, you use the IN (or INP) and OUT instructions with port-mapped I/O, but replace those with PEEK and POKE for memory-mapped I/O. I'll give you examples of both methods when we move into writing UCIS software.

We adapt the UBEC to different control lines with the arrangement of its program jumper wires and with component selection, as covered in Chap. 6. There are also variations in the cable connections from each computer to the UBEC, as you'll see later in this chapter.

Determining the UCIS Starting Address

To interface external hardware with a computer you must find a range of unused addresses, specific locations that can store or send one byte each, in its memory or I/O system. Such blocks of unused addresses are called holes, and the first address in the hole you use for your UCIS is your UCIS starting address.

The minimum acceptable size of the hole is four times the number of I/O cards you plan to use, to a maximum of 256 bytes. This is the maximum because only ad-dress lines A0 through A7 (8-bits) are decoded to select the particular I/O card be-ing addressed. With eight bits you are limited to 256 bytes, providing a maximum of 256/4 or 64 I/O cards. Because each card controls 24 I/O lines, you could handle 1536 external I/O connections with a 64-card system (64 cards × 24 lines per card).

With memory-mapped I/O you'll ordinarily have no problem locating a hole of 256 contiguous bytes. With port-mapped I/O many machines use only the lower eight bits for I/O addresses, and it's usually impossible to use a full 256 bytes for your UCIS because room must be left for other peripherals on the bus. Because few applica-tions need 64 I/O cards, smaller applications should be able to find enough of a hole in most machines.

If you use memory-mapped I/O with smaller computers and you don't have the full memory capacity installed, locating the hole is very simple. For example, if your 64K computer has only 48K of installed memory, the UCIS starting address can be set at 49,152 (from 48 × 1024, because 1 kilobyte = 2^{10} or 1024 bytes). In hex this is C000h.

You might think that it should start at 49,153; however, with 48K of memory in-stalled, the locations used are 0 through 49,151, so the memory-mapped I/O can start at 49,152.

If you have a computer in the IBM PC or XT family, they use 20 address bits, and you can address 2^{20} or 1,048,576 bytes = 1024K directly. As an example, my 286 has 640K of installed memory, thus the size of its potential memory hole is 1024K − 640K, or 384K. That's a 393,216-byte space in which we only need to locate a 256-byte hole for UCIS.

If your machine has its full memory capacity installed, which would be 64K in a machine that can address only 64K, then you must look for a block of 256 bytes of un-used memory in the reserved areas. Typically there are a few kilobytes of address space reserved by the manufacturer for various functions. There's often space for fu-ture expansion in this area, and that's where we'll find our 256 bytes. To find the avail-able space you need a map of your computer's memory layout. Often that's included in your computer's user manual, but if not, check with your dealer or service center.

Also, if you are using the UBEC to interface with any of the PC, XT, 286/AT, 386, 486, or Pentium computers, then Chaps. 3 and 5 provide additional information on selecting between memory-mapped and port-mapped I/O. Chapter 5 also shows how to use computer software enclosed with this second edition to locate a suitable UCIS starting address.

Figure 7–3 shows a memory map that's typical of what you'll find for the smaller hobby-type computers—in this case my H-8. The bottom 8K is defined as reserved by the manufacturer, and contains the ROM chips for the panel monitor (PAM-8) and the Heath disk operating system (HDOS). When I connected my first UCIS I had only 48K of RAM installed, so the lower address of unused memory was 56K (8K reserved + 48K installed). For simplicity I set the starting address for memory-mapped I/O at 60K, midway in the unused space. I later added the full memory and moved the UCIS I/O into the blank reserved area at address 1K, also shown in Fig. 7–3.

Figure 7–4 shows how to determine both the binary code for address 60K and the corresponding DIP switch settings for the UBEC. Try calculating one yourself: say we want the I/O starting address to be 1K (1024), as it is now on my system. The resulting DIP switch positions would be A10 = 1, and all others 0.

Computer Bus Adapter Cards (CBACs)

The CBAC is designed to make the correct computer/UBEC connections for you, so it's easier than hardwiring and offers less chance of error. While cable sockets are always less reliable than good hardwiring, I'd expect that to be a smaller problem for most people than the possibility of crossed connections. In general, the best way to make the computer connections is to use a CBAC and a cable with preassembled connectors, and to install the 50-pin header S11 on the UBEC.

Whenever possible I've specified cables and arranged terminals to provide multiple ground connections. With my 286, for example, the 50-wire cable allows 20 ground connections. The cable conductors are 28-gauge wire, but 20 in parallel give a ground equivalent larger than 18-gauge wire.

Ready-to-assemble CBAC PC cards are available from JLC Enterprises. If you don't care to assemble the card yourself, they are available assembled and tested from ECO Works, EASEE Interfaces, and AIA. See Appendix A for details, including the CBAC artwork for those wishing to fabricate their own cards from scratch.

Computer Connections

Now I'll detail the connections for each computer, but first some general advice.

Warning: *Always be sure that your computer and any other related power supplies are turned off when making any kind of connections. Then double-check to make sure that your ribbon cable and power supply connections are correctly oriented before turning the power on. If you follow the instructions carefully, it's unlikely that anything in the UCIS could damage your computer, but the worst cases would be making incorrect connections or accidental misconnections under power.*

I'll specify the connector pinouts for each computer and their UBEC connections, but don't take my listings blindly. Manufacturers always reserve the right to change these, and may also slip in some new capability that changes the memory mapping. In most cases I've tested the hardware and software, but your machine might be different. Check your computer reference manual to verify cable connections and the available port addresses or memory holes before connecting the UBEC to your machine.

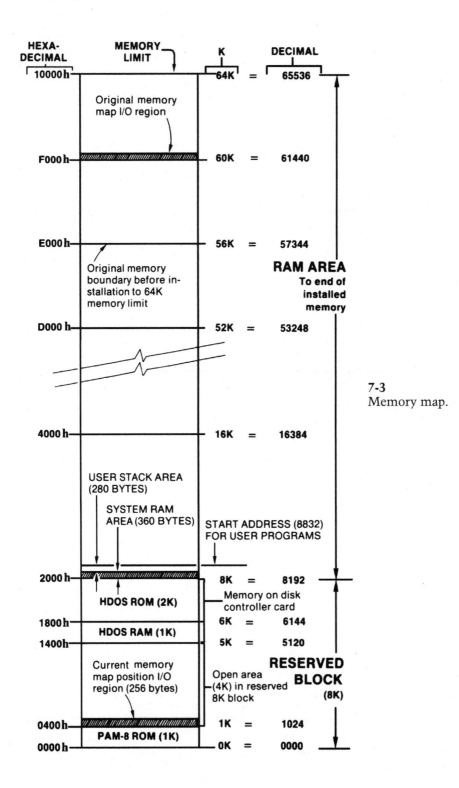

7-3
Memory map.

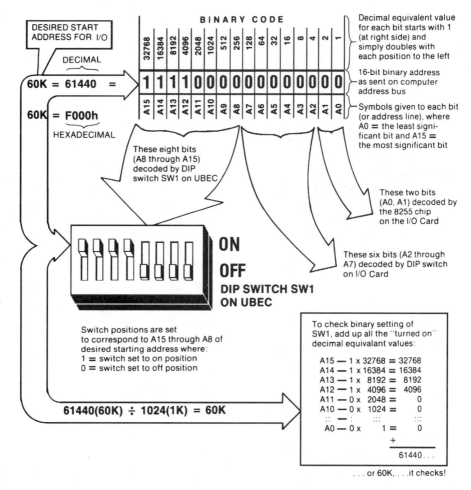

7-4 Setting the UCIS starting address.

I've tried to include as much information about the interface arrangements as I thought might be helpful, but the real basics of each case are shown in Figs. 7–5 through 7–11. Double-check those pinouts, assemble the proper parts the right way, and set the DIP switch(es) as shown, and you'll very likely be home free.

For connecting to UBEC we'll cover the complete gambit of computers from the Pentium back to the earliest Commodore. I'll cover the TRS Color Computer family, the VIC-20, the complete Apple II family, the IBM PC, *x*86 series, and their compatibles, and the IBM System/2 Model 30. I'll even cover some computers that don't even have adapter cards.

Observing how differently each computer interface works is a real eye opener, revealing how remarkable it is that any computer can connect to UCIS. I personally find the idiosyncrasies between the different computer interfaces quite fascinating as well as informative, and I hope you do too.

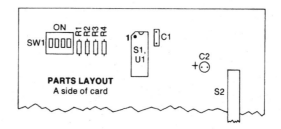

PARTS LAYOUT
A side of card

CBAC SCHEMATIC FOR IBM PC

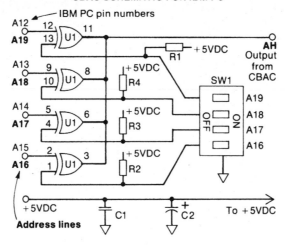

PARTS LIST FOR IBMPC CBAC

Qnty.	Symbol	Description
4	R1-R4	2.2KΩ ¼W resistors [red-red-red]
1	S1	14-pin DIP socket (Digi-Key A9314)
1	S2	50-pin double row header (Jameco 53559)
1	SW1	4-segment DIP switch (Digi-Key CT2064)
1	C1	100000pF, 50V ceramic disk capacitor (Digi-Key P4164)
1	C2	2.2μF, 35V tantalum capacitor (Digi-Key P2059)
1	U1	74LS136 quad EXCLUSIVE-OR gate with open collector (Jameco 46586) or (JDR74LS136)

Author's recommendations for suppliers given in parentheses above with part numbers where applicable.   Equivalent parts may be substituted.

7-5 Special CBAC for IBM PC up through Pentium computers.

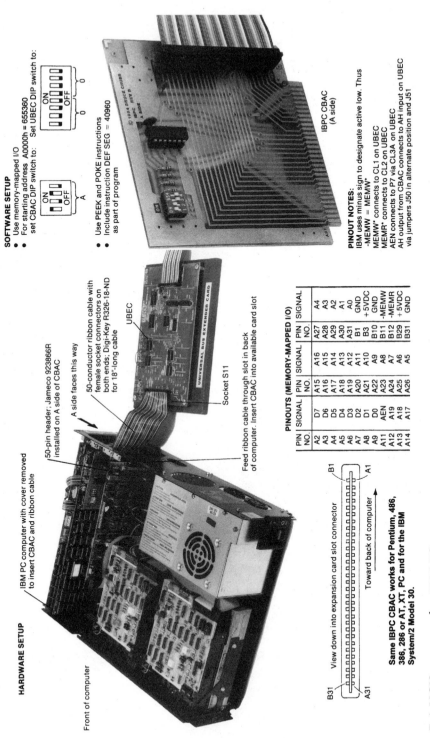

HARDWARE SETUP

Front of computer

IBM PC computer with cover removed to insert CBAC and ribbon cable

50-pin header: Jameco 923866R installed on A side of CBAC

A side faces this way

50-conductor ribbon cable with female socket connectors on both ends; Digi-Key R326-18-ND for 18"-long cable

UBEC

Socket S11

Feed ribbon cable through slot in back of computer. Insert CBAC into available card slot

IBPC CBAC (A side)

SOFTWARE SETUP

- Use memory-mapped I/O
- For starting address A000h = 655360
  set CBAC DIP switch to: Set UBEC DIP switch to:

ON / OFF   A   ON / OFF   0 0

- Use PEEK and POKE instructions
- Include instruction DEF SEG = 40960 as part of program

PINOUT NOTES:

IBM uses minus sign to designate active low. Thus -MEMW = MEMW*

MEMW* connects to CL1 on UBEC
MEMR* connects to CL2 on UBEC
AEN connects to P7 via CL3A on UBEC
AH output from CBAC connects to AH input on UBEC via jumpers J50 in alternate position and J51

PINOUTS (MEMORY-MAPPED I/O)

PIN NO.	SIGNAL	PIN NO.	SIGNAL	PIN NO.	SIGNAL
A2	D7	A15	A16	A27	A4
A3	D6	A16	A15	A28	A3
A4	D5	A17	A14	A29	A2
A5	D4	A18	A13	A30	A1
A6	D3	A19	A12	A31	A0
A7	D2	A20	A11	B1	GND
A8	D1	A21	A10	B3	+5VDC
A9	D0	A22	A9	B10	GND
A11	AEN	A23	A8	B11	-MEMW
A12	A19	A24	A7	B12	-MEMR
A13	A18	A25	A6	B29	+5VDC
A14	A17	A26	A5	B31	GND

View down into expansion card slot connector

B1 A1
B31 A31

Toward back of computer

Same IBPC CBAC works for Pentium, 486, 386, 286 or AT, XT, PC and for the IBM System/2 Model 30.

7-6 UCIS connected to IBM PC.

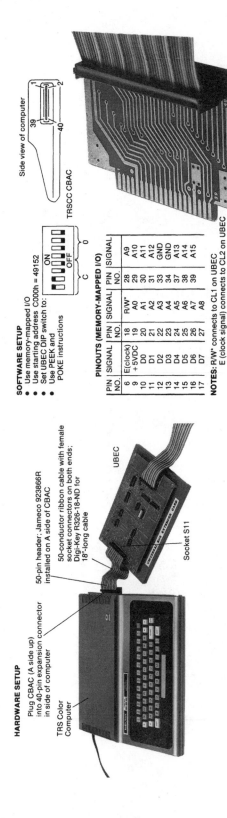

SOFTWARE SETUP

- Use memory-mapped I/O
- Use starting address C000h = 49152
- Set UBEC DIP switch to:
- Use PEEK and
  POKE instructions

PINOUTS (MEMORY-MAPPED I/O)

PIN NO.	SIGNAL	PIN NO.	SIGNAL	PIN NO.	SIGNAL
6	E(clock)	18	R/W*	28	A9
9	+5VDC	19	A0	29	A10
10	D0	20	A1	30	A11
11	D1	21	A2	31	A12
12	D2	22	A3	33	GND
13	D3	23	A4	34	GND
14	D4	24	A5	37	A13
15	D5	25	A6	38	A14
16	D6	26	A7	39	A15
17	D7	27	A8		

NOTES: R/W* connects to CL1 on UBEC
E (clock signal) connects to CL2 on UBEC

HARDWARE SETUP

Plug CBAC (A side up)
into 40-pin expansion connector
in side of computer

TRS Color
Computer

50-pin header; Jameco 923866R
installed on A side of CBAC

50-conductor ribbon cable with female
socket connectors on both ends;
Digi-Key R326-18-ND1 for
18"-long cable

UBEC

Socket S11

7-7 UCIS connected to TRS Color Computer.

I/O select interface slot:	Address range available	UBEC DIP Switch setting
1	49408-49663 C100h-C1FF	
2	49664-49919 C200h-C2FFh	
3	49920-50175 C300h-C3FFh	
4	50176-50431 C400h-C4FFh	
5	50432-50687 C500h-C5FFh	
6	50688-50943 C600h-C6FFh	
7	50944-51199 C700h-C7FFh	

SOFTWARE SETUP

- Use memory-mapped I/O
- For starting address (depending on slot used for CBAC) set UBEC DIP switch as shown;
- Use PEEK and POKE instructions

View down into card slot connector

Toward back of computer

PINOUTS (MEMORY-MAPPED I/O)

PIN NO.	SIGNAL	PIN NO.	SIGNAL	PIN NO.	SIGNAL
2	A0	12	A10	42	D7
3	A1	13	A11	43	D6
4	A2	14	A12	44	D5
5	A3	15	A13	45	D4
6	A4	16	A14	46	D3
7	A5	17	A15	47	D2
8	A6	18	R/W*	48	D1
9	A7	25	+5VDC	49	D0
10	A8	26	GND		
11	A9	40	Φ0(clock)		

NOTES: R/W* connects to CL1 on UBEC
Φ0 (clock signal) connects to CL2 on UBEC

HARDWARE SETUP

50-pin header; Jameco 923866R installed on A side of CBAC

50-conductor ribbon cable with female socket connectors on both ends; Digi-Key R326-18-ND for 18"-long cable

Socket S11

UBEC

APP-II CBAC
A side faces to the right
Feed ribbon cable through any available slot in back of computer

Apple II computer

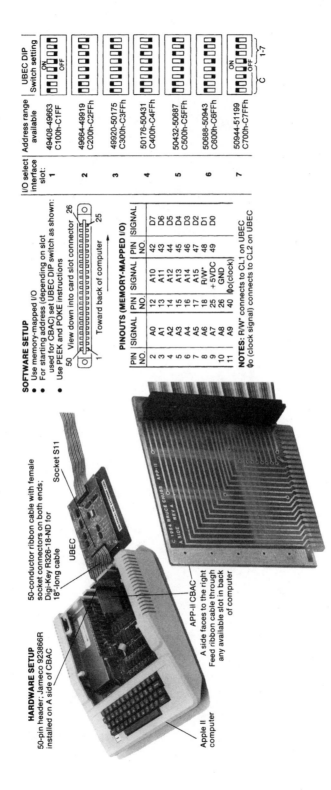

7-8 UCIS connected to Apple II.

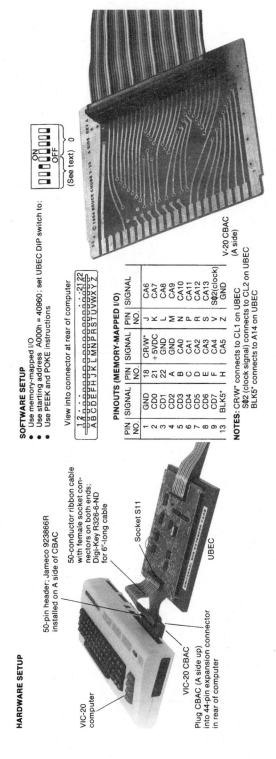

HARDWARE SETUP

VIC-20 computer

50-pin header; Jameco 923866R installed on A side of CBAC

50-conductor ribbon cable with female socket connectors on both ends; Digi-Key R326-6-ND for 6"-long cable

Socket S11

UBEC

VIC-20 CBAC

Plug CBAC (A side up) into 44-pin expansion connector in rear of computer

SOFTWARE SETUP

- Use memory-mapped I/O
- Use starting address A000h = 40960 ; set UBEC DIP switch to:
- Use PEEK and POKE instructions

View into connector at rear of computer

```
  1 2 . . . . . . . . . . . . . . . . . . . . 21 22
  □ □ □ □ □ □ □ □ □ □ □ □ □ □ □ □ □ □ □ □ □ □
  □ □ □ □ □ □ □ □ □ □ □ □ □ □ □ □ □ □ □ □ □ □
  A B C D E F H J K L M N P R S T U V W X Y Z
```

PINOUTS (MEMORY-MAPPED I/O)

PIN NO.	SIGNAL	PIN NO.	SIGNAL	PIN NO.	SIGNAL
1	GND	18	CR/W*	J	CA6
2	CD0	21	+5VDC	K	CA7
3	CD1	22	GND	L	CA8
4	CD2	A	GND	M	CA9
5	CD3	B	CA0	N	CA10
6	CD4	C	CA1	P	CA11
7	CD5	D	CA2	R	CA12
8	CD6	E	CA3	S	CA13
9	CD7	F	CA4	V	SΦ2(clock)
13	BLK5*	H	CA5	Z	GND

V-20 CBAC (A side)

NOTES: CR/W* connects to CL1 on UBEC
SΦ2 (clock signal) connects to CL2 on UBEC
BLK5* connects to A14 on UBEC

7-9 UCIS connected to VIC-20.

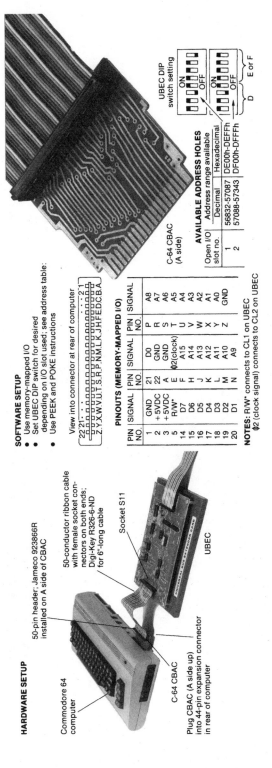

HARDWARE SETUP

Commodore 64 computer

50-pin header; Jameco 923866R installed on A side of CBAC

50-conductor ribbon cable with female socket connectors on both ends; Digi-Key R326-6-ND for 6'-long cable

Socket S11

UBEC

C-64 CBAC

Plug CBAC (A side up) into 44-pin expansion connector in rear of computer

SOFTWARE SETUP

- Use memory-mapped I/O
- Set UBEC DIP switch for desired depending on I/O slot used; see address table:
- Use PEEK and POKE instructions

View into connector at rear of computer

```
22 21 · · · · · · · · · · · · · · · · · · · 2 1
   · · · · · · · · · · · · · · · · · · · · · ·
   Z.Y.X.W.V.U.T.S.R.P.N.M.L.K.J.H.F.E.D.C.B.A
```

PINOUTS (MEMORY-MAPPED I/O)

PIN NO.	SIGNAL	PIN NO.	SIGNAL	PIN NO.	SIGNAL
1	GND	21	D0	P	A8
2	+5VDC	22	GND	R	A7
3	+5VDC	A	GND	S	A6
R/W*		E	Φ2(clock)	T	A5
14	D7	F	A15	U	A4
15	D6	H	A14	V	A3
16	D5	J	A13	W	A2
17	D4	K	A12	X	A1
18	D3	L	A11	Y	A0
19	D2	M	A10	Z	GND
20	D1	N	A9		

NOTES: R/W* connects to CL1 on UBEC
Φ2 (clock signal) connects to CL2 on UBEC

C-64 CBAC
(A side)

AVAILABLE ADDRESS HOLES

Open I/O slot no.	Address range available	
	Decimal	Hexadecimal
1	56832-57087	DE00h-DEFFh
2	57088-57343	DF00h-DFFFh

UBEC DIP switch setting

ON / OFF   D   E or F

7-10 UCIS connected to Commodore 64, 64C, or 128.

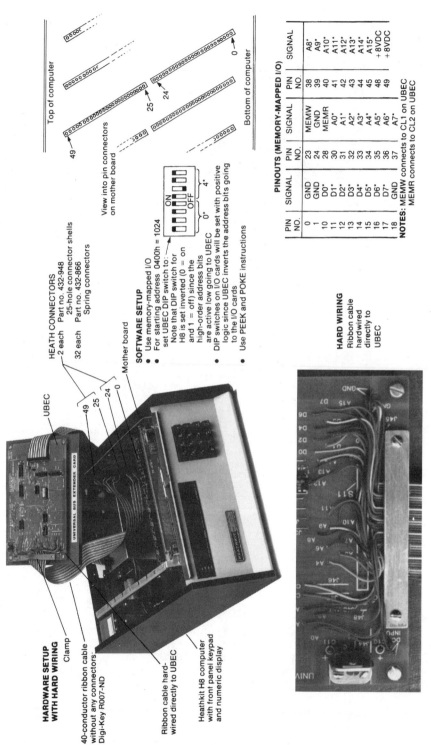

HARDWARE SETUP WITH HARD WIRING

Clamp

40-conductor ribbon cable without any connectors:
Digi-Key R007-ND

Ribbon cable hard-wired directly to UBEC

Heathkit H8 computer with front panel keypad and numeric display

UBEC

HEATH CONNECTORS

2 each   Part no. 432-948   25-hole connector shells

32 each   Part no. 432-866   Spring connectors

Mother board

View into pin connectors on mother board

Top of computer

Bottom of computer

SOFTWARE SETUP

- Use memory-mapped I/O
- For starting address 0400h = 1024 set UBEC DIP switch to:
  Note that DIP switch for H8 is set inverted (0 = on and 1 = off) since the high-order address bits are active low going to UBEC
- DIP switches on I/O cards will be set with positive logic since UBEC inverts the address bits going to the I/O cards
- Use PEEK and POKE instructions

ON
OFF

0*   4*

HARD WIRING

Ribbon cable hardwired directly to UBEC

PINOUTS (MEMORY-MAPPED I/O)

PIN NO.	SIGNAL	PIN NO.	SIGNAL	PIN NO.	SIGNAL
0	GND	23	MEMW	38	A8*
1	GND	24	GND	39	A9*
10	D0*	28	MEMR	40	A10*
11	D1*	30	A0*	41	A11*
12	D2*	31	A1*	42	A12*
13	D3*	32	A2*	43	A13*
14	D4*	33	A3*	44	A14*
15	D5*	34	A4*	45	A15*
16	D6*	35	A5*	48	+8VDC
17	D7*	36	A6*	49	+8VDC
18	GND	37	A7*		

NOTES: MEMW connects to CL1 on UBEC
MEMR connects to CL2 on UBEC

7-11 UCIS direct-wiring for computers without a JLC adapter card.

The chapter's approach is to present the full spectrum of computer hookups. This way, even if your particular computer isn't specifically covered, you will receive sufficient information to see how it can work with UBEC.

Pentium, 486, 386, and 286/AT-Class Computers

Using the IBEC in Chaps. 3 and 5, we covered connecting to Pentium, 486, 386, and 286 or AT-class computers. I've listed these computers here as well, because the UBEC and CBAC can also be used to interface with these machines. Therefore, if you already have a UBEC from a previous computer hookup, you can use it with any one of the computers listed above. You won't have all the options provided by an IBEC, but the system will work just fine.

For example, your UBEC won't provide the address selected test LED. However, that feature can be easily breadboarded and added to the UBEC. The CBAC adapter card you'll need to attach a UBEC to the above computers is the IBPC. It's the same one that's used for the IBM PC and XT and their compatibles, so I'll cover assembling it once we complete discussions on all these machines.

To obtain the port-mapped I/O mode using IBPC, you'll need to cut and jumper a few traces. The important point is that the more popular conventional memory-mapped I/O does work fine with UBEC/IBPC, and it works without any card modification.

IBM PC and XT Computers and Their Compatibles

The address bus of the IBM PC and XT has 20 bits, allowing them to address directly up to 1,048,576 bytes (1024K or 1 Mbyte) of memory, 16 times that provided by other typical eight-bit personal computers. There are five expansion slots on the standard PC and eight on the XT. We can plug our interface into any open slot of the PC, and any except slot 8 of the XT (which is wired differently). There's usually a disk controller card in one slot, a video card takes up another, and, in larger systems, a multifunction I/O-memory-expansion-clock/calendar card can occupy a third. That typically still leaves at least two slots in the PC and five in the XT. The UCIS needs just one.

The IBPC adapter card was originally designed for the PC and XT, and because these computers only activate the lower ten address lines for I/O instructions, their port-mapped I/O space is limited to 2^{10} or 1024 separately addressable ports. This is four times the port I/O of any of the smaller computers we'll consider in this chapter, and four times what the UCIS requires. In spite of these seeming abundant statements, port-mapped I/O isn't attractive for the UCIS with either the PC or the XT. Too many address locations are used by the system, and those remaining are spread out so that it's nearly impossible to find a contiguous block of 256 unused addresses. Therefore, the IBPC wasn't designed to include port-mapped I/O.

Memory-mapped I/O is the better choice for the PC and XT. Because UBEC is not set up to decode the four higher address bits A16–A19, this decoding is placed on IBPC.

Assembling and Installing the IBPC

Figure 7–5 includes the schematic, parts layout, and parts list for this special CBAC. The decoding circuitry is just like the UBEC, but only four bits wide instead of eight. To assemble this CBAC, insert components on the A side of the card and solder them on the B side using the following steps:

☐ *R1-R4.* Bend and insert the leads of each resistor, solder, and trim the leads.

☐ *S1[+].* Be sure to install this DIP socket with pin 1 oriented as shown on the parts layout. Hold the socket tightly against the cards as you solder it in place. *Do not* insert IC U1 at this time, as we'll be doing some testing that is best accomplished without the IC installed.

☐ *S2.* Insert the 50-pin header and hold it tightly against the card as you solder it in place. Solder a couple of pins first near each end, and then when you have verified that it is tight against the card, solder the remaining pins.

☐ *SW1[+].* Install the DIP switch using your VOM to make sure it is oriented so that the contacts are closed when they are set toward the ON label on the CBAC, which is toward the top of the card. Hold the switch tightly against the card as you solder it.

☐ *C1.* Insert this component with the capacitor disk perpendicular to the card, solder it, and trim the leads. Check the photo in Fig. 7–5 carefully and make sure that your C1 is inserted into the top two holes at the location shown and *not* in the two lower feed-through holes that connect to pins 10 and 11 of U1.

☐ *C2[+].* Insert this capacitor with the positive lead in the hole nearest S1, as shown on the parts layout. Incorrect polarity will damage this capacitor. Solder and trim the leads.

If you are using a 286 or AT, 386, 486, or Pentium computer and you would like to modify the IBPC for port-mapped I/O, then continue following these steps; otherwise, skip to cleanup and inspection.

☐ Cut the two traces on the B-side of the card that go to bus connector tabs B11 and B12. Jumper the trace that went to B11 over to B13. Jumper the trace that went to B12 over to B14. This changes the control line inputs to the UBEC from being memory write and read to being I/O write and read.

☐ Either remove U1 or cut the trace line on the A-side of IBPC that leads away from the feed-through hole connecting to U1 pin 11. Either action eliminates the decoding of A16–A19 from impacting the value of AH; in other words, port I/O will only activate the lines A0–A15.

☐ If you would like to be able to switch between conventional memory-mapped I/O and port-mapped I/O, then mount a 3PDT toggle on a small piece of right-angle aluminum or an equivalent, and secure it to the top of the IBPC card. Run wires from each side of the three trace-cuts and connect them to the toggle so that in one direction the toggle makes the needed connections for memory-mapped I/O and in the other direction for port-mapped I/O.

☐ *Cleanup and inspection.*

This completes IBPC, and it's now ready for installation in everything from the original IBM PC up through the latest Pentium. Assuming the standard IBPC configuration, Fig. 7–6 shows the setup for an IBM PC, along with the listing of computer/UBEC pinouts. Note that to set the UCIS starting address you must set both SW1s, the CBAC's as well as the UBEC's.

Note that pin A11 connects to program jumper pad P7 on the UBEC via pad CL3A. With the IBMs, devices other than the CPU can control the address and control lines. When the CPU is in control, the PC pulls its AEN (address enable) line low. Pin A11 is that AEN line, and to keep other devices from inadvertently writing to the UCIS, it is jumped on the UBEC to be gated with MEMW* and MEMR*.

If the example UCIS starting address shown in Fig. 7–6, 655,360 (A0000h), doesn't seem to work on your computer, then follow the procedures described in Chap. 5 for determining the initial memory setting. Remember that for each change you make to the DIP switch settings you also need to make the corresponding change to the DEF SEG statement in your software. Although the Fig. 7–6 example shows the IBPC with a PC, the same installation works for any of the PCs up through the Pentium.

IBM System/2 Model 30

This computer is unique in the System/2 family, because instead of using the IBM proprietary Micro Channel Architecture it retains the AT bus. This makes it a candidate for using either the IBEC or the UBEC/IBPC. It is limited to only three ISA expansion slots. However, the Model 30 has the essential I/O and graphics card features built into the system board, so fitting in the IBEC or IBPC shouldn't be a problem. The slots are horizontal on the System/2 compared to vertical on most of the other machines, but again that doesn't make any difference to our UCIS.

TRS Color Computer Family

Radio Shack's Color Computer, Color Computer II, and Color Computer-64K use the 6809E microprocessor chip, while the Color Computer III uses a somewhat faster 68D09E microprocessor with some 16-bit capability. These compatible chips do not support separate I/O instructions, so we have to use memory-mapped I/O. There is a 40-pin female card-edge connector on each Color Computer's PC board that's handy for attaching our UCIS to the bus. Referred to as the *ROM cartridge port* or *cartridge connector,* this connector is meant for use with cartridge programs such as games, but it provides all the bus signals we need for the UBEC.

The memory map in the reference manual for these computers shows the address range C000h through FEFFh set aside for a plug-in ROM cartridge, a region of 16,128 locations for specialized I/O. With the UCIS plugging into the cartridge connector, this memory range is an excellent spot for our block of I/O addresses. Address C000h corresponds to 49,152 decimal or 48K, and is a good UCIS starting address. Figure 7–7 summarizes the UCIS setup for the Color Computers.

Apple II Family and Compatibles

the Apple II and II+ use the 6502 8-bit microprocessor while the IIe and IIc use the 65C02A, an improved 6502 compatible eight-bit microprocessor. The IIGS uses the 65C816, a 6502-compatible 16-bit microprocessor. None of these chips support separate instructions for port I/O, so they handle all I/O by memory mapping.

Figure 7–8 shows the interfacing setup for the Apples. The 50-wire ribbon cable has a standard socket connector on each end. The IIc, as well as all the Macintoshes except the Macintosh II, do not provide expansion card slots; therefore, they need the USIC for the UCIS as covered in Chap. 4.

The Apple II, II+, IIe, and IIGS are nice machines to interface with UBEC, because they include up to eight 50-pin female card-edge connectors in their cases expressly for system expansion. These slots are numbered 0–7, and the UCIS can use any slot except 0, which is reserved by the manufacturer.

Although slot assignments aren't critical, a typical setup might be:

Slot	Assignment
0	RAM expansion to 64K
1	Printer interface
2	Open
3	Modem and/or 80-column display card
4	Open
5	Extra disk controller for more than two drives or hard disk
6	Disk controller for two drives
7	Open

You can see that there should be plenty of room in the Apple II family for the CBAC. Either slot 4 or 5 is a good choice, because neither one is used for other applications as often as some others. Slot 7 is not as handy; its opening in the back of the case is shorter than those for the other slots, making it more difficult to run the ribbon cable.

The Apple designers made interfacing easy and attractive by providing a separate address decoding line for each of slots 1 through 7. Pin 1 in each of the 50-pin connectors is called I/O SELECT*. The CPU sets this signal low for a given slot to communicate with the address range of that slot. The table in Fig. 7–8 defines the I/O SELECT* address allocations, and you can see that there's a block of 256 bytes for each slot. That corresponds exactly to the maximum 64 I/O cards we can handle with the UCIS.

Because this built-in decoding performs the same function as the U5–U6 circuits on the UBEC, it's tempting to use the Apple I/O SELECT* line to drive the UBEC AH signal line directly. This has three disadvantages. First, I/O SELECT* is inverted logic, and would have to be inverted again to give positive logic for AH. More importantly, using I/O SELECT* makes the UCIS slot-dependent, and there are potential timing problems in that I/O SELECT* is already pregated with the phase zero clock (more about this signal later). For these reasons I recommend that you use the address decoding features of the UBEC rather than the built-in Apple slot decoding features.

Apple combines the read and write control features into a single line labeled R/W* (pin 18). R/W* is brought high to set up a read operation, and low to set up a write. The phase zero clock signal (pin 40) must be gated with the R/W* line to mark the precise times when data on the data bus are valid. Because the hookup shown is not slot-dependent, you are free to use any of the 256-byte address blocks not used by other I/O devices. Corresponding UBEC DIP-switch settings are defined in the Fig. 7–8 address table.

VIC-20

The Commodore VIC-20 uses the same 6502 microprocessor chip as the Apple II; it also uses memory-mapped I/O. The 6502 can address 64K, but to keep initial cost low the VIC-20 typically came with only 29K of its memory space in use. The remaining capacity is available through the memory expansion connector on the rear of the case for expansion of ROM or RAM, or for specialized I/O like the UCIS.

All of the bus lines needed for the UCIS come out to the VIC-20's expansion connector, but they are a bit different from the ones I've explained so far. The eight data lines labeled CD0 through CD7 are pretty standard, but only the 14 low-order address lines, CA0 through CA13, are brought out to the connector.

The VIC-20 manages its memory in eight blocks of 8K each. Only 13 address bits are needed to address 8K, so 14 is one more than required. Each memory block is activated by one of the eight block-select signal lines: BLK1*, BLK2*, BLK3*, BLK5*, RAM1*, RAM2*, RAM3*, I/O1*, and I/O2*.

Block 5 (BLK5*) is often used for ROM programs (such as games) that are plugged into the memory-expansion port, so it's a prime candidate for the UCIS. This block has a total space of 8K from address 40,960 through 49,151. As part of the procedure for addressing this space, the VIC-20 sets block-select line BLK5* low. With BLK5* (pin 13) connected by the CBAC to the A14 address bit on the UBEC, and corresponding segment A14 of DIP-switch SW1 set to OFF, the UBEC can detect the computer addressing both Block 5 and the interface.

The A15 input is not used with the VIC-20, and is jumped to ground on the CBAC. Segment A15 of the UBEC DIP-switch must also be off to achieve a comparison so that AH can go high. The rest of the high-order address bits, A8–A13, operate normally to identify the starting address for the interface I/O.

The VIC-20 memory expansion connector is a 44-pin female card-edge connector with 22 contacts on each side and .156-inch contact spacing. All other computers in this series use .100-in contact spacing. Figure 7–9 shows the interface setup and the computer pinouts.

An aerospace engineering associate of mine programmed a complete aircraft flight management system in a VIC-20, and used it to demonstrate aircraft system performance all around the world. Even in today's world of Pentiums, connecting a UCIS to a VIC-20 can still provide considerable interfacing power.

The Commodore family

The Commodore 64 and 64C use the 6510A microprocessor, and the Commodore 128 uses the 8502 microprocessor. Both these chips have the same basic architecture as the 6502, so these computers also handle all I/O by memory-mapping. The basic 64 and 64C system memory has 64K of RAM and 20K of ROM. This memory allocation includes a built-in BASIC software interpreter, operating system software, a standard display character set, and 4K for I/O. That adds up to a memory capacity of 88K.

How does a computer with a 16-bit address bus address a range of 88K? (Remember that 16 bits can address only 2^{16} or 65,536 locations, otherwise known as 64K.) The technique involves overlaying different memory maps; that is, a given range of addresses might at one time be represented as RAM and at another time as ROM, or I/O. Which condition is in effect at any time depends on the condition of special internal, or external, signal control lines. (Actually this method is much like with the x86 family in terms of overlaying the port addresses and the memory addresses and selecting which is being addressed via the control lines.)

There are two I/O slots that very handily provide memory address holes for the UCIS. These are listed in the Commodore 64 reference manual as reserved for future expansion, and go by the names open I/O slots numbers 1 and 2. Figure 7–10 defines these two slot-address ranges and their corresponding UBEC/SW1, DIP-switch settings. Each range includes 256 bytes, which is the ideal size for the UCIS.

The bus lines come out at the expansion port connector at the rear of the Commodore. This connector, also known as the game connector, is a 44-pin female card-edge connector with 22 contacts on each side and .100-in contact spacing. It provides all the bus signals we need for the UBEC.

Hardwiring Without an Adapter

The CBAC adapter cards from JLC Enterprises are not available for every computer. If yours happens to be one of these, then it's best to give special consideration to the serial form of UCIS using the USIC presented in Chap. 4. However, if you really desire the parallel option, then consider wiring your computer directly into a UBEC.

The original Heathkit H-8, the newer Trionyx H-8, and the newest Trionyx H-16 make good examples of how hardwiring can be accomplished. The processors used by these three machines are the 8080A, Z80, and 8086 respectively, with all three supporting port-mapped and memory-mapped I/O. With port I/O, however, both H-8s limit us to 64 user-available, contiguous port addresses, enough for only 64/4 or 16 I/O cards. It's possible to sneak in more, but there's no point, as all three computers are naturals for memory-mapped I/O. The computers use a 50-wire bus with a 50-pin header on its motherboard. Figure 7–11 shows the interfacing setup and lists the computer to UBEC pinouts.

The first step in hardwiring to any computer is to make and install the UBEC cable clamp shown in Fig. 7–12. Clamp the cable in place so that its last 7 in extends onto the UBEC. Separate the insulated conductors all the way to the cable clamp. Then, after you determine where each wire needs to go, trim it to length to reach into the appropriate hole. Remove about ¼ in of insulation, insert the wire end into the hole, and solder. The UBEC holes involved are A0-A15, D0-D7, CL1-CL3, GND, and DC INPUT. Figure 7–11 shows the location of the cable clamp and the photo insert shows typical wiring termination in the UBEC provided holes.

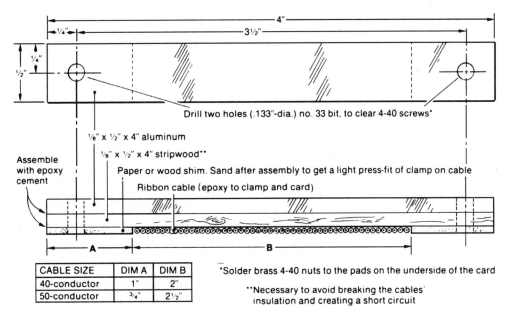

CABLE SIZE	DIM A	DIM B
40-conductor	1"	2"
50-conductor	3/4"	2½"

*Solder brass 4-40 nuts to the pads on the underside of the card

**Necessary to avoid breaking the cables' insulation and creating a short circuit

7-12 UBEC ribbon cable clamp.

Other Computers

If you have a computer that I haven't covered, please don't give up on the UCIS. Study the reference manual for your computer and compare its bus description to examples I've covered. Hopefully I've covered enough different situations whereby you should find enough similarities to be able to tailor the UBEC for your computer.

For example, from a UCIS interfacing point of view, the IBM PC Junior is only slightly different than the basic PC. All required UBEC interface signals are available on the Junior, just as they are on the PC. The Junior employs 60 pinouts on its expansion slot connectors, rather than 62 as on the PC, and the needed signals come out on different pin numbers. To use the IBM PC CBAC you need to file off one pair of edge tabs and add cuts and jumpers to relocate the signal lines.

The main thing to consider in any adaptation is that the computer's data bus, address bus, and control lines must pass through the UBEC to come out as the logic signals shown at the right of Fig. 6–3, the UBEC schematic. Whatever logic signals come out of the computer, and whether they are inverted or noninverted, they must emerge from UBEC with data lines D0–D7 and address lines A0–A7, plus AH, as all noninverted. The RD* and WR* lines, on the other hand, must be inverted. This is essential so that the UBEC interfaces properly with the I/O cards via the IOMB. If you follow these procedures, the UCIS can be readily tailored to fit most any computer.

In the next chapter we'll really start getting the interface working by building the general purpose I/O circuits and the test card.

8
CHAPTER

Digital I/O Circuits and Test Card

In this chapter we'll tackle the I/O end of the UCIS. I'll cover the I/O motherboard and the general-purpose digital I/O cards, and tell you how to build a 5-Vdc power supply for the I/O cards and a test card for checking out the system. The I/O cards are a new design introduced with this second edition. Each digital input (DIN) card reads 24 input lines, and each digital output (DOUT) card controls 24 outputs. One card of each might be enough for a small interface. The average system uses several cards of each, while larger interfaces employ 20 to 40 I/O cards or more.

Being double-sided, the new cards are easier and faster to assemble than before. Using plated-through holes to route circuitry between sides eliminates the need for jumpers. This saves installing the 23 jumpers required with the original output card, and 17 with the input card. Rather than installing the 55 separate resistors required with the original output card, the new design uses resistor networks. Each network contains a group of resistors all packaged together. Lead forming isn't required, and there are fewer pieces to install.

Input noise filtering is a built-in feature on the new input card, making the interface less susceptible to reading *wrong* inputs. By popular request, the new output card is easily configured to activate loads either by connecting them to ground or by driving them with a positive voltage. The first method is called current-sinking and the second current-sourcing. Integrated circuits typically sink higher currents than they can source. Therefore, digital electronics is better off using current-sinking, but sometimes particular loads require current-sourcing.

A possible downside of the new design is the difficulty of fabricating double-sided boards with plated-through holes. Home-type fabrication isn't recommended, and the artwork for the new designs is not included in Appendix A. Also, double-sided boards are more expensive for the professional fabricators and require special equipment to produce. For readers desiring the least expensive approach possible and not minding the extra assembly time, the steps for building the original single-sided cards are retained in Chap. 16, and their artwork is included in Appendix A.

Ready-to-assemble cards are available from JLC Enterprises, including both the new and original versions. If you prefer obtaining your boards fully assembled and tested, they are available from ECO Works, EASEE Interfaces, and AIA. See Appendix A for details. Like all features of UCIS, options are available to best meet everyone's need.

I/O Motherboard

The I/O motherboard (IOMB) is a set of 13 slots or header connectors to let you plug I/O cards into the extended computer bus. Figure 8–1 includes the parts list and shows the parts layout. To fill the card to capacity you need 26 of the 12-pin headers, two for each card slot. The IOMB holds up to 13 I/O cards for a total of 312 discrete wires connected to your external hardware—13 cards × 24 wires per card.

If you need more, you can connect two or more IOMBs together. You simply jump each of the 24 bus traces from one board to the next. Only the first board in the series should have the 2.2-k Ω resistors. These perform as line-terminator pull-up resistors. Here are the IOMB assembly steps:

☐ *Card test.* Use your VOM to check that there are no open circuit traces and no shorts between adjacent traces or pads.

☐ *J1, J2.* Form straight, taut jumpers from #24 solid bare bus wire. Insert, solder, and trim the ends. At this point I recommend that you install the J2 jumper in its "standard" position between trace 10 and the +5 Vdc. Later on we'll revisit this setup, and for certain applications you might desire to move it to its alternate position.

☐ *R1–R19.* Install these resistors as shown in Fig. 8–1. For R13–R19, the resistor leads themselves are too short to reach bus lines 3 through 9. Extend them with bus wire as shown, but slip electrical spaghetti insulation or heat-shrink tubing over the joints to keep adjacent splices from touching. Make sure the jumpers are taut and run straight with no chance of touching each other. For two or more motherboards, install R1–R19 only on the first.

☐ *S1.* If using an IBEC or UBEC, install the 40-pin header. For two or more motherboards, S1 is required only on the first, and S1 is not required on any motherboards if you are using a USIC. It is best to solder only a couple pins first, and then inspect your work to make sure that the connector's bottom surface is pressed flat and tight against the motherboard. If it's not, it's easy to reheat the joints and make it so. Once this direct mating is verified, then solder the remaining pins.

☐ *S2–S27.* Install two 12-pin headers for each I/O card slot, pushing them tight against the card as you solder. It's best to first solder only one pin located near each header end, then place the board and header over a screwdriver clamped in a vice with the blade protruding up. Lay the board bottom side up, with the plastic part of the header resting on the

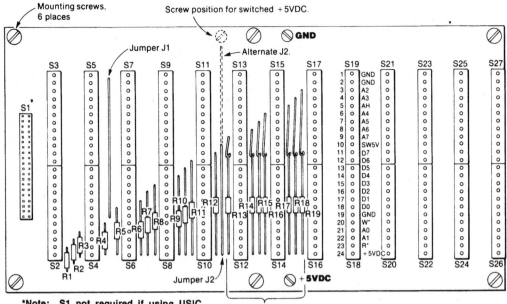

Note: S1 not required if using USIC

R13-R19 leads extended with bus wire

LABELS

GND GND   +5VDC +5VDC

I/O MOTHER BOARD PARTS LIST (In order of assembly)

Qnty.	Symbol	Description
2	J1,J2	Jumpers, make from no. 24 uninsulated bus wire (Belden no. 8022)
19	R1-R19	2.2KΩ ¼W resistor [red-red-red]
1	S1	40-pin double row straight header (Jameco 53532)
26	S2-S27	12-pin headers (Digi-Key WM4410)
3	-	4-40 x ¼" long pan head machine screws (Digi-Key H142)
3	-	4-40 brass nuts

Author's recommendations for suppliers given in parentheses above with part numbers where applicable.
Equivalent parts may be substituted. Resistors are 1/4W, 5 percent and color codes are given in brackets.

8-1 Input/output motherboard parts layout and parts list.

screwdriver blade next to a soldered pin. Reheat the pin while pressing firmly down on the board, and the header will tend to "snap" in place. Repeat for the second pin. Once verified that the header's bottom surface is pressed flat and tight against the board, with the pins straight up, then solder the remaining pins.

☐ *Power terminal screws.* Insert 4-40 screws in the +5-VDC and GND connection holes from the top or component side of the card. Add 4-40 brass nuts on the trace side, tighten them firmly, and solder the nuts to the circuit traces. You can substitute regular nickel-plated nuts, but the brass ones are easier to solder.

☐ *Labels.* Cut out one set of the "+5 Vdc" and "GND" labels from a photocopy of Fig. 8–1, and glue them next to the power terminals as shown. This documentation step is important. A future mistake in hooking up your supply backwards can do severe damage to the electronic parts on your I/O cards.

☐ *Adjacent trace test.* Set your VOM on scale R × 100 and check the resistance between each adjacent pair of pins on any one set of the card slot connectors. Between pins 1 and 2 you should read 0Ω (zero ohms) because these are both ground. Between all the other adjacent pairs—2 and 3, 3 and 4, through 23 and 24—you should read either infinite resistance for an open circuit, close to 4.4 kΩ for two 2.2-kΩ resistors in series, or 2.2 kΩ between lines 23 and 24. A reading close to zero for any of these indicates a solder bridge between adjacent pads somewhere along the two bus lines under test. Locate it by inspection, and reheat the pads to draw the solder away from the gap; then retest to make sure you have solved the problem.

☐ *Ground test.* Keep your VOM on the R × 100 scale, but this time connect one lead to the ground screw (GND). Touch the other lead to each pin in one pair of card slot connectors one at a time. Pins 1, 2, and 19 are grounds and should read 0Ω, but all the rest should be open circuits. If you get different readings, again look for the offending solder bridge and remove it. The most likely spot is among the pads for S1.

☐ *Cleanup and inspection.* For a professional-looking job and to help ensure that your card functions properly, follow the specific steps covered in Chap. 1 regarding cleanup and inspection.

This completes the assembly and stand-alone testing of IOMB. Before constructing the I/O cards, I'll explain how the computer addresses a particular I/O line through UCIS. While an understanding of this could be helpful in later debugging, you may skip ahead to Building the Digital Output Card if you prefer.

Address Decoding

Figure 8–2 shows how the most general 20-bit interface I/O address is decoded in a three-level process. The number of bits that actually need decoding is application-dependent. For example, using an IBEC with the memory-mapped I/O option, you need to decode all 20 bits, while using a UBEC with an Apple II system only requires 16 bits. Regardless of the application, all address bits A8 and higher are decoded by the IBEC (or UBEC plus CBAC). Every decoded high-order bit must match the UCIS starting address, as set on the DIP switches, before signal AH goes high. This is the first level of address decoding.

The same discussion holds if you're using a USIC, except that the computer itself performs the high-level address decoding in selecting the I/O port to which the USIC is attached, and the USIC directly sends out the needed address bits A0–A7. With the USIC, AH is hardwired to high on the USIC card.

Address bits A2–A7 pass directly through the IBEC or UBEC (or are provided as outputs of USIC) to be decoded by the I/O cards, each of which includes a six-segment DIP switch to be set for the individual card's address. If lines A2–A7 match the setting of the I/O card's DIP switch, signal AM* (address middle-order bits) goes low (because it is an active low signal). This is the second level of address decoding.

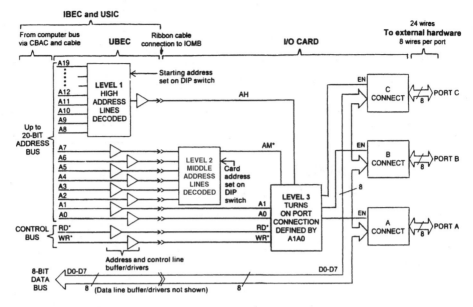

8-2 Address decoding, a three-level process.

Signals AH and AM*, along with either the R* line (for input cards) or W* line (for output cards) are used as enable inputs to the third level of decoding. This level activates one of three ports, A, B, or C, depending upon the values of A1 and A0 as defined in Table 8–1. This way a port is activated only when:

- The interface is being addressed. In this case, address lines A8–A15 (or higher as needed) match the UCIS starting address set on the IBEC (or UBEC and CBAC) DIP switches, causing AH to go high.
- The address lines A2–A7 match the DIP switch settings on the I/O card, the card address, causing AM* to be activated low.
- Either a read or a write operation is taking place, meaning that R* or W* are activated low.

Now let's see how this is all mechanized.

Digital Output Card Schematic

Figure 8–3 is the schematic for the general-purpose digital output card (DOUT). Every output card is interchangeable with every other output card. DIP switch SW1 is used to set each card's unique address. To simplify our understanding of the card, the schematic is divided into five functional areas: card address decode, port select decode, data out buffers, port data latches, and output line drivers.

The address decode circuit uses U4, an eight-bit identity comparator. It works the same as on the IBEC except that unused pins 15 through 18 are grounded, because

Table 8-1. Card Port Selection Based upon A1A0

A1	A0	Port activated	Decimal value of A1A0
0	0	A only	0
0	1	B only	1
1	0	C only	2
1	1	No port activated*	3

*Used to set 8255 control byte on original-design I/O cards and on DAC card.

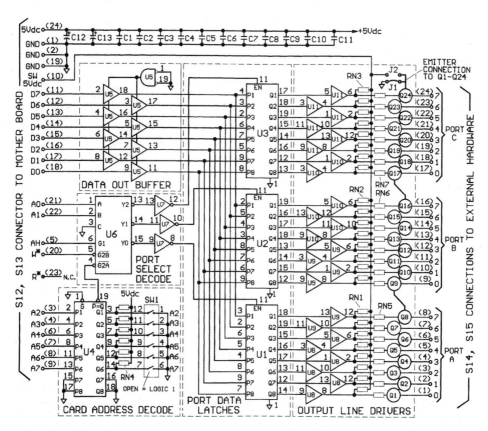

8-3 Digital output card schematic.

only six bits need decoding. This way U4's pin 19 output goes low when the six address inputs match the settings on the DIP switch. This card address compare output, AM* (address middle active low), is connected to pin 4 of U6, the port select decode function.

U6 is a three- to eight-line decoder, but we're using it as a two- to three-line decoder. When enabled, it looks at the two inputs A1 and A0 and sets line Y0 low if A1 and A0 inputs are both low, as indicated by a binary 00. Likewise, for A1A0 inputs at binary 01 line Y1 is set low, and for a 10 line Y2 is set low. Each of these lines perform as a "chip select" to determine which of the three data port latches is activated.

Three gate inputs are required to enable U6, one active high on pin 6 and two active low on pins 4 and 5. This combination works perfectly for our design. AH is connected to pin 6, AM* to pin 4, and the W* to pin 5. Thus, for U6 to be enabled we must satisfy three conditions: (1) AH must be high, indicating that the interface is being addressed; (2) AM* is low, meaning that the specific card within the interface is being addressed; and (3) W* is low, signaling that the computer is performing a write operation. When U6 is not enabled, all three outputs are high.

Each U1–U3 is an octal latch, meaning that it handles eight data lines. For a write operation, the addressed port—A, B, or C—must be latched. This means that the signals on data lines D0–D7 when the latch is selected are transferred to the appropriate port, and remain fixed even after the card is no longer addressed. In fact, the outputs remains latched until either the output port is again selected or the power is turned off. This latching is important, because you don't want the outputs—signal or alarm indications, say, or motor speed settings—to disappear while the computer is doing calculations or communicating with other devices. The outputs stay constant until the CPU sends new data.

To understand latching, we only need to look at a single line's operation, as all the others are the same. Consider pins 7 and 14 of U1, which handle data line D0 for Port A. As long as the gate enable input (EN on pin 11) is high, the D0 output for Port A on pin 14 will be equal to the data line input on pin 7. When the enable input is taken low, the output is held constant (latched) at the last value established before the enable transition. The data remain fixed as long as EN is held low.

Thus, each time we send new data to Port A we must momentarily set EN for U1 high. This action transmits the data being sent out of the computer to the output side of U1. Bringing EN back low latches the data and holds them until new data are transmitted. This latching operation needs to take place right during the short time when Y0 is low, so all we need is an inverter between Y0 and EN. Pins 9 and 8 of U7, a hex inverter, provide this function.

An octal data buffer, U5, is connected between the UCIS's data bus and the parallel connected data inputs to U1, U2, and U3. This way each data line sees only a single load per output card hanging on the bus, thereby reducing bus loading by a factor of 3 to 1.

The outputs from latches U1, U2, and U3 could be used as the direct port outputs from the card. However, IC outputs are quite sensitive and don't provide much drive capability. For example, the 74LS573 used for U1–U3 can sink only 24 mA and source only 2.6 mA. In sink mode you could turn on one LED, but that isn't much drive capability. To increase the card's capability as well as provide options for different forms of output control, I've added an IC buffer and a transistor to each output line.

By substituting different transistors and selecting the jumper position as J1 or J2, it is possible to change the type and level of drive capability. Most applications will do exceptionally well with what I call the standard configuration, using the jumper in J1 position and using the specified 2N3904 NPN output transistor. With this you can drive loads requiring up to 40 Vdc at .3 A. Assuming this configuration meets your needs, you can skip ahead to Building the Digital Output Card if you prefer.

Optional Output Card Configurations

It is easy to vary the output configuration to handle special situations. First off, the 7407 hex noninverting driver provides quite a latitude in system design. It is an open-collector device that can sink 30 mA in the output-low state and can withstand up to 30 Vdc in the output-high state. Because it is open collector, we can override the buffer/driver's output as determined by the latch output, and force the 7407's output low via the pull-up resistor RN1's connection to pin 10 of the IOMB.

Figure 8–4 shows a subschematic detailing the latch, buffer/driver, and output transistor portion of the output card configured for both the standard current-sinking and the optional current-sourcing configurations. Only one output line is illustrated because all are identical. For readers who are into circuit design, many different transistors can be used in either configuration, and the values of the two resistor networks can be tweaked for specific optimizations. However, the four selections illustrated should handle most every need.

I'll cover the standard current-sinking case first, using the 2N3904 NPN output transistor as shown in Fig. 8–4a. Line 10 on the IOMB is normally connected to +5 Vdc. When the output of the latch goes high, the pull-up resistor RN1 pulls the output of the 7407 high. That causes Q1 to conduct, and activate the load. When the latch output is low, the output of the 7407 is low, Q1 is turned off, and the load is deactivated. Software is used to define the latch output, and therefore we have total control of the load's operation.

When any system is powered up, and particularly when you are program debugging or otherwise changing your program, it's very possible and even likely that you will have loads activated when you do not necessarily want them activated. In many cases this doesn't matter, because the program isn't running in normal operation, but in other applications it might make a difference. When it does, such power-up sequencing problems are typically solved by delaying the turn-on of the external hardware until such time as the software is powered up and has all outputs initialized to the desired starting state.

As an alternative approach, we can change the J2 jumper on IOMB so that it connects line 10 to an optional grounding switch as shown in Fig. 8–5. This way, anytime you desire to deactivate the external hardware, that is, turn off all the loads, you simply switch the toggle to the ground position. When line 10 is grounded, it forces the output of the 7407 to ground, causing Q1 to turn off; this switches all the output card signals to the open-circuit state. This action causes all external hardware to de-activate, that is, turn off. For normal operation, turn the toggle to the ON position. It should be

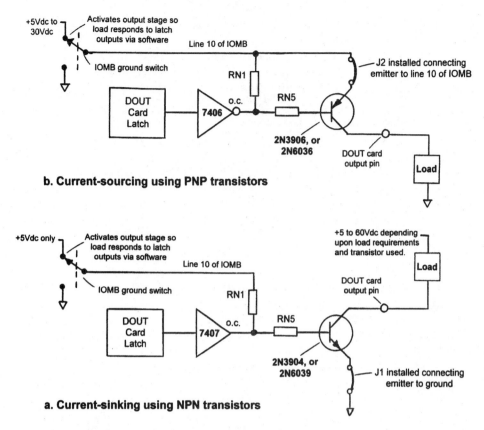

8-4 Current-sinking and current-sourcing configurations.

pointed out that for this standard current-sourcing configuration, the voltage powering the optional grounding toggle switch remains at the +5-Vdc level independent of the voltage applied to the external hardware.

If you need to drive loads with more than the .3 A at 40-Vdc capability, substitute in the 2N6039. This NPN Darlington transistor is a bit more tricky to fit on the card, and it's more expensive, but it increases your drive capability up to about 1 A and 60 Vdc.

The above current-sinking output format should work for most designs. It performs like the computer interface is a large collection of toggle switches, where one side of each switch is connected to ground and the other side is available for connecting to external hardware. Each switch is separately controllable by software, giving you total freedom to turn each circuit on and off as desired. This is the preferred method of using digital electronics to drive external hardware.

You may find situations, however, where you need to connect the interface to external hardware that is all prewired with a common ground, and what you need to do is switch the hot side of the loads. These situations can be handled by replacing the NPN transistor with a PNP, and placing the output card jumper in the J2 position as illustrated in Fig. 8–4b. If you substitute a 7406 inverting buffer/driver in place of the 7407, the circuit retains the same overall logic performance.

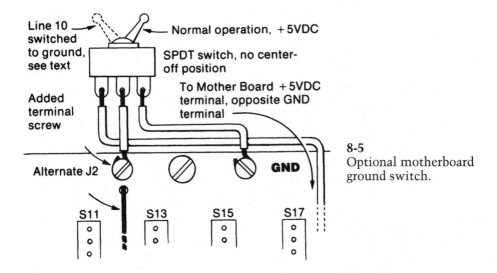

Line 10 switched to ground, see text

Normal operation, +5VDC

SPDT switch, no center-off position

Added terminal screw

To Mother Board +5VDC terminal, opposite GND terminal

Alternate J2

GND

8-5
Optional motherboard ground switch.

S11   S13   S15   S17

Note: Lower end of J2 in alternate position connects to bus line 10 and not to +5VDC.

In this current-sourcing case, Q1's emitter is connected to the IOMB line 10 grounding switch. The voltage applied to the grounding switch must be the same as that powering the loads—but never greater than the 30-Vdc limitation of the 7406. With this switch set for the supply voltage and the output of the latch goes high, the output of the inverter 7406 goes low, causing Q1 to conduct. Having Q1 turned on activates the load. Setting the toggle switch to ground forces the output of the 7406 to low but also grounds the emitter of Q1, effectively keeping the load turned off.

The 2N3906 PNP transistor provides a drive capability of .3 A and 30 Vdc. Changing to the 2N6036 PNP Darlington increases the drive capability to 1 A but still at the 30-Vdc maximum, the limit of the 7406. Because in the current-sourcing case the sum of all the load currents pass through trace 10 of IOMB, and for each DOUT through its smaller traces, it's necessary that you control the size of the loads and the number that are turned on simultaneously.

The current-sinking design is more tolerant of driving heavier loads because the ground traces are wider and there are three ground pins per I/O card. Current-sinking also has the distinct advantage that the IOMB pin 10 trace remains at the +5-Vdc maximum level. When higher voltages are applied, you need to be extremely careful to have each I/O card properly aligned with the IOMB header. Having the connector misaligned by one or more pins will almost certainly result in the higher voltage damaging ICs.

Building the Digital Output Card (DOUT)

Figure 8–6 shows a photo of the digital output card plugged into IOMB. Figure 8–7 is the parts layout, and Table 8–2 gives the parts list. All parts mount on the top or A side of the board and are soldered on the B side. I've indicated steps involving polarity and orientation-sensitive parts with a plus sign in brackets.

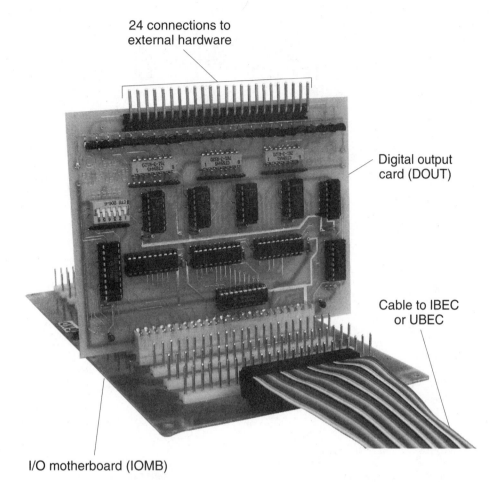

24 connections to
external hardware

Digital output
card (DOUT)

Cable to IBEC
or UBEC

I/O motherboard (IOMB)

8-6 Digital output card plugged into motherboard.

I'll describe how to assemble the output card in the standard configuration, and for the three steps that are different, for the alternate configurations as well. You'll need one output card for every 24 lines of output desired from your interface. Here's how to assemble one:

☐ *Card test.* Use your VOM to check that there are no open-circuit traces and no shorts between adjacent traces or pads.

☐ *J1 or J2.* If you are building the standard current-sinking configuration, insert the jumper in the J1 position. If building the current-sourcing configuration, then insert the jumper in the J2 position. Solder and trim the leads.

☐ *S1–S11[+].* Install the IC sockets with pin 1's orientation as shown in Fig. 8–7.

☐ *RN1–RN4[+].* Install these SIP resistor networks making sure that the pin 1 end, typically marked with a vertical line or dot, is located toward the right

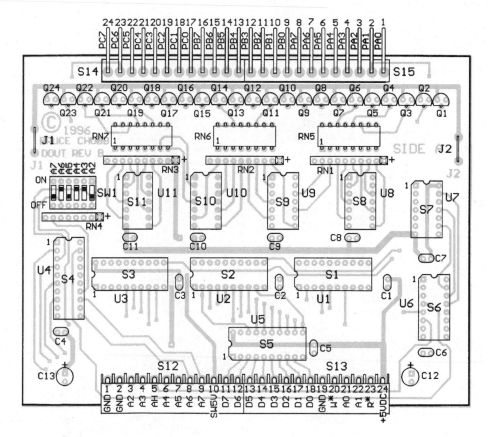

8-7 Digital output card parts layout.

edge of the card as shown in Fig. 8–7. If unsure of the markings, set your VOM on ohms and find the one end of RN that is common to all the others; this is pin 1.

☐ *RN5–RN7.* Install these DIP resistor networks as shown in Fig. 8–7. Orientation isn't important, but for consistency in appearance, it is best to align them just like the ICs. (To reduce cost, you can substitute 24 separate resistors and install them in place of the DIP networks.)

☐ *SW1[+].* Use your VOM to make sure you install this DIP switch so its contacts are closed when thrown toward U4. This may well make the switch read like it is upside down, but this position is absolutely critical to setting correct card addresses. Also, for readers who may have assembled the original-design I/O cards, this closed-switch orientation is opposite.

☐ *S12, S13.* Install the 12-contact side entry connectors by first hooking their nylon retaining fingers over the card edge, then feeding the metal contact pins through the card holes. Make sure all 12 pins of each connector pass through the holes. Hold the connector shell tightly against the card as you solder.

Table 8-2. Digital Output Card Parts List

Qnty.	Symbol	Description
1	J1 **or** J2	Jumper, make from no. 24 uninsulated bus wire (Belden no. 8022) {only install 1 of the 2 jumpers per text}
5	S1-S5	20-pin DIP sockets (Digi-Key A9320)
1	S6	16-pin DIP socket (Digi-Key A9316)
5	S7-S11	14-pin DIP sockets (Digi-Key A9314)
3	RN1-RN3	1.0KΩ 9-element SIP resistor networks (Digi-Key Q9102)
1	RN4	2.2KΩ 7-element SIP resistor network (Digi-Key Q7222)
3	RN5-RN7	120Ω 8-element DIP resistor networks (Digi-Key 761-3-R120)
1	SW1	6-segment DIP switch (Digi-Key CT2066)
2	S12,S13	12-pin Waldom side entry connectors (Digi-Key WM3309)
2	S14,S15	12-pin Waldom right angle headers (Digi-Key WM4510)
11	C1-C11*	.1µF, 50V monolithic capacitors (JDR .1UF-MONO)
2	C12,C13*	2.2µF, 35V tantalum capacitors (JDR 2.2-35)
24	Q1-Q24	2N3904 NPN small signal transistors (Jameco 38359)
3	U1-U3	74HCT573 octal D-type latched flip-flops (Jameco 45090)
1	U4	74HCT688 8-bit Magnitude comparator (Jameco 45129)
1	U5	74LS541 octal buffer/line driver (Jameco 47870)
1	U6	74LS138 3-to-8 decoder (Jameco 46607)
1	U7	74LS04 hex inverter (Jameco 46316)
4	U8-U11	7407 hex buffer/drivers w/open collector output (Jameco 49120)
		Alternate parts for higher current NPN current-sink configuration(see text)
24	Q1-Q24	2N6039 NPN Darlington transistors (Mouser 511-2N6039)
		Alternate parts for standard current PNP current-source configuration(see text)
24	Q1-Q24	2N3906 PNP small signal transistors (Jameco 38375)
4	U8-U11	7406 hex inverting buffer/drivers w/open collector output (Jameco 49091)
		Alternate parts higher current PNP current-source configuration(see text)
24	Q1-Q24	2N6036 PNP Darlington transistors (Mouser 511-2N6036)
4	U8-U11	7406 hex inverting buffer/drivers w/open collector output (Jameco 49091)
		Mating connector for cable
2	—	12-pin Waldom terminal housing (Digi-Key WM2110)
24	—	Crimp terminals (Digi-Key WM2301 for wire sizes 18-24 or WM2301 for wire sizes 22-26)

Author's recommendations for suppliers given in parentheses above with part numbers where applicable. Equivalent parts may be substituted, however if you substitute capacitors, make sure they have .098" (or 2.5mm) lead spacing (as denoted by asterisk above).

☐ *S14, S15.* Install the 12-prong right angle headers and hold them tightly against the card as you solder. Use the same procedure as with the headers on IOMB to ensure that each base is pressed firmly against the card.

Execute only one of the following two steps, depending upon which configuration you are building. Have the board oriented as in Fig. 8–7 and refer to Fig. 8–8 when installing these parts.

☐ *Q1–Q24[+].* These instructions are for the 2N3904 if installed J1 and the 2N3906 if installed J2. Slightly bend the leads of each transistor to fit its holes, and orient each with its flat side facing the resistor networks, with its center (base) lead in the hole nearest the resistors. For a neat installation with Q1–Q24 all lined up in a row, I put a spacer of $\frac{1}{8}$-in^2 stripwood under the curved back sides of the transistors, then press each one down until it just rests on the wood. Once all 24 transistors are soldered in place, I remove the wood and trim the leads.

☐ *Q1–Q24[+]*. These instructions are for the 2N6039 if installed J1 and the 2N6036 if installed J2. Referring to Fig. 8–8, you'll need to bend the leads of each transistor quite extensively to fit its holes, making certain that you have the emitter lead in the emitter hole, the collector lead (the center lead on the transistor) in the collector hole, and the base lead in the base hole,

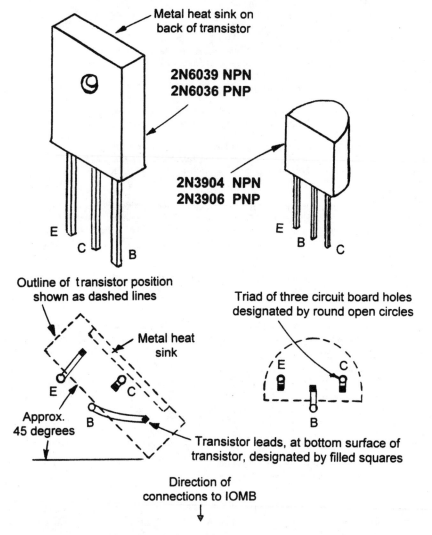

8-8 Installing transistors on the DOUT card.

nearest the resistors. The base lead requires the most bending and forming. The transistors will all end up on about a 45 degree angle, and with the metal tab facing the upper right corner of the board. Solder and trim the leads.

☐ *C1–C11.* Insert, solder, and trim leads.

☐ *C12, C13[+].* Install these capacitors with their plus (+) leads in the plus holes as indicated in Fig. 8–7. Reversed polarity will damage the capacitors.

☐ *U1–U11[+].* See Fig. 1–7 for IC insertion and extraction procedures. Be sure you have the ICs specified, that you put them in the right sockets with correct pin 1 orientation, and that all pins go into the socket. Be sure that for U8–U11 you are using the 7407 if J1 is installed, and 7406 if J2 is installed.

☐ *Cleanup and inspection.* To help ensure that your card functions properly, follow the specific steps covered in Chap. 1 regarding cleanup and inspection. This is a most vital step, so don't cut it short!

☐ *Card power test.* Turn off the +5-Vdc supply and plug the output card into any IOMB slot. Turn the power back on and follow Table 8–3 to see that power reaches each IC. If the voltage is way low, your output card probably has a short circuit. Look for and correct reversed polarity–sensitive capacitors, ICs backward, and/or solder bridges. If necessary, remove ICs one at a time, turning power off and on each time, to help isolate the problem.

☐ *DIP switch test.* Connect your (–) meter lead to ground and move the (+) lead to touch the IC pins listed in Table 8–4. At each position turn the corresponding switch segment on, and you should read +5 Vdc; turning it

Table 8-3. Digital Output Card IC Power Tests

✔	IC	+ METER LEAD ON PIN NO.	- METER LEAD ON PIN NO.
	U1	20	10
	U2	20	10
	U3	20	10
	U4	20	10
	U5	20	10
	U6	16	8
	U7	14	7
	U8	14	7
	U9	14	7
	U10	14	7
	U11	14	7

Each line should read +5Vdc

Table 8-4. Digital Output Card DIP Switch Tests

✔	IC	PIN NO.	DIP SWITCH SEGMENT
	U4	3	A2
	U4	5	A3
	U4	7	A4
	U4	9	A5
	U4	12	A6
	U4	14	A7

With switch segment set to "ON" IC pin should read +5Vdc.
With switch segment set to "OFF" IC pin should read 0Vdc.

off should yield 0 Vdc. If all readings are reversed you have the switch in backward. If only some segments are incorrect, check around for poor soldering.

That completes the output card. I'll discuss the input card's schematic next, but if you wish you may skip ahead to the section on building the digital input card.

Digital Input Card Schematic

Figure 8–9 is the schematic for the digital input card (DIN). The card's address decode section is identical to the output card's; the only difference in the port select decode is that the R* line is connected to pin 5 of U5 instead of the W* line. Also, no latching is required on inputs, because once the computer has performed the READ operation, the data read is stored in memory.

Three 74LS240 octal tri-state inverting bus drivers, U1–U3, are used to connect each of the three input ports to the data bus. The tri-state feature is important so that we don't have multiple chips trying to each connect to the bus at the same time.

The pin 1 and 19 inputs on U1–U3 are normally high, and the devices are held in their tri-state, open circuit condition. Only when pins 1 and 19 for one of the buffer/drivers are brought low by the appropriate output of the port select decode is the input data from the external hardware transferred to the UCIS data bus.

Pull-up resistors, in the form of resistor networks RN1–RN3, are provided to normally hold each of the data input lines high at the 5-Vdc level. To signal an input, the external hardware needs only to pull the input line to ground. I prefer to use the inverter form for buffer/drivers U1–U3 so that when an input line is activated (pulled to ground), the action shows up as a logic 1 in the software. This is the way all the software is written in this book. However, if you prefer the opposite approach, simply replace U1–U3 with their noninverting counterpart, the 74LS244, and rewrite the software accordingly.

The DIN card is designed so that each input line is wired from the external hardware directly into the appropriate pin on U1–U3. This works great for most applications, and is how the original input cards function. However, the interface input lines

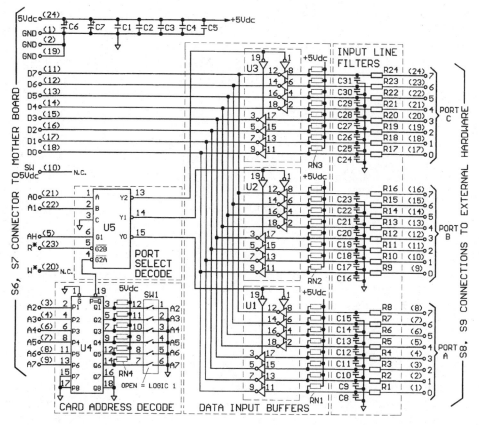

8-9 Digital input card schematic.

emanating from many hardware applications contain extensive electrical noise. Adding a noise filter to each input line can reduce the effects of such noise. If you are not interested in this option, simply skip ahead to the section on building the digital input card.

Optional Input Line Filtering

Adding a low-pass RC (resistor-capacitor) filter circuit to each input line is easy with DIN. Simply cut the thinned-down circuit trace located on the bottom side of the card under each of the resistor locations, and install the corresponding R1–R24 resistors and C8–C31 capacitors. The R and C combination is called a low-pass filter, because they let dc and low-frequency signals pass through with practically zero attenuation while attenuating high-frequency signals.

The product of the resistor and capacitor values (R × C) is called the *filter time constant*. The larger this constant the greater the filtering. The filter's cutoff frequency (in units of cycles per second, or hertz), is calculated as $1/(6.28 \times R \times C)$. AC signals or noise above the cutoff frequency are severely attenuated, while signals below this frequency pass through with little attenuation.

A big question is "How do we go about selecting the best values for R and C?" Well, first off the resistor should never be greater than about 100 Ω. Larger values cause the low input voltage into U1–U3 to rise to a point where it is too close to the .8-Vdc switching threshold of the ICs, thereby actually increasing noise susceptibility.

Going with less isn't good, because lower values do not adequately limit current surges resulting from the capacitor being discharged when the input line is switched to ground by the external hardware. Fix the resistor at 100 Ω and then adjust C to achieve your desired filtering level.

The maximum capacitor that fits on the card is a 470-µF 10-Vdc model (the Digi-Key P6217), which when combined with the 100-Ω resistor produces a filter time constant of .047 s (calculated as .000470 × 100) or 47 ms, and a cutoff frequency of 3.4 Hz. This means that the noise would have to hold the input in the incorrect state for approximately 47 ms before the software would receive an incorrect input. With the 3.4-Hz cutoff, the input line would be very sluggish in responding to valid changes in input. This is extremely heavy filtering, and probably more than is needed for the severest of situations.

Balancing the amount of filtering versus how fast you desire to track valid input changes is important. For example, using QuickBASIC with my Pentium in data acquisition, I read input bytes 200,000 times per second. If you were interested in such fast tracking of inputs and installed the filter time constant of 47 ms, it would take about 9400 samples before you detected an input change, a totally unacceptable situation when you're seeking high-speed data acquisition. Thus, the upper bound on your filter time constant should be about half the time period you expect between reads. In this case the desired maximum capacitor value would be about .02 µF. That's quite a difference.

Large capacitors also have a disadvantage by creating longer-lasting surge currents when card inputs are switched to ground by the external hardware. Using the 100-Ω resistor, this current is 5 V (the normal charge on the capacitor) divided by 100-Ω or 50 mA. Therefore, you wouldn't want to drive this level of filtering if you are using TTL circuits in your external hardware to switch input lines to ground. Larger capacitors make these high-level currents present for longer filter time constants, making the drive requirement worse.

Table 8–5 provides examples for a wide selection of filter capacitor values, along with the resulting filter time constant and cutoff frequency. For example, if your requirement doesn't demand a fast response time for reading inputs and you desire heavy filtering, I would suggest using the 100-Ω resistor with the 10- or 22-µF capacitor. You can adjust these values up and down per Table 8–5 as needed, based upon the speed of your program and the filtering level desired.

One technique is to start with the 100-Ω resistor and the smaller-value capacitors, then gradually increase the capacitor size until any noise problem is solved. Select one size larger, for a design margin, and install the capacitors. Remember, though, always keep your filter time constant under half the time it takes your software to execute one loop, so that you read timely data. Also, for applications where you desire to filter out high-frequency noise, it is better to use tantalum, disk ceramic, or monolithic capacitors in place of the frequently lower-priced electrolytic capacitors.

Table 8-5. Input Filter Time Constant and Capacitor Selection

Capacitor Value (µF)	Filter Time Constant	Filter Break Frequency
.0047	.47 µsec	2 MHz
.01	1 µsec	160 kHz
.047	4.7 µsec	213 kHz
.1	10 µsec	16 kHz
.47	47 µsec	3.4 kHz
1	100 µsec	1.6 kHz
4.7	470 µsec	340 Hz
10	1 msec	159 Hz
22	2.2 msec	72 Hz
47	4.7 msec	34 Hz
100	10 msec	16 Hz
470	47 msec	3.4 Hz

If you are unsure of the level of noise filtering you need, or for that matter if you need it at all, then a good approach is to assemble the cards without installing R1–R24 and C8–C31. Also, this reduces parts cost and reduces assembly time. In the future, if you ever do find that you desire to add filtering, its a simple task to cut the trace on the bottom side of the board and insert the desired parts.

In place of using hardware filtering, you can perform software filtering in your application program. The software simply checks to see that the input has remained stable for two or more samplings before it accepts any change in input. A disadvantage of software filtering is that it requires extra processing time, therefore slowing down your real-time loop. Its advantage is that extra hardware isn't required, and it's easy to vary the effective filter time constant. Examples of software filtering are included on the 3.5-in disk enclosed with this second edition.

Building the Digital Input Card (DIN)

You'll need one input card for every 24 lines of desired input from your external hardware. Figure 8–10 shows a digital input card plugged into IOMB. Figure 8–11 gives its parts layout and Table 8–6 the parts list. Here's how to build one:

☐ *Card test.* Use your VOM to check that there are no open circuit traces and no shorts between adjacent traces or pads.

Skip the next three steps unless you are including input line filtering.

☐ Cut the traces on the bottom side of the board in the necked-down thin-area right under the center of each resistor location. Use an X-Acto knife or a cutoff wheel in a Dremel®-type hand power tool. The cut should be at least $1/32$ in wide, and need not be any deeper than the trace itself.

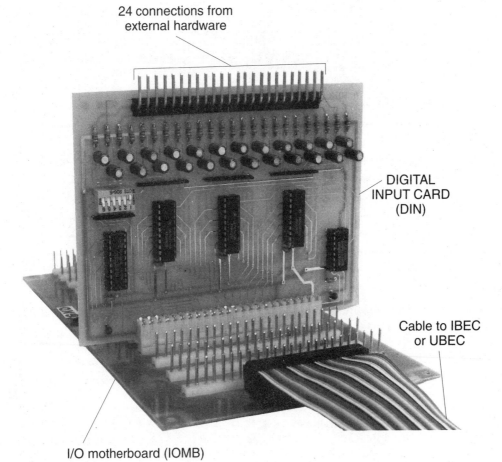

24 connections from
external hardware

DIGITAL
INPUT CARD
(DIN)

Cable to IBEC
or UBEC

I/O motherboard (IOMB)

8-10 Digital input card plugged into motherboard.

☐ Use your VOM to make sure that you have an open circuit across each of
the cuts.

☐ *R1–R24.* Install, solder, and trim leads. Hold off installing the capacitors,
because it's easier to install shorter components first.

☐ *S1–S5[+].* Install the IC sockets with pin 1 oriented as shown in Fig. 8–11.

☐ *RN1–RN4[+].* Install these SIP resistor networks, being careful to have the
correct pin 1 orientation. Solder and trim leads.

☐ *SW1[+].* Use your VOM to make sure you install this DIP switch so that its
contacts are closed when thrown toward U4. This may well make the switch
read like it is upside down, but this position is absolutely critical to setting
correct card addresses. Also, for readers who may have assembled the
original-design I/O cards, this switch orientation is opposite.

☐ *S6, S7.* Install the 12-contact side entry connectors by first hooking their
nylon retaining fingers over the card edge, then feeding the metal contact
pins through the card holes. Make sure all 12 pins of each connector pass

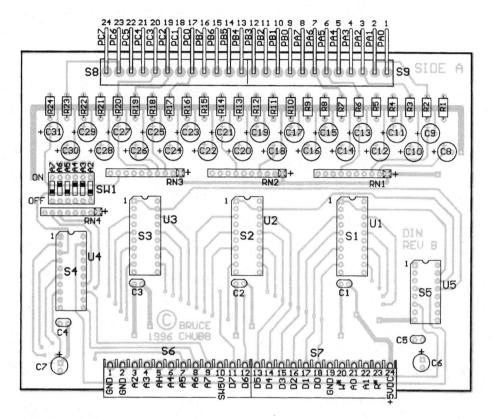

8-11 Digital input card parts layout.

through the holes. Hold the connector shell tightly against the card as you solder.

☐ *S8, S9.* Install the 12-pin right-angle headers, and hold them tightly against the card as you solder. Use the same procedure as with the connectors on the IOMB to ensure that each base is pressed firmly against the card.

☐ *C1–C5.* Insert these components, solder, and trim leads.

☐ *C6–C7[+].* Install these capacitors with their plus (+) leads in the plus holes. Reversed polarity will damage the capacitors.

Skip the next step unless you are including input line filtering.

☐ *C8–C31[+].* Install with the value of your choice per Table 8–5 and corresponding text discussion. For those that are polarity sensitive (tantalum and electrolytic) make sure their plus (+) leads go into the plus holes.

☐ *U1–U5[+].* Be sure that you have the ICs specified, that you put them in the right sockets with the correct orientation, and that all pins go into the socket.

☐ *Cleanup and inspection.* To help ensure that your card functions properly, follow the specific steps covered in Chap. 1 on cleanup and inspection. This is a most vital step, so don't cut it short!

Table 8-6. Digital Input Card Parts List

Qnty.	Symbol	Description
4	S1-S4	20-pin DIP sockets (Digi-Key A9320)
1	S5	16-pin DIP socket (Digi-Key A9316)
3	RN1-RN3	2.2KΩ 9-element resistor networks (Digi-Key Q9222)
1	RN4	2.2KΩ 7-element resistor network (Digi-Key Q7222)
1	SW1	6-segment DIP switch (Digi-Key CT2066)
2	S6,S7	12-pin Waldom side entry connectors (Digi-Key WM3309)
2	S8,S9	12-pin Waldom right angle headers (Digi-Key WM4510)
5	C1-C5*	.1μF, 50V monolithic capacitors (JDR .1UF-MONO)
2	C6,C7*	2.2μ, 35V tantalum capacitors (JDR 2.2-35)
3	U1-U3	74LS240 octal inverting buffer/drivers (Jameco 47141)
1	U4	74HCT688 8-bit Magnitude comparator (Jameco 45129)
1	U5	74LS138 3-to-8 decoder (Jameco 46607)

Fixed resistor and most popular alternate capacitors for different levels of filtering (see text)

24	R1-R24	100Ω resistor [brown-black-brown]
24	C8-C31	.1μF, 35V tantalum capacitor (Jameco 33486)
24	C8-C31	.1μF, 50V monolithic capacitors (JDR .1UF-MONO)
24	C8-C31	1μF, 35V tantalum capacitor (Jameco 33662)
24	C8-C31	10μF, 16V tantalum capacitor (Jameco 94060)
24	C8-C31	22μF, 16V radial lead electrolytic capacitor (Digi-Key P6224)
24	C8-C31	47μF, 16V radial lead electrolytic capacitor (Digi-Key P6226)
24	C8-C31	100μF, 16V radial lead electrolytic capacitor (Digi-Key P6227)

Note: Lead spacing for capacitors should be between .079"(2mm) and .1" (2.5mm)

Mating connector for cable

2	—	12-pin Waldom terminal housing (Digi-Key WM2110)
24	—	Crimp terminals (Digi-Key WM2301 for wire sizes 18-24 or WM2301 for wire sizes 22-26)

Author's recommendations for suppliers given in parentheses above with part numbers where applicable. Equivalent parts may be substituted, however if you substitute capacitors, make sure they have .098" (or 2.5mm) lead spacing (as denoted by asterisk above). Resistors are ¼W, 5 percent and color codes are given in brackets.

☐ *Card power test.* Turn off the +5-Vdc supply and plug the output card into any IOMB slot. Turn the power back on and follow Table 8–7 to see that power and ground reach each IC. If the voltage is way low, your output card probably has a short circuit. Look for and correct reversed-polarity-sensitive capacitors, backward ICs, and/or solder bridges. If necessary, remove the ICs one at a time, turning the power off and on each time, to help isolate the problem.

☐ *DIP switch test.* Connect your (–) meter lead to ground and move the (+) lead to touch the IC pins listed in Table 8–4, the same as for DOUT. At each position turn the corresponding switch segment on, and you should read +5 Vdc; in the off position you should measure 0 Vdc.

This completes the assembly of the general-purpose digital input card.

Table 8-7. Digital Input Card IC Power Tests

✓	IC	+ METER LEAD ON PIN NO.	- METER LEAD ON PIN NO.
	U1	20	10
	U2	20	10
	U3	20	10
	U4	20	10
	U5	16	8

Each line should read +5Vdc

Building a 5-Vdc Power Supply

You can purchase industrial/laboratory type 5-Vdc power supplies from electronic suppliers, but another alternative is to purchase a surplus computer power supply. For example, Jameco continually lists a variety of computer power supplies with +5-Vdc ratings ranging from 5 to 30+ A , with prices ranging from $10 to $60. Additionally, these supplies typically provide ±12 Vdc. The +5 Vdc can drive the UCIS logic level requirements, including the I/O cards and system LEDs and so forth, and the ±12 Vdc can power other hardware such as relays and lamps. For model railroad applications it can power the Optimized Detectors presented in Chap. 17.

It's hard to beat the surplus prices even when you build your own supply from scratch. If you do buy one, make sure you obtain a hookup wiring diagram and/or a color-code listing showing which output wire is which voltage, ground, power good, on-off control, and so on. Often the supplies require that one of the wires be connected to ground to turn on the supply. Also, computer supplies frequently require a minimum load, such as 10 percent of their rated output, before they will turn on. In these cases you may need to temporarily add a power resistor or an automotive lamp across the supply's 5-Vdc output until your interface is providing enough of a load.

Current requirements vary with the application, so it's a good idea to rough out what yours will be. Simply total the items you expect to drive with the 5-Vdc supply. Each I/O card draws about .25 A, each LED turned on draws about .02 A, and from there you add any other loads you desire. Here's a sample breakout of the 5-Vdc current requirements for one of my particular applications:

Load	Current, amp
1500 LEDs @ .02 A (for my application I could assume no more than ⅓ turned at a time)	10
24 I/O cards @ .25 A	6
Additional decoding/encoding circuitry	6
Miscellaneous allocation	2
Total	24

Even if you need a lot less, it's still a good idea to purchase or build a hefty supply to allow for future expansion.

If you are building your supply from scratch, the voltage regulators I presented in the first edition (the ones with 5-A and 10-A capacities) are no longer manufactured. The best choice now is the LM323K, with 3-A capacity. It is possible to use the LM338K adjustable voltage regulator that has a 5-A capacity, but it costs about twice as much, plus it requires some external circuitry to set the voltage. JLC Enterprises may come out with a small circuit card that does this for you, but at this point it's easier to use the LM323K.

For efficient operation and to prevent the regulator and its heat sink from having to dissipate too much heat, the input voltage should be held reasonably close to the maximum desired output voltage plus about 3.5 V. You can get by with a larger differential voltage if your current drain is less than the specified maximum. Having the input much lower than 8.5 Vdc (5 + 3.5) causes the regulator to stop regulating.

Figure 8–12 shows a power supply schematic that can be tailored to meet your UCIS requirements. Table 8–8 lists the necessary parts. The power transformer T1 and the large filter capacitor C1 can be costly, but you can use low-priced surplus parts for these. I have found good sources to be Fair Radio Sales, P.O. Box 1105, 1016 E. Eureka St. Lima, OH 45802, and All Electronics, P.O. Box 567, 14928 Oxnard St., Van Nuys, CA 91408.

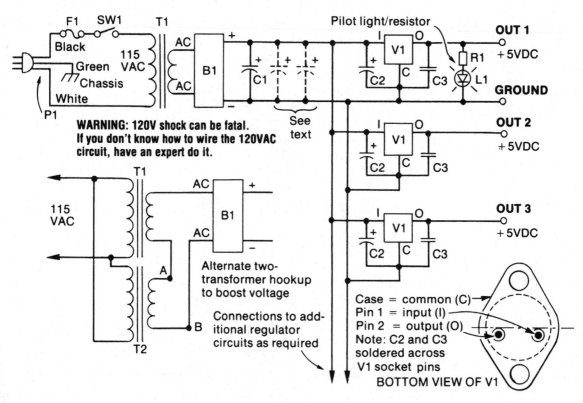

8-12 Schematic of 5-Vdc power supply.

Table 8-8. 5-Vdc Power Supply Parts

Qnty.	Symbol	Description
1	P1	Three-conductor power cord w/plug (Jameco 37997)
1	—	Strain relief (Radio Shack 278-1636)
1	F1	Panel-mount fuse holder (Jameco 18702)
1	—	2A slow-blow fuse (Jameco 25646)
1	SW1	Toggle switch (Jameco 76241)
1	T1	Power transformer (see text)
1	B1	30A, 200V Bridge rectifier MDA999-3 (Jameco 25574)
1	C1	Filter capacitor(s) (see text)
as reqd.	V1	LM323K 5Vdc fixed positive 3A regulator (Jameco 23667)
as reqd.	—	TO-3 mounting kit with socket (Mouser 534-4732)
as reqd.	C2	1μF, 35V tantalum capacitor (Jameco 33662)
as reqd.	C3	100000pF, 50V ceramic disk capacitor (Digi-Key P4164)
as reqd.	H1	Fin-type heat sink for B1 and all V1s (Jameco 16512)
1	R1	330Ω resistor [orange-orange-brown]
1	L1	Diffused red LED (Digi-Key P300)
1	—	Terminal strip (Mouser 504-TB100-02)
as reqd.	—	Grommets (Radio Shack 64-3025)

Author's recommendations for suppliers given in parentheses above with part numbers where applicable. Equivalent parts may be substituted. Resistors are ¼W, 5 percent and color codes are given in brackets.

The transformer output should be at least 6.3 Vac, but not over 7.3 Vac. Some transformers have taps on the primary for power line variations, for example 107–125 Vac. If so, wiring your line-cord input to the lower primary helps raise the output voltage and vice versa. If in doubt about what current capacity to buy, choose the larger value. Also consider weight as a factor, because it indicates how much iron is in the core laminations. For a given current capacity, more iron gives better regulation, thus providing a more constant output voltage under changing loads. Heavier transformers also tend to run cooler.

A rule of thumb for the size of filter capacitors like C1 is to allow 5000 μF per amp of output. You needn't try to find all the capacitance you need in a single capacitor. In Fig. 8–12, the dotted additional capacitors in parallel with C1 show how to add as many of these parts as it takes to get what you need. For example, four 50,000-μF capacitors would be good for 20 A. C1's voltage rating isn't important as long as it's 15 Vdc or greater. When you order large filter capacitors, make sure you also purchase mounting clamps for easy installation.

Figure 8–12 shows multiple V1 regulators, assuming that you're using the LM323K. With an adequate heat sink, each V1 will deliver up to 3 A. To meet my 24-A requirement, for example, you would want at least eight regulators. Ten would be better so they would run cooler. Its nice to divide up your system requirements between the different regulators so that currents stay well below maximum ratings, and a short in one function doesn't affect others.

Figure 8–13 shows a 5-Vdc, 3-A power supply that's adequate for smaller interface systems. The basic structure is a 7 × 13 × 2-in aluminum chassis (Bud #AC-409) with a ¹⁄₁₆-in thick aluminum sheet used as the component-mounting surface on top. For safety you should always enclose the high-voltage connections to prevent accidental shock.

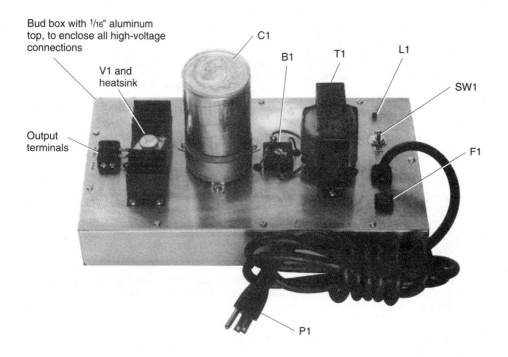

Bud box with ¹⁄₁₆" aluminum top, to enclose all high-voltage connections

V1 and heatsink

Output terminals

C1

B1

T1

L1

SW1

F1

P1

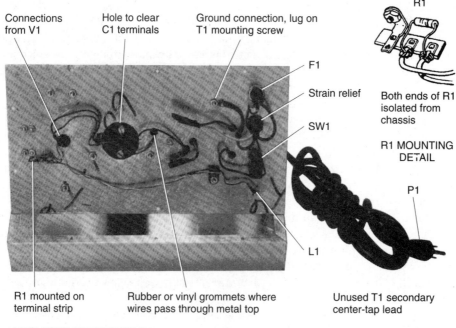

Connections from V1

Hole to clear C1 terminals

Ground connection, lug on T1 mounting screw

F1

Strain relief

SW1

L1

R1

Both ends of R1 isolated from chassis

R1 MOUNTING DETAIL

P1

R1 mounted on terminal strip

Rubber or vinyl grommets where wires pass through metal top

Unused T1 secondary center-tap lead

ENCLOSED CONNECTIONS

8-13 Power supply example.

All parts should be mounted before you begin wiring except V1 and its heat sink. How you lay out your power supply isn't critical, but use the checklist below so you won't forget to include a necessary item.

☐ Fuse holder F1 and toggle switch SW1
☐ Transformer T1
☐ Bridge rectifier B1, with heat sink if required
☐ Filter capacitor C1, or two or more C1s in parallel
☐ Voltage regulator V1, or two or more V1s, with a heat sink for each
☐ Pilot light LED L1 and current-limiting 330-Ω, ¼-W resistor
☐ Output screw terminals, including at least one ground terminal and one +5-Vdc terminal for each V1 regulator.

V1 must be mounted on a heat sink, and for currents greater than 5 A, B1 should have a heat sink too. B1's case is typically insulated and V1's case is the common or ground connection, so the heat sinks don't need to be insulated from each other.

High-voltage wiring

Install the high-voltage wiring as follows:

Warning: *If you are unfamiliar with electrical safety precautions, you should not make the 115-Vac connections. These connections must be made properly to avoid the possibility of a dangerous, potentially fatal, electrical shock.*

☐ *Power line cord.* Trim about 6 in of the outer insulation casing off the end of the cord. Separate its three wires, then trim about ¼-inch of insulation from each.
☐ *Green wire (ground).* Solder to a lug mounted under any one of the transformer mounting screws. This grounds both the metal frame of the transformer and the metal chassis, if your power outlet wiring is properly grounded.
☐ *White wire/T1.* Solder the wire to one of the transformer primary lugs. If exposed, insulate this connection with heat-shrink tubing or electrical tape.
☐ *Black wire/F1.* Solder the wire to the end tab of the fuse holder.
☐ *F1/SW1.* Run a piece of #18 stranded and insulated hookup wire from the side tab of the fuse holder to one tab of the toggle, and solder both connections.
☐ *SW1/T1.* Run a second piece of #18 hookup wire from the other tab of the toggle to the other transformer primary lug. If the latter connection is exposed, insulate it with heat-shrink tubing or electrical tape.
☐ *Inspection.* Examine all connections to make sure that the high-voltage wiring is as shown in Fig. 8–13, and that any exposed connections are insulated.

☐ *Voltage test.* Set your VOM to measure a voltage of around 10 Vac. Place a 2-A fuse in the fuse holder, plug in the line cord, turn the toggle on, and measure the voltage at the transformer secondary. It should be just a bit higher than the rated value for your transformer, for example, 6.5–7.5 Vac for a 6.3-V transformer without a load. Make a note of your reading. If it isn't between 6 and 8 Vac, you either have the transformer wired incorrectly or have an unsuitable transformer. Unplug the line cord and correct this before proceeding.

Low-Voltage Wiring

Make sure the power cord is unplugged before proceeding with the low-voltage connections.

☐ *T1/B1 (One of two).* Run a length of #18 or larger insulated hookup wire from either of the transformer secondary lugs to either of the B1's AC input lugs, and solder them.

☐ *T1/B1 (2 of 2).* Run a length of #18 or larger insulated hookup wire from the other transformer secondary lug to the other B1 AC input lug, and solder it.

☐ *B1/C1 (positive).* Run a #18 or larger insulated hookup wire from the positive (+) lug of B1 to the positive terminal of C1, or to the positive terminal of each C1 if you use more than one filter capacitor.

☐ *B1/C1 (negative).* Repeat the above step, but connecting all the negative (–) terminals.

☐ *Voltage test.* Plug in the line cord and set your VOM to measure dc volts. Touch the meter's plus (+) and minus (–) leads to the plus and minus terminals of C1, and make a note of your measurement. If it isn't between 8 and 10 Vdc, you have made a wiring error, a measurement error, or have a faulty B1 or C1(s).

Warning: *Until your power supply is finished and connected to a load, voltage will be stored in the capacitor even after you turn off the SW1 power switch and unplug the ac line cord. To avoid shocks, you should hook up a 100-Ω 1-W resistor across the capacitor terminals for about 20 s. Also, connect your VOM across the capacitor and watch the voltage drop as the resistor discharges it to a safe level.*

☐ *V1[+].* Attach the regulator to its heat sink. If you mount the heat sink on the same metal chassis you've grounded through the green wire of the power line cord, you must electrically isolate V1 from its heat sink. Figure 8–14 shows how to isolate V1 using a TO-3 transistor socket and a TO-3 insulated washer with heat sink grease (Radio Shack 276-1372) for heat transfer.

Apply a film of heat sink grease to the bottom of V1 and fit the mica insulator washer—the grease will hold it in place. Make sure the holes line

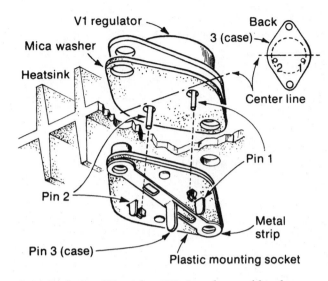

8-14 Isolating V1 with a TO-3 socket and hardware.

up, then apply grease to the bottom of the washer. Fit the socket to the heat sink from the bottom, making sure the socket's plastic bushings project into the mounting holes and that the socket connections for the two regulator pins line up with the corresponding clearance holes in the heat sink. If things don't line up or fit correctly, you have the socket backward or the wrong heat sink.

Hold the socket in place and plug in V1, making sure the washer's screw holes stay in line with the regulator's holes. Then insert the two mounting screws and tighten until the socket is tight against the bottom of the heat sink, with the plastic bushings in their holes.

If your assembly doesn't require isolation, you can eliminate the mica washer or even eliminate the socket by bolting V1 directly to the heat sink and soldering the connecting wires directly to the regulator pins and a solder lug attached to one of the mounting screws.

☐ *Isolation test.* With your VOM set on scale R × 1, connect one VOM lead to the heat sink. Touch the second lead to the heat sink too, and adjust your meter to read 0 Ω. Then touch the second lead to all three lugs of the socket. The two side or pin lugs should read very high resistance, nearly an open circuit. For an isolated assembly using the mica washer, the case connection should also read a very high resistance; with a nonisolated assembly it should read 0 Ω. If you get 0 on a lug where you should find a high resistance, that lug is touching the heat sink and is not isolated. Rework your assembly to get the necessary isolation.

☐ *V1 connections.* Solder a different color #18 or larger insulated wire to each regulator socket pin, and note your colors. Write them down for future reference.

☐ *C2[+].* Wrap the plus (+) lead around V1 socket pin 1, and the minus (–) lead around V1 socket pin 3, keeping both exposed leads as short as possible so they won't touch anything else. Solder the connection at pin 1, being careful not to unsolder the wire previously attached.

☐ *C3.* Wrap one lead around V1 socket pin 2 and the other around V1 socket pin 3, keeping both exposed leads as short as possible so they won't touch anything else. Solder both connections, being careful not to unsolder the previous connections on these pins.

☐ *Heat sink installation.* Attach the heat sink to your power-supply chassis, making sure that you maintain V1's isolation by not letting bare wires or pins touch the chassis or heat sink.

☐ *V1/C1 (negative).* Connect the wire from V1 pin 3 to the minus (–) terminal of C1.

☐ *V1/C1 (positive).* Connect the wire from V1 pin 1 to the plus (+) terminal of C1.

☐ *Labels.* Cut out the second set of labels from Fig. 8–1 (or a photocopy) and glue them beside the power supply output terminals, to mark one screw terminal as the +5-Vdc output and the other as ground.

☐ *Output connection.* Connect the wire from V1 pin 2 to the +5-Vdc output terminal.

☐ *Ground connection.* Run a #18 or larger wire from the ground output terminal to the negative (–) terminal of C1.

☐ *L1(pilot light)[+].* This LED lights up to indicate when the power supply is turned on. Use a small, solid wire to connect it, along with the 330Ω ¼W resistor as shown in Fig. 8–13.

☐ *Final test.* Plug in the power cord and turn on the toggle; the pilot light should come on. Use your VOM to measure the output of the supply and make a note of your measurement. If everything is working correctly your output reading will be between +4.9 and +5.1 Vdc, the tolerance limits of the regulator. This completes the +5-Vdc power supply.

You can also use these instructions for building supplies with greater current capacity. Simply use a larger chassis box, use a higher-current transformer, mount the bridge rectifier on a heat sink, mount larger (and/or more) filter capacitors in parallel, use heavier wire, and allow space for more V1 regulator/heat sinks using the same circuit shown in Fig. 8–12.

Building the Digital Output Test Panel

In later chapters we'll be using software to test our interface. The first set of instructions, or program, will be a short one to show that a message can be sent from the computer to the output card's output terminals. This will test the computer's ability to write to each of 24 output lines. To do that efficiently we'll need a digital output test panel (DOTEST) as shown in Fig. 8–15. Ready-to-assemble DOTEST cards are available from JLC Enterprises, and fully assembled and tested panels are available from ECO Works, EASEE, and AIA. See Appendix A for

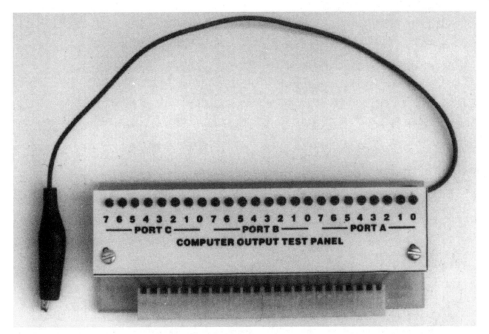

8-15 Digital output test panel DOTEST.

details, including the artwork for those wishing to fabricate their own DOTEST card from scratch.

The test panel's LEDs display the status of each of the 24 output lines at any time. This makes the initial checkout a snap and can also come in handy any time you need to debug the UCIS or confirm that it's operating properly.

There's an extra set of pads on the DOTEST card marked A1, A2, and so on. If the corresponding pads are connected using insulated jumpers and the underside traces cut, the card tests the original-design output card covered in Chap. 16. For this chapter we can simply ignore these extra pads.

Figure 8–16 includes the schematic and parts layout for the Test Panel. Table 8–9 is the parts list. I'll just name those assembly steps like those you've already performed on other cards, and give details on anything new.

☐ *Card inspection.*
☐ *R1–R24.*
☐ *S1, S2.*
☐ *Panel plate.* Make the LED display panel as shown in Fig. 8–17. You can lay a photocopy of the panel face, Fig. 8–18, over the aluminum and punch through the paper to mark the metal for drilling. If you can't find aluminum handy, a piece of .06- or .08-in styrene sheet (available at most hobby shops) will work fine.
☐ *L1–L24, positioning[+].* Install the LEDs with the positive leads in the holes next to the top of the card and the negative leads in the holes next to the resistors. Figure 8–17 shows how to identify LED leads. DO NOT solder the LEDs at this time.

SCHEMATIC

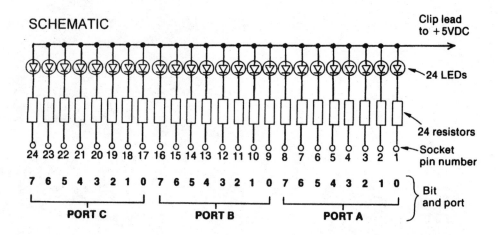

PARTS LAYOUT

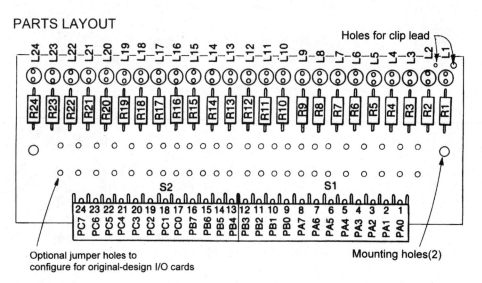

Optional jumper holes to
configure for original-design I/O cards

Mounting holes(2)

8-16 Digital output test panel schematic and parts layout.

- ☐ *Panel assembly.* Mount the panel on the card as in Fig. 8–17, carefully working each LED into its panel hole.
- ☐ *L1–L24, soldering.* One at a time, push the lip of each LED firmly against the panel, then solder and trim the leads. Solder the two 4-40 brass nuts to their circuit board pads.
- ☐ *Clip lead.* Cut a clip lead in two, leaving one end about 12 in long. Strip off ⅛ in of insulation from the cut end and install it as shown in Fig. 8–17.
- ☐ *Cleanup and inspection.*
- ☐ *Panel face.* Cut the panel face out of Fig. 8–18 or a photocopy. Dismount the panel and use spray adhesive to cement the face to it with the dots centered over the holes. Use an X-Acto knife with a #11 blade to trim excess

Table 8-9. Digital Output Test Panel Parts List

Qnty.	Symbol	Description
24	R1-R24	330Ω resistors [orange-orange-brown]
2	S1,S2	12-pin Waldom side entry connectors (Digi-Key WM3309)
24	L1-L24	Red diffused T1 size LEDs (Digi-key P363)
2	—	No. 4 x ¼"-long standoffs (Digi-Key J167)
2	—	4-40 x ½"-long pan-head machine screws (Digi-Key H146)
2	—	4-40 brass nuts
1	—	Clip lead (Jameco 10444)
1	—	1/16" x 1¼" x 5" aluminum (or styrene) LED panel

Optional jumpers if building card for testing original-design I/O cards (COUT24 and CIN24)

24	J1-J24	Jumpers, make from no. 24 insulated bus wire

Author's recommendations for suppliers given in parentheses above with part numbers where applicable. Equivalent parts may be substituted. Resistors are ¼W, 5 percent and color codes are given in brackets.

paper off the edges, then work from the face side to trim out the 26 holes. Remount the panel.

☐ *Card test.* Attach the card's clip lead to the +5-Vdc terminal of your power supply. Attach another clip lead to the ground terminal and, with the supply turned on, touch its other end to each of the 24 connector inputs. Only one LED should light for each input, and their order should exactly correspond to the connector pinout. If more than one lights you have a solder bridge. If one doesn't light you have a poor solder joint, a bad LED, or a reversed LED. Debug until all 24 LEDs work correctly.

If you need to test a number of DOUTs configured as current sources, you can build up a second test panel with the LEDs installed the opposite way. When you conduct the test, you'll connect the clip lead to ground rather than +5 Vdc.

This completes the assembly and test of the test panel. We need to assemble one more item, a special cable, to be all set for system testing.

Assembling the Wraparound Test Cable

For our system testing we will be wrapping the outputs from an output card back in as inputs to an input card. For that we need to construct the wraparound test cable shown in Fig. 8–19. Each pin of one connector is simply wired straight to the corresponding pin of the other connector. Figure 8–20 shows how to use conventional tools to manually crimp and solder the terminals to the wires and insert them into the connector shells.

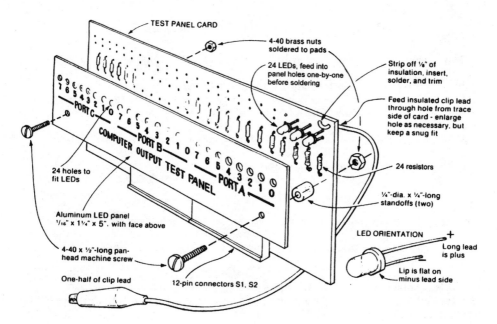

8-17 Digital output test panel assembly.

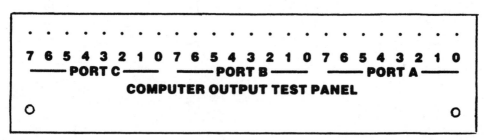

Actual size panel face

8-18 Digital output test panel artwork.

8-19 Wraparound test cable.

Because you will need to repeat this process later to connect external hardware devices to your I/O cards, you might want to invest in a special crimping tool designed specifically to crimp these particular terminals. Once you get the hang of using it, crimping goes much easier and faster. Also, because the connections made by the tool are so tight, you can typically skip the soldering requirement.

The particular tool I use is available from Digi-Key as WM9903. It is manufactured by Waldom, and their part number is W-HT-1919. Instructions for using the crimping tool are printed on the back of the package. Don't try to use a general-purpose crimping tool, as they are not designed to wrap the tabs around the wire and around the insulation. You will not obtain good connections. Without the tool designed specifically for this task, you are better off following the steps in Fig. 8–20.

If you are constructing a UCIS using an IBEC, proceed to the next chapter; if you are using the serial USIC, skip to Chap. 10, and if you are using a UBEC/CBAC skip to Chap. 12.

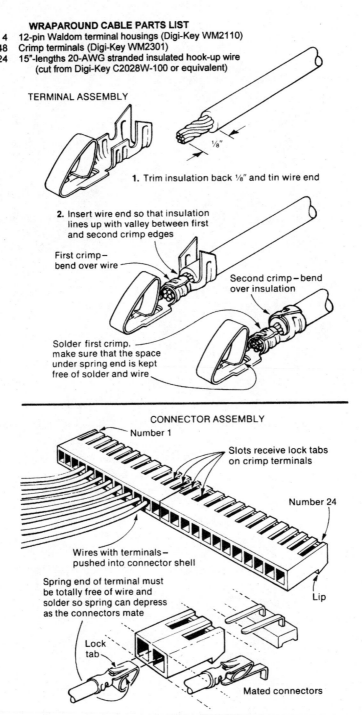

WRAPAROUND CABLE PARTS LIST
4   12-pin Waldom terminal housings (Digi-Key WM2110)
48   Crimp terminals (Digi-Key WM2301)
24   15"-lengths 20-AWG stranded insulated hook-up wire
      (cut from Digi-Key C2028W-100 or equivalent)

TERMINAL ASSEMBLY

1. Trim insulation back ⅛" and tin wire end

2. Insert wire end so that insulation lines up with valley between first and second crimp edges

First crimp – bend over wire

Second crimp – bend over insulation

Solder first crimp, make sure that the space under spring end is kept free of solder and wire

CONNECTOR ASSEMBLY

Number 1

Slots receive lock tabs on crimp terminals

Number 24

Wires with terminals – pushed into connector shell

Lip

Spring end of terminal must be totally free of wire and solder so spring can depress as the connectors mate

Lock tab

Mated connectors

8-20 Making cable connections without the special crimping tool.

9
CHAPTER

Testing the Parallel (IBEC-Based) System

If you have been building along with me, you've now built all the hardware for your IBEC-based UCIS. You can be proud of that, and soon you'll be even prouder when you see it at work with your external hardware. In this chapter we'll test the system to see that everything is assembled and connected correctly. I'll explain how to test the UCIS hardware shown in Fig. 9–1.

We'll start out with initial checks of your UCIS without the I/O cards. Then I'll show how to use the test panel and wraparound test cable along with software to verify the operation of your entire UCIS, including the I/O cards.

Initial System Testing

Systematic checking takes time, but it also goes a long way toward ensuring that your UCIS will operate correctly when you first turn it on. The most likely sources of trouble will be poor soldering, incorrect parts, and/or incorrect part insertion. DIP switch settings that are inconsistent with each other and/or with the software is another common fault. Check each of the following boxes only after you get the correct results.

☐ *Verify correct computer operation.* At this point I'm assuming that you have proven to yourself that with IBEC installed, everything is operating fine with your computer. If you do suspect any problems you need to go back and repeat the steps in Chap. 5. Once you are satisfied with the operation, then proceed with this chapter.

☐ *IOMB power test.* Connect your power supply to the +5-VDC and GND terminals. With the motherboard NOT connected to IBEC, turn the supply on and use your VOM to check for the correct voltage and to make sure the polarity is correct. Also make sure that you measure +5 Vdc between pin 10 on any one of the card slots and GND.

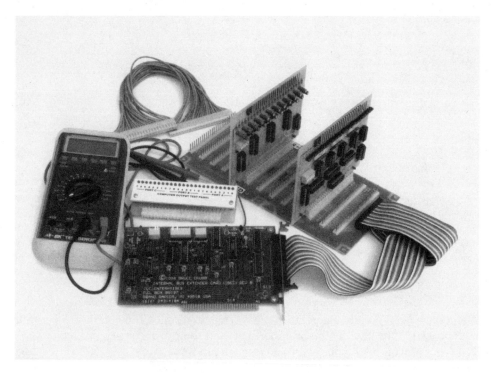

9-1 UCIS parallel (IBEC-based) hardware setup for testing.

☐ *Examine the ribbon cable.* With the ribbon cable laid out flat without any twists, both connector sockets should be facing the same direction (meaning that if one points upward, the other should point upward as well).

Warning: *Always make sure your computer is turned off whenever you insert or remove your IBEC and whenever you attach or remove the ribbon cable between IBEC and IOMB.*

☐ *Connect the ribbon cable.* With both your computer and your 5-Vdc supply turned off, connect your 40-wire ribbon cable from your IBEC to your IOMB. It is important that both ends of the cable be connected with the exact orientation as illustrated in Fig. 9–1. The 20 ground pins on the IBEC must connect to the 20 ground pins on the IOMB. Either end wrong will reverse all power and signal lines with all the ground connections, resulting in possible damage to the IBEC and to your computer.

☐ *RRON sensor test.* Once you are totally convinced that the ribbon cable is connected correctly, turn on your computer and make sure it is operating correctly. Then watch the green LED on IBEC and turn on your 5-Vdc supply. This action should light L2. If not, turn off your 5-Vdc supply and recheck your ribbon cable, and if required use your VOM to determine where you have a wrong connection. Do not continue until you obtain correct operation of the green LED.

☐ *Power on-off test.* During this test, the IBEC's red LED should remain off and the green LED should correspond to your power supply being off or on. Turn your 5-Vdc supply on and off several times as you are using your computer for other functions. You should notice no effects on computer operation.

If this stops the program, it might be causing a line transient in the power outlet feeding your computer. Try plugging the computer and power supply into outlets on different circuits and/or installing a power strip with power surge and noise protection to power your computer; these might correct the problem. If this doesn't appear to be the problem, repeat the test with U3 and U4 removed. If these are OK, then most likely the pin 19 enable inputs are being held low when they shouldn't be, and this is loading down the computer's data bus. Check for soldering problems around U9, U3, and U4, or possibly a bad IC. Correct any problems before continuing.

Warning: *Turn off the 5-Vdc power supply whenever you add or remove an I/O card on IOMB. Turn off both the computer and the 5-Vdc supply when plugging in or unplugging the IBEC or its ribbon cable from the IBEC or IOMB. You need not turn off your computer or your 5-Vdc supply when making connections or disconnections with the I/O cards.*

☐ *Plug in output card.* With your 5-Vdc supply turned off, plug one of your output cards (either a DOUT configured with the J1 jumper, or a COUT24 card) into one of the IOMB card slots. Sets its DIP switch to all segments off, meaning with all segments in position toward the IOMB.

☐ *Attach the test panel.* Plug the appropriate test panel onto your output card and attach its clip lead to the +5-VDC terminal of IOMB. Turn on your power supply. The test panel's LEDs should light in a random pattern, which can include all on or all off but most likely will be some on and some off if you are using COUT24. With DOUT the LEDs will most likely be all off. If none light, use your VOM to make sure you still have 5 Vdc at one of the ICs on the output card. If you installed the optional switch for controlling the IOMB's bus line 10, make sure it's set for +5 Vdc as measured between bus line 10 and ground.

Output Card Test Program

You are now ready to take a major step forward in UCIS testing by using the test program listed in Fig. 9-2. This program will demonstrate the complete output capabilities of your interface using the test setup shown in Fig. 9-3. This listing is in BASIC, and assumes conventional memory-mapped I/O. Once we have it understood, I'll list the changes needed for handling the port-mapped I/O mode.

The program is written using very basic code, so it should run with any version of BASIC including BASICA, GW-Basic, QBasic and QuickBASIC. It should also be straightforward to convert it to your favorite language. The source code for this

VALUE OF N IN LOOP	VALUE OF DISPLAY			DISPLAY BIT "ON"
	Decimal	Binary	Hexadecimal	
0	1	00000001	01	0
1	2	00000010	02	1
2	4	00000100	04	2
3	8	00001000	08	3
4	16	00010000	10	4
5	32	00100000	20	5
6	64	01000000	40	6
7	128	10000000	80	7

BIT PATTERN FOR RELATIONSHIP $2 \uparrow N$

```
10   DIM PN$(2)
20   PN$(0) = "A": PN$(1) = "B": PN$(2) = "C"
30 REM DEFINE 8255 CONTROL BYTE
40   CO = 128
50 REM DEFINE PORT A ADDRESS FOR CARD UNDER TEST
60 REM THIS STEP IS APPLICATION DEPENDENT
70 REM FOLLOWING ASSUMES IBM PC THRU PENTIUM WITH
80   DEF SEG = 40960 'MEMORY-MAPPED I/O WITH SA = 640KB
90   PRINT "OUTPUT CARD LED DRIVER TEST PROGRAM"
100 REM INCREMENT PORT TO BE TESTED IN A LOOP
110   FOR P = 0 TO 2
120 REM OUTPUT 8255 CONTROL BYTE
130   POKE 3, CO
140 REM INCREMENT DISPLAYED BIT NUMBER IN A LOOP
150   FOR N = 0 TO 7
160 REM OUTPUT TEST STATUS TO MONITOR
170   PRINT "PORT IS", PN$(P), "BIT NUMBER IS", N,
180 REM OUTPUT LED DISPLAY DATA TO CARD
190   D = 2 ^ N
200   POKE P, D
210 REM WAIT FOR CARRIAGE RETURN TO INCREMENT
220   INPUT "ENTER RETURN TO CONTINUE TEST", CR
230 REM COMPLETE LOOPS
240   NEXT N
250   POKE P, 0
260   NEXT P
270   GOTO 90 'REPEAT TEST TILL HALT PROGRAM
```

Changes if exponent not available
```
15   DIM D(7)
25   D(0) = 1: D(1) = 2: D(2) = 4: D(3) = 8
26   D(4) = 16: D(5) = 32: D(6) = 64: D(7) = 128
190   D = D(N)
```

Changes for port-mapped I/O
```
70 REM FOLLOWING ASSUMES IBM PC THRU PENTIUM
80   SA = 1280 'PORT-MAPPED I/O STARTING ADDRESS
130   OUT SA + 3, CO
200   OUT SA + P, D
250   OUT SA + P, 0
```

9-2 Output card LED-driver test program using the IBEC.

9-3 Test setup for running an output card test program.

program, as well as equivalent programs in Pascal and C, are contained on the 3.5-in disk enclosed with this second edition. The program uses five variables, which are defined as follows:

PN$( ) This is a three-position array used to print out the port being tested; *PN$(0)* = "A," *PN$(1)* = "B," and *PN$(2)* = "C."

CO This 8255 output control byte is set at 128. It is required for the COUT24 card.

P This numeric variable identifies the I/O card's port to be tested; 0 = Port A, 1 = Port B, and 2 = Port C.

N This variable designates the bit number (LED) to be lit, and will range from 0 to 7.

D This variable is the data output byte to be sent to the LEDs on the test circuit.

Because this is the first test program in this book, I'll briefly explain each line of code. Line 10 sets aside three locations in memory and labels them *PN$(0)*, *PN$(1)*, and *PN$(2)*. Line 20 stores the symbol "A" in *PN$(0)*, "B" in *PN$(1)*, and "C" in *PN$(2)*.

BASIC lines beginning with REM (meaning remark) are not instructions to be carried out. Their function is simply to carry comments that help readers understand the program. Thus line 40 explains what line 50 does: it gives control-byte variable *CO* the numeric value 128.

For this example, I've selected a UCIS starting address of 640K, which is 655360 and A0000h as defined by Table 5–5. Line 80 defines the corresponding segment register equal to the UCIS starting address divided by 16, in this case 655360/16 or 40960. The required format is DEF SEG = 40960. This value is also provided in Table 5–5. If your computer requires a different UCIS starting address, then simply use Table 5–5 and change the DEF SEG value accordingly.

Line 90 displays the program title. Line 110 begins a test loop by incrementing variable *P* from 0 to 1 to 2, to test all three ports. Line 130 sends the output control byte to initialize the 8255 ports for output. Note that if you never intend to use any of the original-design COUT24 output cards or a DAC3 card, then this statement can be deleted, but it doesn't hurt to keep it in.

Line 150 sets up an inner loop whereby the number of LED to be lit (*N*) is incremented from 0 to 7. Line 170 provides a monitor display of the port under test and the bit number to be lit.

The next step is to calculate and send the bit patterns to the port under test, so we can see if they get through correctly to light the proper LED on the test panel. Line 190 does the calculation by setting *D* equal to 2 raised to the power *N*. This arithmetical statement sets the bit pattern relationships shown at the top of Fig. 9–2 for each value of *N* from 0 through 7. Line 200 transmits *D* to the output card to light bit number *N* of Port *P*.

Line 220 is a monitor prompt message asking for input from the operator if the test is to continue. Line 240 loops back to increment the bit pattern. Line 250 turns off all 7 LEDs for the port that has just been tested prior to executing line 260, which loops back to begin testing the next port. Once the program has sent an output to all eight bits in all three ports, it reaches line 270, which repeats the test by branching back to line 90.

As I said earlier, there can be subtle differences between one dialect of BASIC and another. For example, your particular BASIC might use a symbol other than the colon I used for including more than one statement on a line, as in line 20.

Figure 9–2 includes some special-case instructions, too. If your BASIC doesn't have the exponent function used in line 190, you can get the same result by adding or changing the four lines as indicated.

For port-mapped I/O, you will need to replace the DEF SEG statement with a direct definition of the UCIS starting address; see the revised line 80 for port-mapped I/O and then use OUT instructions instead of POKE, as noted by the revised lines

130, 200, and 250. For the port-mapped case I assumed a UCIS starting address of 1280 as defined by Table 5–7. If your computer requires a different value, then simply change line 80 accordingly.

Conducting the Test

To perform the test, copy the Fig. 9–2 program from the enclosed disk and make any adjustments in the copy required by your particular computer and your selected I/O mode. Equivalent programs are also enclosed in Pascal and C. Run the program and compare the port and bit combination on the monitor with the LED that lights on the test panel. If they agree, press RETURN to increment the display. Every time you do this the program will loop, and the next LED to the left should come on.

If the LEDs don't agree with the monitor display, you have some debugging to do. First make sure you are using the correct test card for your output card. This should be an unmodified DOTEST panel if you are using DOUT or OUTEST (or the jumper-modified DOTEST) for COUT24. Make sure you are using the Fig. 9–2 software. It's always tempting to change a program to suit one's personal taste. That's fine once we have proven that the hardware is perfect, but until you reach that point, I highly recommend that you use the specific software listed in Fig. 9–2.

Once one port works correctly, any remaining problem is almost certainly on the output card itself. Check the card for bad solder joints, solder bridges, reversed ICs, and so on. By knowing which bits fail and using your schematic, you can tell exactly where in the circuit card to look for the problem.

Success with the test program proves not only that the output card works correctly, but also that the address decoding and data out portions of IBEC are working along with the IOMB and your software. It's an effective visual test of UCIS output. Once you get one output card working on all three ports, the hardest part of the UCIS project is behind you. Total interface success is just a small step ahead; you need only to test the input card and the input portions of IBEC. Because you know everything else checks out, this will be much easier.

In Case of Difficulty

If these instructions, which are tailored for your selected I/O mode, UCIS starting address, and card address, don't control the LEDs on your test panel, your software isn't communicating with your output card. You have what's known as an "interface communication problem." The question is, "How do I fix it?" The solution will probably be very simple once you find it, but first you have to find it!

The first thing to take notice of is "Does the red Address Selected test LED blink on each time the program loops?" If necessary, adjust R7 as your program is operating. Clockwise should make the LED tend to stay on longer. When you stop your program, the LED should stay off.

If you can't get the red LED to come on, then your UCIS starting address as defined by software doesn't agree with the DIP switch settings on SW2 and SW3, or the I/O mode your software is using doesn't agree with the settings of SW1, or there is an assembly problem with IBEC in the address decode or U12 area. Review Chap. 5 and make sure you are using the correct starting address, the right value in your DEF SEG statement, and that everything matches your DIP switch settings for I/O mode and starting address. If the switch settings aren't exactly correct, and if they don't correspond exactly to the address being used by your DEF SEG and in your POKE or OUT instruction, then you won't get any response from the red LED nor on the test panel.

If this fails to find an error, then the best procedure is to totally recheck your IBEC per Chaps. 3 and 5 and the first part of this chapter. Are the ribbon cable connectors mounted correctly on the ribbon cable and plugged in correctly on both ends? Is every IC used in IBEC in the correct socket with the correct pin 1 orientation? Are all the pins properly pushed into their corresponding socket holes? And so on, checking everything methodically, including going over every single solder connection.

Assuming your red test LED is functioning but your test panel isn't responding, then your problem is either in the output card itself, or in IBEC—excluding the area of high-order address decode. If you've built two output cards you can try again with the second one, but set its DIP switch to the same address so you're making the same test. This doesn't guarantee 100 percent that the problem isn't in the output cards, because it's possible to make the same assembly mistake on both cards, but it certainly points to the problem being on IBEC. In any event, go back over the output cards and the IBEC. Re-examine each card carefully with a magnifying glass under a bright light to see that all parts are the correct part and inserted with correct orientation, that all solder joints are good, and that there are no solder bridges between adjacent pins or traces.

Remember, most problems can be traced to faulty soldering, parts insertion, or cabling. Keep searching and retesting until you find the error, correct it, and achieve successful operation.

If you still can't find the trouble, get a friend (preferably one with some electronics background) to go through all the checks with you. Often someone else will find something wrong where the person who made the error passes right over it. As you perform the checks, have your friend challenge you with questions like, "How do you know that is the correct part?" and, "Are all the wires in that connector making contact?" As you try to explain what you're doing at each step, you will very likely find your problem. In industry we call this a "design walk-through."

If you can't find any errors, and plugging in new ICs doesn't help, try to get help from a computer expert in your area. In many computer clubs you'll find hobbyists just looking for tough problems to solve. The technician at your computer service center might be able to help you on the side.

If you live near a college or school that teaches electronics, check out the digital electronics classes. A student might be able to help you and earn course credit for doing it, plus a payment from you for helping out. An experienced person with the aid of an oscilloscope and/or a logic analyzer should be able to find most any interface problem with minimum effort.

The most important requirement for problem-solving is persistence. The reward will be worth it.

Automated Wraparound Test Program

We could test the input card by connecting a DIP switch to the card inputs, using a test program to read the switch settings, and displaying the results on the computer screen. It's much easier to use the computer to perform the whole test automatically. Such levels of automated testing and diagnostics are some of the advantages of having a computer equipped with external hardware.

Because you've already tested your first output card, you can use it to test all the input cards you need. In the computer industry this kind of I/O check is called a "wraparound test," because the computer's outputs are simply wrapped around and connected to its inputs. This wraparound capability is achieved using the wraparound test cable described in Chap. 8.

Figure 9–4 shows the setup for performing the automated wraparound test. The test program in Fig. 9–5 writes a bit pattern to the output card and then reads the input card. It then compares what it reads with what it wrote, and if they don't agree it has found an I/O error and displays an error message on the monitor. Because you've already proven the output card, the fault will most likely be in the input card or in the portions of the IBEC that operate during input operations but not during output operations.

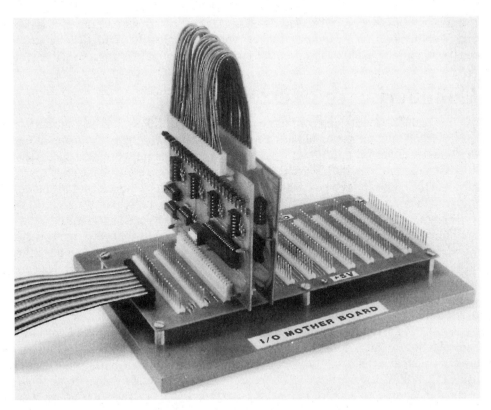

9-4 Test setup for running the automated wraparound test.

If the program reads exactly what it wrote, it changes the bit pattern and repeats the write/read cycle. Once all bit patterns pass the test, both cards in the test loop are proven good.

To successfully run the wraparound test, you need a matched set for the input and output card types, for example a DIN with a DOUT or a CIN24 with a COUT24. The DOUT must also be configured in the standard current-sinking configuration.

Because the wraparound test program closely parallels the LED-driver test program we just used to test the output card, I'll explain only the significant differences. Line 50 defines the 8255 control byte for input, and line 160 transmits it. Line 200 reads the input card. Lines 220 through 230 convert output-bit pattern D and input-bit pattern C to hexadecimal.

If your BASIC doesn't have a decimal-to-hexadecimal conversion function, the first change at the bottom of the listing in Fig. 9–5 will print out *D* and *C* directly. That's not as handy, because you'll have to do conversions by hand to locate bad bit(s), but it still gets the job done. The next change simplifies that process, so if you need to make the first, you'll probably want to make the second.

The other changes at the bottom of Fig. 9–5 cover port-mapped I/O instead of memory-mapped I/O. This program as well as similar programs in Pascal and C are located on the disk enclosed with this second edition.

The important comparison test is the IF, THEN statement in line 270. If *C* equals *D* there is no error, and the program branches to line 320 to increment to the next bit pattern *D* and repeat the cycle. If *C* doesn't equal *D* there's an error, and a message is displayed on the monitor. To continue the test you simply press RETURN. The program runs continuously, looping through all 256 combinations of bit patterns for each port.

Conducting the Test

To conduct the automated wraparound test, set the DIP switch on your output card to all off, and for the input card all off except leave the rightmost segment on. Plug the cards into the IOMB and connect the wraparound test cable. Copy the Fig. 9–5 program from the disk enclosed with this second edition. Make any adjustments in the copy required by your particular computer and your selected I/O mode, then run the program.

Figure 9–6 includes a few lines of sample program output as seen on the monitor. I forced that error message by grounding one of the input connector pins on the input card with a clip lead. If you do get errors, first check to see that you have the correct addresses set in the DIP switches on each card and that these settings correspond to the addresses in the program. If, for example, you accidentally interchanged the addresses, you would be trying to write to the input card and read from the output card, and nothing would work.

Seeing which bits fail can help locate where on the input or output card you have bad soldering, a part incorrectly inserted, or a faulty part. To find which bit or bits are faulty, expand the hex representations of C and D into their binary equivalents using Fig. 1–6. Whichever bits don't compare define the area of trouble. Probe those parts of the card with your VOM, looking for 0 V or +5 Vdc on the appropriate lines to pinpoint the problem.

```
10      DIM PN$(2)
20      PN$(0) = "A": PN$(1) = "B": PN$(2) = "C"
30 REM DEFINE 8255 CONTROL BYTE
40      CO = 128
50      CI = 155
60 REM DEFINE UCIS STARTING ADDRESS
70 REM THIS STEP IS APPLICATION DEPENDENT
80      DEF SEG = 40960
90      PRINT "WRAPAROUND TEST PROGRAM"
100 REM INCREMENT PORT TO BE TESTED IN A LOOP
110     FOR P = 0 TO 2
120 REM INCREMENT TEST BIT PATTERN IN A LOOP
130     FOR D = 0 TO 255
140 REM OUTPUT CONTROL BYTES TO 8255s
150     POKE 3, CO
160     POKE 7, CI
170 REM OUTPUT BIT PATTERN TO OUTPUT CARD
180     POKE P, D
190 REM READ BIT PATTERN ON INPUT CARD
200     C = PEEK(4 + P)
210 REM CONVERT OUTPUT AND INPUTS TO HEX
220     HO$ = HEX$(D)
230     HI$ = HEX$(C)
240 REM PRINT TEST STATUS ON MONITOR
250 PRINT "PORT IS", PN$(P), "HEX OUTPUT IS", HO$, "HEX INPUT IS", HI$
260 REM COMPARE INPUT TO OUTPUT AND BRANCH IF NO ERROR FOUND
270     IF C = D THEN GOTO 320
280 REM ERROR FOUND SO PRINT TO MONITOR
290 PRINT "ERROR FOUND IN ABOVE I/O, ENTER RETURN TO CONTINUE"
300     INPUT CR
310 REM COMPLETE LOOPS
320     NEXT D
330     POKE P, 0
340     NEXT P
350     GOTO 40  'START TEST OVER AGAIN
```

Changes if hexadecimal conversions not available
```
220 **DELETE LINE**
230 **DELETE LINE**
250 PRINT "PORT IS", PN$(P), "DEC OUTPUT IS", D, "DEC INPUT IS", C
```

Changes to simplify manual hex conversions
```
130     FOR N = 0 TO 7
175     D = 2^ N
320     NEXT N
```

Changes for port-mapped I/O
```
80      SA = 1280  'ASSUMED PC PORT-MAPPED STARTING ADDRESS
150     OUT SA + 3, CO
160     OUT SA + 7, CI
180     OUT SA + P, D
200     C = INP(SA + 4 + P)
330     OUT SA + P, 0
```

9-5 Automated wraparound test program using IBEC.

NO ERROR

PORT IS	A	HEX OUTPUT IS	Ø	HEX INPUT IS	Ø
PORT IS	A	HEX OUTPUT IS	1	HEX INPUT IS	1
PORT IS	A	HEX OUTPUT IS	2	HEX INPUT IS	2
PORT IS	A	HEX OUTPUT IS	3	HEX INPUT IS	3
PORT IS	A	HEX OUTPUT IS	4	HEX INPUT IS	4
PORT IS	A	HEX OUTPUT IS	5	HEX INPUT IS	5
PORT IS	A	HEX OUTPUT IS	6	HEX INPUT IS	6
PORT IS	A	HEX OUTPUT IS	7	HEX INPUT IS	7
PORT IS	A	HEX OUTPUT IS	8	HEX INPUT IS	8
PORT IS	A	HEX OUTPUT IS	9	HEX INPUT IS	9
PORT IS	A	HEX OUTPUT IS	A	HEX INPUT IS	A
PORT IS	A	HEX OUTPUT IS	B	HEX INPUT IS	B
PORT IS	A	HEX OUTPUT IS	C	HEX INPUT IS	C
PORT IS	A	HEX OUTPUT IS	D	HEX INPUT IS	D
PORT IS	A	HEX OUTPUT IS	E	HEX INPUT IS	E
PORT IS	A	HEX OUTPUT IS	F	HEX INPUT IS	F
PORT IS	A	HEX OUTPUT IS	1Ø	HEX INPUT IS	1Ø
PORT IS	A	HEX OUTPUT IS	11	HEX INPUT IS	11
:	:	:		:	

ERROR INDUCED

PORT IS	A	HEX OUTPUT IS	Ø	HEX INPUT IS	1

ERROR FOUND IN ABOVE I/O, ENTER RETURN TO CONTINUE TEST
?

PORT IS	A	HEX OUTPUT IS	1	HEX INPUT IS	1
PORT IS	A	HEX OUTPUT IS	2	HEX INPUT IS	3

ERROR FOUND IN ABOVE I/O, ENTER RETURN TO CONTINUE TEST
?

PORT IS	A	HEX OUTPUT IS	3	HEX INPUT IS	3
PORT IS	A	HEX OUTPUT IS	4	HEX INPUT IS	5
:	:	:		:	

9-6 Automated wraparound test program sample output.

Once you have a proven input card, you can use it to test the rest of your output cards automatically, just as your first proven output card can be used to test input cards. Keep your test programs handy for use if you ever suspect faulty operation in your UCIS.

To make your wraparound test more interesting, and to help provide faster debugging, you can place a second output card on the IOMB with the same DIP switch setting as the first output card. Install the output test panel on the second card. When the computer sends output bytes it drives both cards simultaneously with the same data. This way the second card shows the particular bit pattern being tested for each port, ranging from decimal 0 to 255 (00h to FFh). If the program halts due to an error, you instantly see which bit on which port is causing the problem. Checking the input card schematic for the routing of this bit shows you exactly where to look on the input card to correct the problem.

At this point you can feel very proud. Your UCIS is fully operational, and you've written and executed two interface programs. Assuming your UCIS is only IBEC based, you may skip to Chap. 13, where I'll show how to connect a variety of external hardware devices to your interface.

10
CHAPTER

Serial Interface Software

In Chap. 4 we built the universal serial interface card or USIC, and in this chapter I'll show you the special software required to make a USIC functional.

The USIC's main software requirement is a protocol to allow sending card addresses and data in series over the same wire. The protocol software will automatically form packets of information to be transmitted. It also inserts special control characters to mark the beginning and end of transmissions, and to define which bytes are addresses and which are pure data. This might sound complicated, but software engineer Bruce Wahl and I have done the hard work for you by preparing standardized subroutines for the USIC protocol.

The protocol uses four types of messages: initialization, transmit data, poll request, and receive data. I'll describe the general message format first, then we'll look at the details of each type.

General Message Format

Each message consists of framing and data. In serial I/O we use data as a general term for both the I/O card addresses and the information being communicated to or from their ports. In serial I/O, addresses and information are all just signals sent over the wires. Figure 10–1 depicts a typical serial I/O message, bit by binary bit: a mark in a given bit period represents a logic 1, while a space represents a logic 0; in each byte the LSB (*least significant bit*) is sent first.

Framing separates one message from another and defines the type of data in each byte. Three standard control characters are used in framing the packets: STX for start-of-text, ETX for end-of-text, and DLE for data-link-escape.

To understand the differences in how the serial channel handles control characters versus data, it's important to know how alpha characters (letters) are stored in a computer. In computer memory, numbers, letters, and special characters are all stored as binary codes made up of ones and zeros.

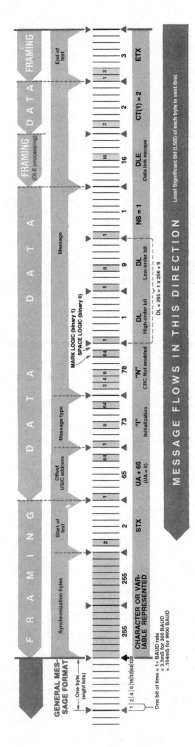

10-1 Serial I/O message packet.

Figure 10–2 is the *American Standard Code for Information Interchange,* or ASCII. It defines standard relationships between bytes of eight-bit binary code and their multiple representations in decimal, hexadecimal, and symbolic character notation. Symbolic character notation includes function codes or control characters and other character types, including uppercase and lowercase alpha characters, numeric characters (numerals) 0 through 9, and special characters such as punctuation marks.

ASCII defines function and character codes only for decimal 0 through 127, leaving 128 through 255 undefined except for the mathematical relationship between binary, decimal, and hexadecimal codes. Figure 10–2 is an important reference to keep handy as you do more and more UCIS application programming. You might not have realized it, but we've already been using ASCII codes.

For example, in the output card test program in Fig. 9–2, when the computer's BASIC interpreter read the instruction PN$(0) = "A" it accepted the letter A as an ASCII symbol and stored the binary code 01000001 in the memory location for variable *PN$(0)*. With an arithmetic statement such as $G = 65$, the interpreter would place the same 01000001 code in the memory location for variable G. Note that the same binary code can represent decimal integers (numbers) and either alpha or control characters. It's up to the program to keep track of the correct interpretation for any given application. You'll see how this works as you learn more about message formats and protocol subroutines.

Returning to Fig. 10–1, each message format begins with two bytes of all ones to synchronize the hardware. Next comes an STX to show that the text of the message begins with the next byte. The text is a series of bytes with each bit of each byte sent in serial, one bit after another, and each new byte following right behind its predecessor.

The first byte after the STX is the offset address *UA* of the USIC being commanded. It is formed by adding decimal 65 to the USIC's DIP switch setting, written as *UA* + 65, to keep *UA* from conflicting with any control characters. That makes the assigned values of *UA* + 65 for the 16 possible USICs correspond to alpha characters A through P.

Following the address byte is an alpha character defining the message type as an "I" for initialize, "T" for transmit data, "P" for poll-request, or "R" for receive data. The details of the remaining message bytes depend on the message type, and I'll describe them separately.

The DLE character is used when it's necessary to include binary codes in the data for decimal numbers 2, 3, and 16, which are the same codes as for control characters STX, ETX, and DLE. In the Fig. 10–1 example, a decimal 2 is sent by first sending a DLE byte, decimal 16, and following that with a byte of 2. An ETX character marks the end of each message.

Now turn to Fig. 10–3 as we take a closer look at each message format. You can refer to Table 10–1 for a handy reminder of the programming acronyms being used.

Initialization message

This message tells the USIC how many input and output cards are connected to its motherboard, and which types are at which address. Once the USIC's MC68701S processor receives this information, it automatically initializes the 8255s on the I/O cards.

BINARY CODE	DECIMAL CODE	HEXADECIMAL CODE	FUNCTION CODE	MEANING
00000000	0	00	NUL	All zero character
00000001	1	01	SOH	Start of heading
00000010	2	02	STX	Start of text
00000011	3	03	ETX	End of text
00000100	4	04	EOT	End of transmission
00000101	5	05	ENQ	Enquiry
00000110	6	06	ACK	Acknowledge
00000111	7	07	BEL	Bell
00001000	8	08	BS	Backspace
00001001	9	09	HT	Horizontal tabulation
00001010	10	0A	LF	Linefeed
00001011	11	0B	VT	Vertical tabulation
00001100	12	0C	FF	Form feed
00001101	13	0D	CR	Carriage return
00001110	14	0E	SO	Shift out
00001111	15	0F	SI	Shift in
00010000	16	10	DLE	Data link escape
00010001	17	11	DC1	Device control 1
00010010	18	12	DC2	Device control 2
00010011	19	13	DC3	Device control 3
00010100	20	14	DC4	Device control 4
00010101	21	15	NAK	Negative acknowledge
00010110	22	16	SYN	Synchronous idle
00010111	23	17	ETB	End of transmission block
00011000	24	18	CAN	Cancel
00011001	25	19	EM	End of medium
00011010	26	1A	SUB	Substitute
00011011	27	1B	ESC	Escape
00011100	28	1C	FS	File separator
00011101	29	1D	GS	Group separator
00011110	30	1E	RS	Record separator
00011111	31	1F	US	Unit separator
00100000	32	20	SP	Space

BINARY CODE	DECIMAL CODE	HEXADECIMAL CODE	CHARACTER
00100001	33	21	!
00100010	34	22	"
00100011	35	23	#
00100100	36	24	$
00100101	37	25	%
00100110	38	26	&
00100111	39	27	'
00101000	40	28	(
00101001	41	29	)
00101010	42	2A	*
00101011	43	2B	+
00101100	44	2C	,
00101101	45	2D	-
00101110	46	2E	.
00101111	47	2F	/
00110000	48	30	0
00110001	49	31	1
00110010	50	32	2
00110011	51	33	3
00110100	52	34	4
00110101	53	35	5
00110110	54	36	6
00110111	55	37	7
00111000	56	38	8
00111001	57	39	9
00111010	58	3A	:
00111011	59	3B	;
00111100	60	3C	<
00111101	61	3D	=
00111110	62	3E	>
00111111	63	3F	?
01000000	64	40	@
01000001	65	41	A

BINARY CODE	DECIMAL CODE	HEXADECIMAL CODE	CHARACTER
01000010	66	42	B
01000011	67	43	C
01000100	68	44	D
01000101	69	45	E
01000110	70	46	F
01000111	71	47	G
01001000	72	48	H
01001001	73	49	I
01001010	74	4A	J
01001011	75	4B	K
01001100	76	4C	L
01001101	77	4D	M
01001110	78	4E	N
01001111	79	4F	O
01010000	80	50	P
01010001	81	51	Q
01010010	82	52	R
01010011	83	53	S
01010100	84	54	T
01010101	85	55	U
01010110	86	56	V
01010111	87	57	W
01011000	88	58	X
01011001	89	59	Y
01011010	90	5A	Z
01011011	91	5B	[
01011100	92	5C	\
01011101	93	5D	]
01011110	94	5E	^
01011111	95	5F	_
01100000	96	60	`
01100001	97	61	a
01100010	98	62	b

BINARY CODE	DECIMAL CODE	HEXADECIMAL CODE	CHARACTER
01100011	99	63	c
01100100	100	64	d
01100101	101	65	e
01100110	102	66	f
01100111	103	67	g
01101000	104	68	h
01101001	105	69	i
01101010	106	6A	j
01101011	107	6B	k
01101100	108	6C	l
01101101	109	6D	m
01101110	110	6E	n
01101111	111	6F	o
01110000	112	70	p
01110001	113	71	q
01110010	114	72	r
01110011	115	73	s
01110100	116	74	t
01110101	117	75	u
01110110	118	76	v
01110111	119	77	w
01111000	120	78	x
01111001	121	79	y
01111010	122	7A	z
01111011	123	7B	{
01111100	124	7C		
01111101	125	7D	}
01111110	126	7E	~
01111111	127	7F	DEL
11111110	254	FE	.
11111111	255	FF	.

10-2 American Standard Code for Information Interchange (ASCII).

a. INITIALIZATION, "I"
255, 255, STX, UA + 65, "I", "N", DL
high-order bit, DL low-order bit, NS,
CT(1),... CT(NS), ETX

b. TRANSMIT DATA, "T"
255, 255, STX, UA + 65, "T", OB(1),....
OB(NO), ETX. (OB is output data array.)

c. POLL REQUEST, "P"
255, 255, STX, UA + 65, "P", ETX

d. RECEIVE DATA, "R"
255, 255, STX, UA + 65, "R", IB(1),...
IB(NI), ETX. (IB is input data array.)

10-3 Message formats.

Table 10-1. Summary of Programming Acronyms

Acronym	Function	Acronym	Function
CRC	Cyclic Redundancy Check	OB	Output Byte
DL	USIC Delay	PA	Port Address
IB	Input Byte	PN	Port Number
LM	Length of message	TB	Transmit Buffer
NI	Number of Input Ports	TP	Transmit Pointer
NO	Number of Output Ports	UA	USIC Address
NS	Number of Card Sets		

The format of the initialization message is given in Fig. 10–3a, with *N* signifying CRC not enabled, *DL* for USIC delay, and variable *NS* giving the number of card sets of four. These will be explained further with the initialization protocol subroutine.

Card types are defined by a CT( ) byte for each group of four cards. A byte allows two bits to define each card, and we'll use 0 = no card, 1 = input card, 2 = output card, and 3 = reserved for expansion. Figure 10–4 shows how to calculate CT( ) bytes.

Steps to calculate each CT( ) card set code

1. Enter card type and type numbers for each set of 4 cards (slot 1 location always closest to USIC)
2. Multiply each card type number by the card position constant 1, 4, 16, or 64
3. Sum the products to get the card set code
4. Repeat process for each card set of 4

Example:

$$
\text{1st card set of 4}
\begin{bmatrix}
\text{I} & 1 \times 1 = 1 \\
\text{O} & 2 \times 4 = 8 \\
\text{O} & 2 \times 16 = 32 \\
\text{I} & 1 \times 64 = \underline{64} \\
& CT(1) = 105
\end{bmatrix}
\qquad
\text{2nd card set of 4}
\begin{bmatrix}
\text{I} & 1 \times 1 = 1 \\
\text{I} & 1 \times 4 = 4 \\
\text{O} & 2 \times 16 = 32 \\
\text{none} & 0 \times 64 = \underline{0} \\
& CT(2) = 37
\end{bmatrix}
$$

SLOT LOCATION IN SET OF 4	CARD		MULTIPLICATION CONSTANT	PRODUCT OF CARD TYPE TIMES CONSTANT
	TYPE	TYPE NO.		
1			1	
2			4	
3			16	
4			64	

No card = 0
Input card = 1
Output card = 2

CARD SET CODE   Sum of above numbers goes here

Note: All occupied card positions must be in a single contiguous block, i.e. no blank slots are permitted with USIC. The only time that No card = 0, is permitted is at the end of the last card set of four, i.e. to fill out to the end of the last card set.

10-4 Calculating card-type-definition array variables.

Transmit-data message

For this message (Fig. 10–3b) the byte immediately following the address is a "T." The data section includes three binary bytes for each output card on the motherboard. The first byte corresponds to port A of the first card, followed by ports B and C, followed by the same data format for the second output card, the third, and so forth. The DLE processing described above is used in the data transmission to allow transmission of binary code equivalents of the control characters.

Poll-request and receive-data messages

Two messages are needed for the computer to receive data. First the computer sends a poll-request or "P" message (Fig. 10–3c) to tell the USIC to send back the state of the input card input lines connected to its motherboard. When the USIC receives the "P" message, it reads its input cards and forms the receive-data or "R" message (Fig. 10–3d), which carries the information back to the computer. The data section of the "R" message contains three bytes for each input card on the motherboard, and again it uses DLE processing.

Protocol Subroutines

Figure 10–5 is a listing in BASIC of the protocol subroutines used for USIC communications. Similar code for Pascal and C is included on the software disk enclosed with this book. Lines 200 through 1970 contain the source code and are a standard block you put at the top of every BASIC application program for the USIC serial interface.

The protocol code is "ready-to-use" and essentially application-independent, so I'll just cover the high points to give you a general understanding of how it works. I've used MicroSoft BASIC and avoided exotic codes, so most of it should work on any computer with little change. This particular code is tailored to the IBM PC-to-Pentium family of computers. If your machine isn't one of these, I'll point out the few lines that are computer-dependent and which might need to be changed to fit your computer.

There are three user-invoked subroutines for transmit data (T), receive data (R), and initialize (I), but first we need to cover a couple of utility subroutines that are called by T, R, and I. Including the two utilities, the five protocol subroutines are:

Receive one input byte

Lines 310 through 350 are a utility subroutine that receives a single input byte (IB) over the UCIS via port address *PA*. I've made it a separate subroutine because this function is used over and over again. Once you have your protocol package fully debugged, you might want to embed duplicates of these lines at each location where they're called, to increase operating speed.

Lines 320 and 330 are user-dependent. They check that the computer I/O port connected to the UCIS is operating correctly to receive input, and that a byte has been properly accumulated and is ready to be placed on the computer's data bus.

I've set up the subroutines based on serial hardware using an 8250 Asynchronous Communication Element (ACE) chip, which is the baseline for many makes of computers. If your computer uses the same device or the newer 16550 to control its RS-232 port, the two statements will be similar if not identical. For other machines you might have to study your reference and/or operations manuals to derive similar tests, and alter the code accordingly to provide the same functions.

```
200  REM***STANDARD USIC PROTOCOL SUBROUTINES***
210  REM**REVISION 16-MAR-86
310  REM**RECEIVE ONE INPUT BYTE**
320  IF (INP(PA + 5) AND 2) <> 0 THEN PRINT "OVERRUN"
330  IF (INP(PA + 5) AND 1) = 0 THEN GOTO 320
340  IB = INP(PA) 'received byte from port PA
350  RETURN
400  REM**FORM PACKET AND SEND
410  TB(1) = 255 'set 1st start byte to all 1's
420  TB(2) = 255 'set 2nd start byte to all 1's
430  TB(3) = 2  'define start-of-test
440  TB(4) = UA + 65 'add 65 offset to USIC address
450  TB(5) = MT 'define message type
460  TP = 6 'next position for transmit pointer
470  IF MT = 80 THEN GOTO 550 'poll request; end message
480  FOR I = 1 TO LM 'loop to setup output data
490  IF OB(I) = 2 THEN TB(TP) = 16: TP = TP + 1: GOTO 520
500  IF OB(I) = 3 THEN TB(TP) = 16: TP = TP + 1: GOTO 520
510  IF OB(I) = 16 THEN TB(TP) = 16: TP = TP + 1
520  TB(TP) = OB(I) 'move byte to transmit buffer
530  TP = TP + 1 ' increment to next byte position
540  NEXT I
550  TB(TP) = 3 'set end-of-text
630  REM**send packet
640  FOR I = 1 TO TP 'loop through transmit buffer
650  IF (INP(PA + 5) AND 32) = 0 THEN GOTO 650
660  OUT PA, TB(I) 'byte transmitted
670  NEXT I
680  RETURN
800  REM**TRANSMIT DATA**
810  MT = 84 'message type = "T"
820  GOSUB 410 'form packet and send
830  RETURN
1000 REM**RECEIVE DATA**
1010 MT = 80 'send out poll request "P" for input data
1020 GOSUB 410 'form packet and send

1060 REM**receive data back from USIC
1070 GOSUB 320 'receive input byte
1080 IF IB <> 2 THEN GOTO 1070 'loop till get start-of-text
1090 GOSUB 320 'receive input byte as USIC address
1100 IB = IB - 65 'subtract offset for address check
1110 IF IB <> UA THEN PRINT "ERROR RECEIVED BAD UA": GOTO 1010
1120 GOSUB 320 'receive input byte as "R"
1130 IF IB <> 82 THEN PRINT "ERROR RECEIVED NOT = R": GOTO 1010
1140 FOR I = 1 TO NI 'loop through number of input ports
1150 GOSUB 320 'receive input byte
1160 IF IB = 2 THEN PRINT "ERROR: NO DLE AHEAD OF 2": GOTO 1010
1170 IF IB = 3 THEN PRINT "ERROR: NO DLE AHEAD OF 3": GOTO 1010
1180 IF IB = 16 THEN GOSUB 320 'DLE so next byte read
1190 IB(I) = IB 'store as valid data byte
1200 NEXT I
1210 GOSUB 320 'receive input byte as end-of-test
1220 IF IB <> 3 THEN PRINT "ERROR: ETX NOT PROPERLY RECEIVED"
1230 RETURN
1800 REM**INITIALIZATION**
1810 OUT PA + 3, 128 'turn CR bit 7 on to access BAUD rate
1820 OUT PA, 12 'set LS latch for 9600 BAUD rate
1830 OUT PA + 1, 0 'set MS latch to 9600 BAUD rate
1840 OUT PA + 3, 3 'set up for 8 data bits and 1 stop bit
1850 MT = 73 'message type = "I"
1860 OB(1) = 78 'CRC setup not used = "N"
1870 OB(2) = INT(DL / 256)'set USIC delay high order byte
1880 OB(3) = DL - (OB(2) * 256) 'set USIC delay low order byte
1890 OB(4) = NS 'number card sets of 4
1900 LM = 4 'establish length of message
1910 FOR I = 1 TO NS 'loop through number of card sets
1920 LM = LM + 1 'accumulate message length
1930 OB(LM) = CT(I) 'define card type array
1940 NEXT I
1950 GOSUB 410 'form packet and send
1960 LM = NO 'initialize length message for "T" type
1970 RETURN
```

10-5 USIC protocol subroutines.

With the 8250, program line 320 examines port *PA*'s line status register, equivalent to a special memory byte at port address *PA* + 5. This register lets the program determine the status of the port's I/O lines at a given moment.

For example bit 1, "overrun error," is set to one to signify that data is arriving faster than it can be processed. To check for this condition, line 320 logically adds the contents of *PA* + 5 with decimal 2, a binary 00000010. This strips out all bits except bit 1, and checks for it being a one. (For Apple BASICs, replace the ANDs with equivalent arithmetic operations to strip out and test appropriate bits.)

If the result is a one, line 320 prints "OVERRUN" on the monitor as an alert that this is happening. If you get such an error message, you can increase the USIC delay variable *DL* to purposely slow down the USIC and give your computer more time to process input.

Line 330 performs a similar test of the line status register to check when port PA has received a complete byte of input data and has converted it to parallel format so it can be placed on the computer's data bus. This is accomplished by examining bit 0, the "data ready bit" of port *PA*'s line status register.

If bit 0 is a zero, data aren't available, so the program loops back through line 320 until bit 0 becomes a one, showing that data are available. Once data are available, the program drops through the IF statement to execute line 340, which inputs the byte and returns control to the calling program.

Form packet and send

Lines 400–680 are a utility subroutine to form a message packet and send it over the UCIS. It's a separate subroutine, because it is used by the transmit-data, receive-data, and initialization subroutines. Once more, for increased speed you can embed copies of this code where it is needed once your program is operational.

The array variable *TB( )*, the transmit buffer, contains the data bytes in sequence for the ports of each output card to be transmitted to the USIC for the transmit-data message. *TB( )* also holds data for the initialization message.

As we've seen, the first and second bytes are set to all 1s, decimal 255, by lines 410 and 420. Line 430 sets the third byte to STX (decimal 2), 440 sets the fourth byte to *UA* + 65, and 450 sets the fifth byte to the message type. These five bytes use five positions in the *TB* array, so line 460 sets the transmit pointer *TP* to 6, the next available buffer position.

Line 470 then checks to see if the message type is a "P" or poll request. If it is, then all necessary data are in the buffer except for the ETX (decimal 3), which is then added by the branch to 550. If the message is not a poll request but is either a "T" or an "I," the program falls through the test at 470 to execute line 480.

The loop in lines 480–540 sets up the data part of the message packet by transferring the desired output byte data array *OB( )* to the transmit buffer. The next available TB variable, beginning with *TB(6)*, is set equal to the next OB variable beginning with *OB(1)*, except for the cases when a particular *OB( )* equals a 2, 3, or 16.

In these cases the data to be sent is in the same code as a control character, and per our protocol an extra *TB( )* value of 16, a DLE character, is inserted ahead of the 2, 3, or 16 byte. *TP( )* is incremented by one, and the 2, 3, or 16 data byte is placed after the DLE. Once all *OB( )* values are in *TB( )*, line 550 defines the end of the data by setting *TB(TP)* equal to decimal 3 for ETX.

With the formulated message packet in the transmit buffer, the loop in lines 640–670 transmits the message over the serial channel a byte at a time. Line 650 is user-dependent and checks that the computer I/O port connected to the UCIS is free to transmit a byte. In my serial card, which uses the 8250 ACE, this is accomplished by looking at port PA's line status register bit 5, the transmitter holding register empty bit.

If this bit is a one, the port is ready to accept a parallel data byte for transmission over the serial port. If the bit is a zero the port isn't ready, and the program continues to branch back on line 650 until the bit becomes a one, showing that the port has become available. Then line 660 is executed to output the *I*th byte *TB(I)* to the UCIS. Once all bytes are transmitted, line 680 returns control to the calling program.

Transmit data

Lines 800–830 are the subroutine used to transmit data to a set of output cards connected to a USIC. Line 810 sets the message type variable *MT* to decimal 84, ASCII "T." With this definition complete and the output byte array *OB( )* established by the application program, all that's needed is to invoke the form-packet-and-send subroutine with the GOSUB 410 in line 820.

Receive data

Lines 1000–1230 are the subroutine used to receive data from a set of input cards connected to a USIC. As I've explained, the program first sends a "P" message to the USIC with lines 1010 and 1020. It then loops on, reading input bytes via the GOSUB 320 in line 1070 and the test in line 1080, until it receives a byte of decimal 2, STX, from the USIC.

Once an STX is received, line 1090's GOSUB 320 pulls in the next input byte, the offset USIC address *UA* + 65. Line 1100 removes the offset and line 1110 checks to make sure the address received is correct. If not, the transmission is assumed to be garbled. An error message is printed on the monitor, and the program branches to line 1010 and tries again.

If the addresses do match, line 1120 is executed to bring in the next input byte. This should be the message type sent by the MC68701S processor, a decimal 82 for "R." If the byte is not an 82 the transmission is again assumed garbled. Another error message is printed, and the program branches back to 1010 and tries again.

If both these inputs are received correctly, the program proceeds to pull in the data from the input cards. Lines 1140 through 1200 loop through each input port 1 to *NI*. In line 1150 GOSUB 320 is invoked to read in the input byte *IB*. If line 1160 finds that it's a 2, STX, the transmission must be garbled or out of synchronization and the program prints another error message and branches back to 1010 to try once more. If line 1170 finds that *IB* is a 3, ETX, it again must be garbled, resulting in another error message and another retry.

If *IB* is a 16, DLE, the program skips that byte and invokes GOSUB 320 to read the next *IB*, the actual data byte following the data-link-escape character. The value read in as *IB* is then placed in its proper array position *IB(I)*, and the program increments loop variable *I* to process the next input port.

Once the program has received the input bytes for all the *NI* input ports, it reads the next *IB* with line 1210 and checks in 1220 to make sure it is ETX, decimal 3. If not a 3, the program prints an error message on the monitor and returns control back to the user program in line 1230.

Initialization

Lines 1800–1970 are the subroutine that initializes the USIC and its I/O cards. It starts by setting the baud rate for the host computer—in my serial interface the 8250 ACE puts the baud rate and basic serial format under software control. Many other computers also use the 8250, making my code applicable to them, while for others you might need to add a jumper wire or set a DIP switch in your computer's hardware to set the baud rate.

In the case of the 8250, line 1810 sets a software-controlled switch so that when the computer writes out to the port it actually writes to the 8250's internal baud rate generator; it does this to latch in a divisor to be used for generating the baud rate from the computer's basic clock frequency, very much as the CD4040 counter chip and DIP switch do on the USIC.

Once the latch is made accessible by line 1810, lines 1820 and 1830 set the baud rate. The constants in the program set it for 9600. Figure 10–6 defines the programming constants required for each of the USIC DIP switch settings. Your computer might possibly be different, so if these don't work for you, consult your operations manual for the computer's serial port to find the procedures for setting the baud rate with your machine. Similarly, line 1840 formats the 8250 to handle eight data bits, no parity, and one stop bit, which are also requirements for proper USIC operation. Line 1850 defines the *MT* variable as "I," decimal 73.

I've chosen not to use the cyclic redundancy check feature provided by the Chesapeake Group Inc.'s MC68701S program. CRC helps ensure a high probability

Baud Rate	MS	LS
150	3	0
300	1	128
600	0	192
1200	0	96
2400	0	48
4800	0	24
9600	0	12
19200	0	6

Programming statements:

1820 OUT PA, LS

1830 OUT PA + 1, MS

Where: PA = port address

Equation for Baud rate:

$$\text{Baud} = \frac{1,843,200/16}{(MS \times 256) + LS}$$

10-6 Software baud rate settings.

that data bits aren't being lost or garbled, but it takes processing time. It also requires additional lines in the protocol subroutine to handle CRC codes, and limits the MC68701S to processing no more than 16 output cards at a time unless you add other additional protocol lines. All that takes still more processing time, but in BASIC real-time application programs for serial I/O there is NO spare time for CRC error checking.

Besides, I've operated the USIC in RS-232, RS-422, and 20-mA current-loop modes for hours on end, with a long transmission cable interwoven with external hardware wiring, all without a noted error. Also, as you've seen, I've included enough error messages in the protocol subroutines so that if you have problems with electrical noise, or with something like a baud-rate generator set incorrectly, you will soon see error messages on your monitor.

The program in the MC68701S expects to receive CRC codes, and will generate and transmit them if the CRC function is enabled. Line 1860 disables it by setting $OB(1) = 78$, an ASCII "N" for no CRC checking.

Lines 1870 and 1880 set the USIC delay variable DL to be sent to the USIC. The value of this variable causes the USIC software to build in a delay of about 10 microseconds per unit of DL. A value of $DL = 0$ yields no added delay, and $DL = 6000$ adds a delay of 60 ms per transmitted byte.

This delay keeps the USIC from sending data faster than your computer can receive and process it. Because DL values can easily go over 255, the maximum decimal value we can store in one byte, it's necessary to use one each of high- and low-order bytes, set by lines 1870 and 1880.

In line 1890, variable $OB(4)$ is equated to the number-of-card-sets-of-four variable, NS. A loop then fills following OB array positions with the card type (CT) variables for each set of four cards. With all card types defined, the line 1950 GOSUB 410 is invoked to form the complete initialization packet and transmit it to the USIC. The resulting transmission initializes the USIC's MC68701S, and it in turn automatically sets the control bytes for the 8255s on each of the attached I/O cards.

With its initialization work complete, the subroutine executes line 1960 to set message length variable LM equal to NO, the number of output card ports. LM will be used by the form-packet-and-send subroutine during repetitive calls from the transmit-data subroutine. With this initialization complete, control returns back to the user program in line 1970. That's all of the protocol subroutines, so now let's see how to put them to use in application programs.

Initialization of Variables

The protocol subroutines do all the I/O work for us, but for them to work properly certain variables must be "handed off" between your main program and the subroutines. For example, before invoking the transmit-data subroutine you must define the output byte data array, $OB( )$, to be transmitted to the output cards. Before invoking the initialization subroutine you must define the USIC address UA and the $CT( )$ arrays to specify what I/O cards are plugged into the USIC motherboard.

In short, all variables except the $IB( )$ array must be defined before invoking the protocol subroutines, and each of the variable symbols $CT( )$, $IB( )$, NI, NO, NS,

OB( ), PA, and *UA* must be written with exactly these alphabetical notations. Also, certain variables used internally in and/or between the protocol subroutines should be treated as "reserved variables" not to be used in an application program. These include *ML, IB, MT, TB( ),* and *TP.*

The loop index variable *I* is used internally in the protocol subroutines, but "I" can be used in your application programs. Just remember that its value will most likely be changed any time you invoke and return from a protocol subroutine.

The numeric values, of course, have to be adjusted to fit your own UCIS and whatever application programs you write. For example, you must allow one element in the *CT( )* array for each group of four I/O card address slots, and for any group of less than four, used in a given application, and set *NS* equal to the largest subscript needed to define your *CT( )* array. When you invoke the receive-data subroutine, it returns with the input byte data array *IB( )* as read from the input cards. The main program must be ready to unpack data in this form.

The IBM PC, XT, AT or 286, 386, 486, and Pentium support four serial ports, known as COM1 through COM4; quite typically only COM1 and COM2 are installed as part of the original computer configuration. In these cases COM3 and COM4 can be added by installing an expansion card. The corresponding PA (port starting address) constants needed to initialize your use of the protocol subroutines are defined in Table 10–2.

Table 10-2. COM Port Addresses for PC Up Through Pentium Computers

Port Configuration	Hexadecimal Address Range	Decimal Port Starting Address (PA Value)
COM1	3F8-3FF	1016
COM2	2F8-2FF	760
COM3	3E8-3EF	1000
COM4	2E8-2EF	744

Application Program Format

To put everything in perspective, I'll describe a hypothetical application program and see how it interfaces with the protocol subroutines. To make it more general I'll use an application with two USICs, each with a different assortment of I/O cards as shown in Fig. 10–7.

The first step is to calculate the *CT( )* array variables to let the USICs know what I/O cards are in which address slots. The five *CT( )* values shown in Fig. 10–7 were calculated following the procedure in Fig. 10–4. Try calculating them based on the card complements in this example to make sure you understand the method and that you get the same values.

The general outline for an application program using the hardware of Fig. 10–7 is shown in Fig. 10–8. It's quite straightforward, but I'll walk you through the high points.

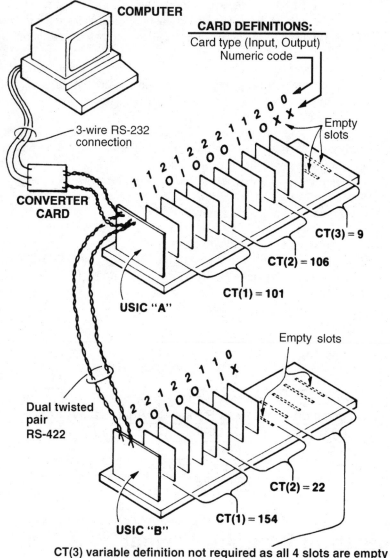

10-7 Serial UCIS with two USICs.

The line 10 through 190 area is set up for your insertion of DECLARE SUB and COM-MON statements for readers using CALLs to subprograms and for DEFINT and DIM statements. Many compilers, QuickBASIC included, require such statements to be at the front of your program. The last statement in this section must be the GOTO 3000 that branches around the block of protocol subroutines, lines 200–1970. Except for the few top-level statements placed before line 190, the main part of your application program (whether it's a test/diagnostic program, a program controlling the environment in a family of greenhouses, or something else altogether) always starts at line 3000.

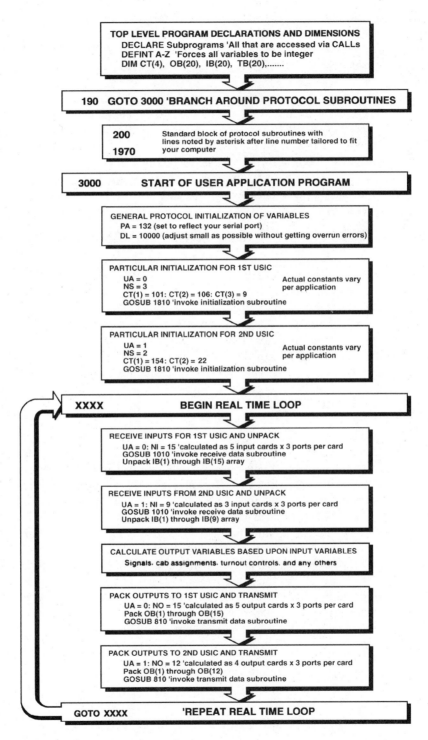

TOP LEVEL PROGRAM DECLARATIONS AND DIMENSIONS
DECLARE Subprograms 'All that are accessed via CALLs
DEFINT A-Z 'Forces all variables to be integer
DIM CT(4), OB(20), IB(20), TB(20),.......

190   GOTO 3000 'BRANCH AROUND PROTOCOL SUBROUTINES

200
         Standard block of protocol subroutines with
         lines noted by asterisk after line number tailored to fit
1970   your computer

3000   **START OF USER APPLICATION PROGRAM**

GENERAL PROTOCOL INITIALIZATION OF VARIABLES
PA = 132 (set to reflect your serial port)
DL = 10000 (adjust small as possible without getting overrun errors)

PARTICULAR INITIALIZATION FOR 1ST USIC
UA = 0   Actual constants vary
NS = 3   per application
CT(1) = 101: CT(2) = 106: CT(3) = 9
GOSUB 1810 'invoke initialization subroutine

PARTICULAR INITIALIZATION FOR 2ND USIC
UA = 1   Actual constants vary
NS = 2   per application
CT(1) = 154: CT(2) = 22
GOSUB 1810 'invoke initialization subroutine

XXXX   **BEGIN REAL TIME LOOP**

RECEIVE INPUTS FOR 1ST USIC AND UNPACK
UA = 0: NI = 15 'calculated as 5 input cards x 3 ports per card
GOSUB 1010 'invoke receive data subroutine
Unpack IB(1) through IB(15) array

RECEIVE INPUTS FROM 2ND USIC AND UNPACK
UA = 1: NI = 9 'calculated as 3 input cards x 3 ports per card
GOSUB 1010 'invoke receive data subroutine
Unpack IB(1) through IB(9) array

CALCULATE OUTPUT VARIABLES BASED UPON INPUT VARIABLES
Signals. cab assignments. turnout controls. and any others

PACK OUTPUTS TO 1ST USIC AND TRANSMIT
UA = 0: NO = 15 'calculated as 5 output cards x 3 ports per card
Pack OB(1) through OB(15)
GOSUB 810 'invoke transmit data subroutine

PACK OUTPUTS TO 2ND USIC AND TRANSMIT
UA = 1: NO = 12 'calculated as 4 output cards x 3 ports per card
Pack OB(1) through OB(12)
GOSUB 810 'invoke transmit data subroutine

GOTO XXXX   **'REPEAT REAL TIME LOOP**

10-8 Application program format.

The initialization section first defines the general protocol variables, those independent of the particular USIC. The *PA* = 132 statement needs to be altered to reflect your serial hookup. For example, if you are using an IBM PC up through Pentium, then set *PA* = 1016 if you're using the COM1 serial port, and *PA* = 760 if using COM2.

Next, the specific initialization for the first USIC is defined, followed by GOSUB 1810 and a repeat of the initialization procedure for the second USIC.

The real-time loop is set to receive UCIS inputs and unpack, calculate output variables from input variables, pack outputs, and transmit to the UCIS. (I'll describe the unpack and pack operations in Chap. 15.) The only difference with more than one USIC is that the receive and unpack operations are grouped together for USIC A, then repeated for USIC B, and the same is true for the pack and transmit operations.

Program Timing Considerations

One point to keep in mind with the UCIS serial interface is that operational speed will be slower than the equivalent program using IBEC or UBEC. For example, at 300 baud we are sending 300 bits per second. A six-I/O-card interface at three ports each needs to transmit 18 bytes, 144 bits, during each real-time loop. Thus it would take 480 ms or almost half a second just to transmit the data. Going to 9600 baud would speed that up to 15 ms, and to 7.5 ms at 19,200 baud.

On the other hand, an I/O setup with 23 I/O cards, 69 bytes, and 552 bits would have transmission times of 1.8 s at 300 baud, 57 ms at 9600 baud, and 28.5 ms at 19,200 baud. For medium-to-large systems where you want fast response times, you definitely need the higher baud rates.

Besides transmission time, the major slowness of the serial interface comes from the machine cycle overhead of the protocol subroutines. It's hard to give meaningful times for this, as it varies greatly dependent upon the language used and from machine to machine. Even though using noncompiled BASIC might be too slow for your final operational program, it's still a good way to get started on developing your system.

The easiest way to make a significant speedup is to use a compiled language like Microsoft's QuickBASIC, Pascal, or C. This conversion has a double benefit, as it speeds up your whole program and not just the protocol subroutines. Much faster operation can also be achieved by converting the protocol subroutines to your computer's assembly language (AL). Using AL your serial I/O speeds will approach the times of parallel I/O, except for the small overhead factor associated with formatting messages and the address bytes sent along with the data (plus, of course, the actual serial transmission time).

11
CHAPTER

Testing the Serial (USIC-Based) System

If you have been building along with me, you've now built all the hardware for your USIC-based UCIS. You can be proud of that, and soon you'll be even prouder when you see it at work with your external hardware. In this chapter we'll test the serial system to see that everything is assembled and connected correctly. I'll explain how to test the UCIS hardware, shown in Fig. 11–1 with the computer output test panel and the other test equipment that you will need. The computer is included because I'll present programs that turn it into an interface tester.

We'll start out with initial checks of your USIC, less the I/O cards. Then I'll show how to use the test panel and the wraparound test cable along with software to verify the operation of your entire UCIS, including the I/O cards.

The most likely sources of trouble, if you have any, will be poor soldering and incorrect parts insertion. The next thing to look for is cable trouble between the USIC and the serial port on your computer, or the way in which the protocol subroutines interact with your computer's serial I/O port when connected to the USIC. Systematic checking takes time, but it also goes a long way toward ensuring that your UCIS operates correctly.

USIC Testing

Before we start actual testing, it's important to understand how the 68701S signals its operational status via the USIC's status indicator LED. It's programmed to use this LED to display the following three different status indications:

1. 68701S is running, but not receiving messages:
   Green LED on 1.8 s, off .2 s.
2. 68701S is running, and receiving data without errors:
   Green LED on 1 s, off 1 s.

11-1 UCIS serial hardware setup for testing.

3. 68701S is running, but receiving data with errors:
   Green LED on .05 s, off .05 s.

Watching for these three different indications as your USIC is operating can significantly help get your initial system operating, as well as help locate and solve future problems if they should develop.

The following steps check that the correct power is reaching each USIC IC, U1 is correctly initialized when powered up, and its internal software is working. Check each of the following boxes only after you complete the requested action and, where applicable, get the correct results.

☐ *Set SW1.* All four segments are off.

☐ *Set SW2.* Any one segment on and all others off. Which baud setting you use doesn't matter when the USIC isn't connected to the computer.

☐ *Power-up test.* Turn off the IOMB's 5-Vdc supply and remove any I/O cards; then plug the USIC into the first slot of IOMB. The USIC should not be connected to your computer. Now turn on the 5-Vdc power to IOMB; the green LED should start blinking about once every two seconds as indicated by the status 1 code noted above. It's controlled by a real-time loop counter in U1's internal software and shows that the chip is working properly. The red and amber LEDs should be off unless you're using 20-mA current loop I/O, in which case red should be off and amber on. Even if the LED indications aren't correct, go on to the next test.

☐ *USIC IC power tests.* Set your VOM to measure dc voltage, and make the tests listed in Table 11–1 to see that each IC installed on USIC is receiving the proper voltage. If you find one not getting power, work back along the circuit path to locate and correct the open circuit. A low voltage, less than 4.8 Vdc, most likely means a short circuit on your card caused by a solder bridge or a reversed part. Turn off the power and re-examine the card to find and correct the error.

If necessary, remove ICs one at a time to isolate the problem. If you do trace a problem to an IC, replace it with a new part. It's handy to keep at least one spare of each IC type used in your UCIS for debugging and possible replacement.

For RS-232 you'll need a ±12 Vdc supply connected to S15 to check the U12 voltage. Figure 11–2 shows a schematic and parts list for building such a supply.

☐ *Recheck operation.* If IC power is correct in all tests, the green LED should blink continuously according to the code 1 noted above when +5 Vdc is applied to the USIC. If the green LED still doesn't blink correctly, look for a bad solder joint, a solder bridge, or an incorrectly inserted or faulty part. For example, make certain that you have the correct capacitor values for C12 and C13. Values other than the specified 22 pF can cause the crystal to not oscillate correctly and prevent U1 from operating. Be persistent in searching for such problems. If necessary to isolate the fault, you can remove U2 through U6 and U9 through U13, as they should have no effect

Table 11-1. USIC IC Power Tests

✔	IC	+ METER LEAD ON PIN NO.	− METER LEAD ON PIN NO.	VOLTAGE READING
	U1	6	1	+ 5VDC*
	U1	7	1	+ 5VDC
	U1	8	1	+ 5VDC*
	U1	21	1	+ 5VDC
	U2	20	10	+ 5VDC
	U3	20	10	+ 5VDC
	U4	20	10	+ 5VDC
	U5	20	10	+ 5VDC
	U6	14	7	+ 5VDC
	U7	14	7	+ 5VDC
	U8	16	8	+ 5VDC
	U9	16	8	+ 5VDC
	U10	16	8	+ 5VDC
	U11	14	7	+ 5VDC
	U12	14	7	+ 12VDC
	U12	7	1	+ 12VDC

No application uses all ICs.
You should read indicated voltage on all
that are installed.

*Reading may be slightly less than
other + 5VDC readings

on the green LED's indication. The cost of U1 includes program verification
and full testing in a functional USIC, so it should be the very last thing to
suspect in debugging your card.

Do not proceed with the rest of the testing until you have the green
LED blinking. The most likely sources of trouble will be poor soldering and
incorrect parts insertion on USIC. If you need additional ideas for help,
temporarily jump ahead in this chapter and read through the initial parts
of the section "In Case of Difficulty", but do not proceed with testing
beyond this point until you have the green LED blinking correctly.

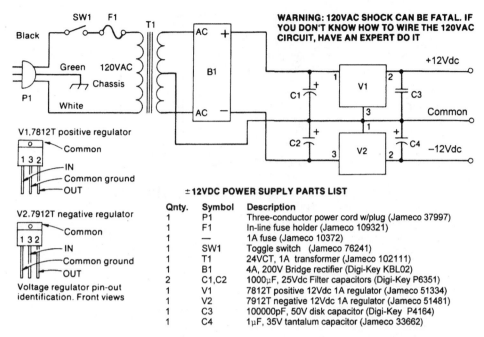

WARNING: 120VAC SHOCK CAN BE FATAL. IF YOU DON'T KNOW HOW TO WIRE THE 120VAC CIRCUIT, HAVE AN EXPERT DO IT

V1,7812T positive regulator

Common
1 3 2
IN
Common ground
OUT

V2,7912T negative regulator

Common
1 3 2
IN
Common ground
OUT

Voltage regulator pin-out identification. Front views

±12VDC POWER SUPPLY PARTS LIST

Qnty.	Symbol	Description
1	P1	Three-conductor power cord w/plug (Jameco 37997)
1	F1	In-line fuse holder (Jameco 109321)
1	—	1A fuse (Jameco 10372)
1	SW1	Toggle switch (Jameco 76241)
1	T1	24VCT, 1A transformer (Jameco 102111)
1	B1	4A, 200V Bridge rectifier (Digi-Key KBL02)
2	C1,C2	1000μF, 25Vdc Filter capacitors (Digi-Key P6351)
1	V1	7812T positive 12Vdc 1A regulator (Jameco 51334)
1	V2	7912T negative 12Vdc 1A regulator (Jameco 51481)
1	C3	100000pF, 50V disk capacitor (Digi-Key P4164)
1	C4	1μF, 35V tantalum capacitor (Jameco 33662)

Author's recommendations for suppliers given in parentheses above with part numbers where applicable. Equivalent parts may be substituted.

11-2 ±12-Vdc RS-232 power supply.

Output Card Test Program

The previous testing established that the ICs are receiving proper power and that the program in the MC68701S is executing. Now I'll take you one step further with a BASIC program, listed in Fig. 11–3, that will demonstrate the complete output capabilities of your interface. The source code for this program is contained on the 3.5-in disk along with equivalent programs in Pascal and C.

The program uses ten variables, defined as follows:

PN\$( )   This is a three-position array used to print out the port being tested; *PN\$(0)*="A," *PN\$(1)* = "B," and *PN\$(2)* = "C."

UA   This variable is the USIC address, and it equals 0.

PA   This is the computer's port address that is being used for your UCIS (set for your computer configuration).

DL   This is the USIC delay, and is set to 5000. (Adjust it to larger or smaller values as required for your computer; instructions follow.)

| VALUE OF | VALUE OF DISPLAY | | | DISPLAY |
N IN LOOP	Decimal	Binary	Hexadecimal	BIT "ON"
0	1	00000001	01	0
1	2	00000010	02	1
2	4	00000100	04	2
3	8	00001000	08	3
4	16	00010000	10	4
5	32	00100000	20	5
6	64	01000000	40	6
7	128	10000000	80	7

BIT PATTERN FOR RELATIONSHIP $2\uparrow N$

```
10    REM**OUTPUT CARD TEST PROGRAM USING USIC**
20    REM**DEFINE VARIABLES
30    DIM PN$(3), CT(9), OB(60), IB(60), TB(80)
40    PN$(1) = "A": PN$(2) = "B": PN$(3) = "C"
50    GOTO 3000      'BRANCH TO USER PROGRAM


3000  REM*******************************************
3010  REM**BEGIN USER PROGRAM TO TEST OUTPUT CARD**
3020  REM*******************************************
3030  PRINT "OUTPUT CARD LED DRIVER TEST PROGRAM"
3100  REM**INITIALIZE USIC**
3110  UA = 0      'USIC ADDRESS
3120  PA = 760    'FOR IBM PC (FOR COM1 USE PA = 1016, COM2 PA = 760)
3130  DL = 5000   'USIC DELAY
3140  NS = 1      'NUMBER OF CARD SETS OF 4
3150  NI = 0      'NUMBER OF INPUT CARD PORTS
3160  NO = 3      'NUMBER OF OUTPUT CARD PORTS
3170  CT(1) = 2  'CARD TYPE DEFINITION (ONE OUTPUT CARD IN SLOT 1)
3190  GOSUB 1810 'INVOKE INITIALIZATION SUBROUTINE
4000  REM**INITIALIZE ALL LEDS TO OFF
4010  OB(1) = 0: OB(2) = O: OB(3) = 0
5000  REM**INCREMENT PORT TO BE TESTED IN A LOOP
5010  FOR PN = 1 TO 3
5020  REM**INCREMENT DISPLAYED BIT NUMBER IN A LOOP
5030  FOR N = 0 TO 7
5040  REM**OUTPUT TEST STATUS TO CRT
5060  PRINT "PORT IS", PN$(PN), "BIT NUMBER IS", N
5070  REM**SETUP TEST LED PATTERN
5080  OB(PN) = 2 ^ N
5090  REM**OUTPUT LED DISPLAY DATA TO CARD
5100  GOSUB 810   'INVOKE TRANSMIT DATA SUBROUTINE
5110  FOR IJ = 1 TO 5000: NEXT IJ
5120  NEXT N
5130  OB(PN) = 0
5140  NEXT PN
5150  REM**REPEAT TEST
5160  GOTO 3110
```

11-3 Output card LED-driver test program using a USIC.

NI   This is the number of input card ports, and is set to 0.

NO   This is the number of output card ports, and is set to 3.

CT(1)   Card type definition for card in slot address 1. Setting this to 2 signifies that we have only a single output card, and it is in slot 1—see Fig. 10–4 in Chap. 10.

OB( )   A three-position array used to store the three data output bytes to be sent to the LEDs on the test panel.

PN   This is a numeric variable identifying the port being tested; 0 = port A, 1 = port B, and 2 = port C.

N   This variable is the bit number of the LED to be lit; the range is 0–7.

Because for the serial case this is our first program, I'll briefly explain each line of code. BASIC lines beginning with REM, for remark, are not instructions to be carried out. Their function is simply to carry comments that help readers understand the program. Line 10 is a remark defining the program title. Line 30 sets aside memory locations and labels them *PN$(1)*, *PN$(2)*, and *PN$(3)*, plus nine similar locations for variable *CT( )*, 60 each for output byte *OB( )* and input byte *IB( )*, and 80 for *TB( )*. Line 40 stores the symbol "A" in *PN$(1)*, "B" in *PN$(2)*, and "C" in *PN$(3)*. Line 50 is the standard GOTO 3000 statement, defined in Fig. 10–8, required to branch around the protocol subroutines (for example, code in Fig.10–5) to be inserted right after line 50.

Because our test uses only one USIC, line 3110 establishes the USIC address as zero. Line 3120 defines the computer's port address being assigned to USIC as 760. Remember, this is an application-dependent variable that you must set to reflect the port address to which you have attached your USIC. Line 3130 sets the USIC delay variable at 5000, which is again a variable you will want to experiment with and set according to the speed of your computer's processing.

The next three lines set the variables to reflect our use of one set of four cards (actually just one card, but as defined in Fig. 10–4 subsets of four count as a full set); zero input card ports; and 3 output card ports. Line 3170 defines the card type definition for the "set of four" to be 2, which following the procedure outlined in Fig. 10–4 defines one output card in the first slot. Follow the procedure defined in Fig. 10–4 to make sure you obtain the same answer.

Line 3190 follows the general application program format established in Fig. 10–8 to invoke the initialization subroutine. Line 4010 sets the three output card port output bytes to zero corresponding to an initialization with LEDs off.

Line 5010 begins a test loop by incrementing variable *PN* from 1 to 2 to 3, to test all three ports. Line 5030 sets up an inner loop whereby the bit number of the LED to be lit (*N*) is incremented from 0 to 7. To keep track of what's being tested, line 5060 displays the port under test and the bit number of the LED to be lit.

The next step is to calculate and send the bit patterns to the port under test, so we can see if they get through correctly to light the proper LED on the test panel. Line 5080 does the calculation by setting *OB(PN)* equal to 2 raised to the power *N*. This arithmetical statement sets the bit pattern relationships shown at the top of Fig. 11–3. If your BASIC doesn't have the exponent function, you can get the same result by following the alternate procedure I provided in Fig. 9–2.

Line 5100, by invoking the transmit data subroutine, transmits all three *OB( )*s to the output card to light bit number *N* of port *PN* while keeping the LEDs for the other two ports dark.

Line 5120 loops back to 5030 to increment the LED bit number. Once bit 7 is executed for a given port, the program executes line 5130 to reset the port LEDs to off by setting *OB(PN)* = 0. Then line 5140 increments *PN* to start testing the next port. With all three ports tested, line 5160 is executed to start the whole test cycle over again by branching back to line 3110.

☐ *Connect your USIC to your computer.* Follow Fig. 4–1 using the appropriate subfigure for your choice of 20-mA current loop, RS-232, or RS-422 standards.

☐ *Add a test card.* Plug an output card (either a DOUT with J1 configuration or a COUT24) into your motherboard. Plug the appropriate output test panel onto the output card and connect the test panel's clip lead to +5 VDC on IOMB as shown in Fig. 9–3.

☐ *Reconfigure your protocol subroutines.* Copy the Fig. 10–5 program from the enclosed disk and make any adjustments required to fit your computer. Once you have your protocol subroutines operational, they will be your standard USIC protocol for all applications, so be sure to save it in a secure location.

Depending on your computer and the particular serial hardware it uses, you may alternatively need to make DIP switch settings to set for serial operation with the desired baud rate, eight data bits, one stop bit, and no parity as required for operation with USIC.

To perform the test, enter the program with adjustments for your application, then run the program. The operation should be identical to that described in Chap. 9, except I've eliminated the need to enter the carriage return. Every time the program loops, the LED that is lit should move one position to the left. Compare the port and bit combination on the monitor with the LED that lights up on the test panel. If you find the program executes too fast for you to track the test panel LEDs, you can slow the process down by adding the requirement for a carriage return input after each update as I did in Fig. 9–2, or alternatively add a loop delay line such as: 5035 FOR IJK = 1 TO 5000: NEXT IJK. Pick different numbers for the loop length (the maximum value of *IJK*) to control the program's speed.

If the LEDs don't agree with the screen display, you have some debugging to do. First make sure your program is correct by checking it against Figs. 11–3 and 10–5. Once one port works correctly, any remaining problem is almost certainly on the output card itself. Check the card for bad solder joints, solder bridges, reversed ICs, and so on. The debug procedures defined in Chap. 9 should help here as well.

Success with the test program proves not only that the output card works correctly, but also that most of the USIC and cabling from USIC to the computer serial port, address decoding, and software are okay. It's an effective visual test of UCIS output. Once you get one output card working on all three ports, the hardest part of the UCIS project is behind you. Total interface success is just a small step ahead—testing the input card and the ability of USIC to send input card received data back to your computer. Because you know everything else checks out, this will be much easier.

If everything works the first time, you have earned the right to be elated! In all probability there will be at least one thing wrong, and you'll have some rechecking and debugging to do.

In Case of Difficulty

Remember, the most important requirement for problem solving is persistence, and reward will be worth the effort. I'll cover general things to look for first, and then lead you through suggested debugging steps, but once you have a successful output card test you can skip directly to the section on automated wraparound test programs.

If these instructions (when they are tailored for your own computer and card address) don't control the LEDs on your test panel, your software isn't communicating with your output card. You have what's known as an "interface communication problem." The question is, "How do I fix it?" The solution will probably be very simple once you find it, but first you have to find it!

The first thing to look for is cable trouble, like serial connections that are not correct for your particular computer. If you haven't already done it, get out your computer reference manual and make sure that all the serial connections are correct for your computer.

If you can't find a cabling problem, then the next best place to start is to recheck your entire USIC assembly patiently and carefully per Chap. 4. Have you used the correct part number ICs in the USIC? Did you follow the correct subsection for the serial standard you chose: 20-mA current loop, RS-232 or RS-422?

Reread the sections of your computer's BASIC manual that cover the BASIC instructions you are using, to be doubly sure your use of the BASIC commands are correct. If you've built two output cards, you can try again with the second one, but set its DIP switch to the same address so you're making the same test.

Make sure you have the correct USIC DIP switch settings, and the correct card-slot setting on the output card's DIP switch. Are all the switches installed with correct on/off orientation? If all switch settings aren't correct for your UCIS configuration, and if they don't correspond exactly to the address and baud rate you are using in your program and computer, then you won't get any response on the test panel.

If everything above is correct and you still don't get any response, then re-examine each card carefully to see that all parts are inserted correctly, that all solder joints are good, and that there are no solder bridges between adjacent pins or traces. In checking soldering, first make sure your card has been cleaned of all soldering flux, and then perform your inspection under a bright light and using a magnifying glass. Look at every single joint and from different angles. If you are positive all else is good, then replace ICs as you might have a faulty one. Leave U1 on the USIC till last, as it was thoroughly tested before it was shipped to you.

Remember, most problems can be traced to faulty soldering, parts insertion, or cabling. Keep searching and retesting until you find the error, correct it, and achieve successful operation.

If you still can't find the trouble, follow the general procedures I outline in the last portion of Chap. 9. Having someone else (preferably one with some electronics background) go through all the checks with you can often find the problem with minimum total effort. If you still have problems, here are some added steps you can take.

Get the Yellow LED Flickering

The flickering of the LED when you run the test program shows that data is reaching the USIC. If it doesn't flicker at all, you have a basic interface problem in the setup and initialization of your computer's serial port, in the cabling between the port and the USIC, or in the U9, U11, U13, and U6 circuit areas on the USIC itself.

Restudy your computer manuals to make sure of your connections and wiring. If using RS-232, make sure you pass the proper ±12-Vdc readings on USIC's U12 chip. Also try interchanging the connections to pins 2 and 3 of the serial port connector, to make sure you haven't reversed the receive and transmit lines. Make sure the computer-dependent lines of code in the protocol subroutines are correct, and that you haven't made a mistake keyboarding any of your modifications.

If your program appears to be hung up, do a CTRL-C or CTRL-BREAK to check where in the loop your program is stuck; most likely it will be at line 650, showing that you have a problem in the way you are handling the line status register. Also, you can insert a temporary line: 665 Print TB(I) to display what's actually transmitted each time a data byte is sent on the serial channel. This type of temporary test intermediate printout comes in very handy for all types of interface and application program checkout.

At this point, baud rate settings aren't a factor. The yellow LED simply shows that data is being sent over the transmission line and through the USIC's I/O-unique circuitry, because this diode is located before the MC68701S processor. Passing this step is more a function of having a correct setup at the computer rather than a test of the USIC itself. Once you have the yellow LED flickering, check the test panel LEDs for proper operation. Once they work, skip to the wraparound test. If not, do some more checking.

Check Port Initialization and Baud Rate

Recheck your manuals. Have someone else read and go over what you have done to be certain that you have your computer's serial port set up correctly. Check any jumpers you've added, all DIP switch settings, the program code, and so on. If your computer serial port isn't set up to talk correctly, the USIC won't hear correctly. Note that the speed at which the yellow LED flickers is a direct function of baud rate.

Once you are certain everything at the computer end is correct, make sure the two USIC DIP switches are set correctly. Make sure you have only the one output card, and that its DIP switch is set with all six segments off to match the CT(1) = 2 statement in your test program.

If everything is set correctly, then you most likely have a hardware problem on the USIC, motherboard, or output card. Shut down your system and repeat your inspection. Again, go over your assemblies using a magnifying glass under a bright light. Look, from different angles, at each solder joint for poor soldering or solder bridges, and look at each part for correct insertion with all pins in sockets, and so on. Review Chap. 1 on PC card soldering. It's a good idea to have a second output card ready to substitute to help eliminate the first card as the problem.

Once you have a successful output card test, total interface success is just a small step ahead. That step is testing the input card and the USIC's ability to transmit information back to your computer.

Automated Wraparound Test Program

We could test the input card by connecting a DIP switch to the card inputs, using a test program to read the switch settings, and displaying the results on the computer screen. Or without wiring a DIP switch, you could use clip leads to selectively ground input pins. It's much easier to use the computer to perform the whole test automatically. Automated testing and diagnostics is one of the advantages of having a computer connected to your external hardware.

Because you've already tested your first output card, you can use it to test all the input cards you need. In the computer industry this kind of I/O check is called a wraparound test, because the computer outputs are simply wrapped around and connected to its inputs. This wraparound capability is achieved using the wraparound test cable we assembled in Chap. 8.

Figure 11–4 shows the test program for performing the automated wraparound test. The source code for this program is contained on the 3.5-in disk, along with equivalent programs in Pascal and C.

The program writes a bit pattern to the output card and then reads the input card. It then compares what it reads with what it wrote, and if they don't agree, it has found an I/O error and displays an error message. Because you've already proven the output card, the fault will most likely be in the input card or in the input section of the USIC.

If the program reads exactly what it wrote, it changes the bit pattern and repeats the write/read cycle. Once all bit patterns pass the test, the USIC as well as both I/O cards in the test loop are proven good.

Because the wraparound test program closely parallels the LED-driver test program we just used to test the USIC's output capability and the output card, I'll explain only the significant differences. Line 3130 has defined a different USIC delay, but its value has a very wide acceptable range. Once your programs are operational, you make it as small as possible to speed up your program. Line 3150 is altered to

```
40       DIM PN$(3), CT(9), OB(60), IB(60), TB(80)
50       PN$(1) = "A": PN$(2) = "B": PN$(3) = "C"
60       GOTO 3000      'BRANCH TO USER PROGRAM


3000 REM****************************************
3010 REM**BEGIN USER WRAPAROUND TEST PROGRAM**
3020 REM****************************************
3030 PRINT "WRAPAROUND TEST PROGRAM FOR I/O CARDS"
3100 REM**INITIALIZE USIC**
3110    UA = 0         'USIC ADDRESS
3120    PA = 760  'COM1 FOR IBM PC  (USE 760 FOR COM2 or 1016 FOR COM1)
3130    DL = 2000  'USIC DELAY
3140    NS = 1     'NUMBER OF CARD SETS OF 4
3150    NI = 3     'NUMBER OF INPUT CARD PORTS
3160    NO = 3     'NUMBER OF OUTPUT CARD PORTS
3170    CT(1) = 6  'CARD TYPE DEFINITION (OUTPUT IN 1, INPUT IN 2)
3190    GOSUB 1810  'INVOKE INITIALIZATION SUBROUTINE
3200 REM**INITIALIZE ALL OUTPUTS TO ZERO
3210    FOR I = 1 TO 3: OB(I) = 0: NEXT I
5000 REM**INCREMENT PORT TO BE TESTED IN A LOOP
5010    FOR PN = 1 TO 3
5020 REM**INCREMENT TEST PATTERN IN A LOOP
5030    FOR D = 0 TO 255
5040    OB(PN) = D
5050 REM**OUTPUT BIT PATTERN TO OUTPUT CARD
5060    GOSUB 810  'INVOKE TRANSMIT SUBROUTINE
5070 REM**READ BIT PATTERN FROM INPUT CARD
5080    GOSUB 1010  'INVOKE RECEIVE SUBROUTINE
5090 REM**OUTPUT TEST STATUS TO CRT
5100 PRINT "PORT IS "; PN$(PN); " DEC OUT IS "; OB(PN); " DEC IN IS "; IB(PN)
5110 REM**COMPARE INPUT TO OUTPUT AND BRANCH IF NO ERROR FOUND
5120    IF IB(PN) = OB(PN) THEN GOTO 5210
5130 REM**ERROR FOUND SO PRINT TO CRT
5140 PRINT "ERROR FOUND IN ABOVE I/O, ENTER RETURN TO CONTINUE TEST"
5150    INPUT CR
5200 REM**COMPLETE LOOPS
5210    NEXT D
5220    OB(PN) = 0
5230    NEXT PN
5240    GOTO 3210   'REPEAT TEST
```

11-4 Automated wraparound test program using a USIC.

reflect picking up the input card; likewise the altered value for *CT(1)* in line 3170 defines to the USIC that we have an output card in slot 1 and an input card in slot 2, calculated following the procedure of Fig. 10–4.

Because in this test we are exercising every possible bit combination in every port, line 5040 defines *OB(PN)* (the output byte being tested) equal to *D*, thereby covering the whole range 0 to 255. Line 5060, by invoking the transmit data subroutine, transmits all three $OB(\ )$s to the output card with the binary pattern corresponding to the decimal number D for port PN, while keeping the other two ports at off.

Line 5080 reads all three ports of the input card by invoking the "receive data" subroutine. Line 5100 prints the port number along with the data sent out and the data received back.

The important comparison test is the IF, THEN statement in line 5120. If $IB(PN)$ = $OB(PN)$ there is no error, and the program branches to line 5210 to increment to the next bit pattern D and repeat the cycle. If $IB(PN)$ doesn't equal $OB(PN)$, there's an error and a message is displayed by line 5140. To continue the test after an error, you simply press return. The program runs continuously, looping through all combinations of bit patterns for each port.

If your BASIC has a decimal-to-hexadecimal conversion function, you can follow the guidelines set up in Fig. 9–5 to convert the $OB(PN)$ and $IB(PN)$ values to hexadecimal before the printout in line 5100. That makes it easier to determine the offending bits if you do run into some errors, or you can add a second output card with its test panel as I described in Chap. 9's wrap test

Conducting the Test

As with the output card test program, copy the Fig. 11–4 program from the enclosed disk and make any required changes. Then set up the hardware with one input and one output card joined by the wraparound cable discussed in Chap. 8. The two cards must be a compatible set, meaning that both are new or both are the original design.

Make sure you have the correct value of PA for your port in line 3120. To match the $CT(1)$ = 6 definition in line 3170, check that the output card DIP switch segments are all off and that the input cards are off except the rightmost segment, which should be turned on. Save the program so you'll have it to use again, and then run the program. Depending on your version of BASIC, you might have to alter the format of some instructions to get the program to run correctly.

If you get the overrun message, try increasing values of $DL,$ and if this doesn't correct the problem, look for possible coding errors. The USIC's red LED should flicker after the yellow one does, showing that the USIC is receiving transmissions, getting the requested data from the input card, and sending it back to the computer. If not, look for improper cabling from the USIC to your computer or a coding error in your modifying the protocol subroutines that receive one input byte or receive data.

Again, if your program appears hung up, do a CTRL-C or CTRL-BREAK to check where your program is stuck. Most likely it will be at lines 320 or 330, showing that you have a problem in the way you are handling the line status register. You can add a temporary line: 345 PRINT IB to observe what data bytes your computer is receiving from the USIC.

Figure 9–6 shows an example of the type of program output seen on the computer monitor. I forced that error message by grounding one of the input connector pins on the input card with a clip lead. If you do get errors, first check to see that you have the correct addresses set in the DIP switches on each card, and that these settings correspond to the addresses in the program. If, for example, you accidentally interchanged the addresses, you would be trying to write to the input card and read from the output card, and nothing would work.

Seeing which bits fail can help locate where on the input or output card you have bad soldering, a part incorrectly inserted, or a faulty part. To find which bit or bits are faulty, expand the hex representations of C and D into their binary equivalents using Fig. 1–6. Whichever bits don't compare define the area of trouble. Probe those parts of the card with your VOM, looking for 0 Vdc or +5 Vdc on the appropriate lines to pinpoint the problem.

Once you have a proven input card, you can use it to test the rest of your output cards automatically, just as your first proven output card can be used to test input cards. Keep your test programs handy for use if you ever suspect faulty operation in your UCIS.

At this point you can feel very proud. Your UCIS is fully operational, and you've written and executed two interface programs. Once you have a successful wrap-around test, you have a fully operational serial UCIS with distributed I/O capability. Skip to Chap. 13 where we'll begin hooking up your external hardware.

12
CHAPTER

Testing the Parallel (UBEC-Based) System

If you have been building along with me, you've now built all the hardware for your UBEC-based UCIS. You can be proud of that, and soon you'll be even prouder when you see it at work with your external hardware. In this chapter we'll test the system to see that everything is assembled and connected correctly. I'll explain how to test the UCIS hardware, shown in Fig. 12–1 with the computer output test panel and the other test equipment that you will need. The computer is included because I'll present programs that turn it into an interface tester.

We'll start out with initial VOM checks of your UCIS without the I/O cards. Then I'll show how to use the test panel and wraparound test cable along with software to verify the operation of your entire UCIS including the I/O cards.

Initial System Testing

The following tests check that your computer operates properly with the UBEC attached, that the UBEC is receiving power from your computer, that it is reaching each of the ICs, and that the DIP switch is installed correctly. Systematic checking takes time, but it also goes a long way toward ensuring that your UCIS will operate correctly.

The most likely sources of trouble will be poor soldering and incorrect parts insertion. The next thing to look for is cable trouble, like CBAC connections that are not correct for your particular computer, or an open or poor contact at one of the ribbon-cable connector pins.

If you haven't already done it, get out your computer reference manual and make sure that all the connections are correct for your computer. If the computer pinouts aren't listed in the manuals you have, then at least double check against the tables for your machine in Chap. 7. Check each of the following boxes only after you get the correct results.

217

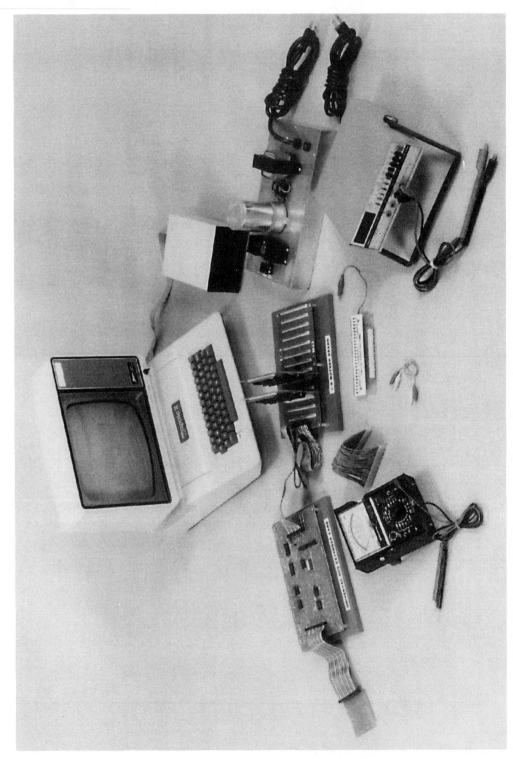

12-1 UCIS parallel (UBEC-based) hardware setup for testing.

☐ *Computer power test.* Set your VOM to measure +5 Vdc, then clip the (−) lead to ground on the UBEC and the (+) lead to the +5-V line via J21.

Warning: *Always make sure your computer is turned off whenever you connect or disconnect the UCIS, and whenever you install or remove ICs on the UBEC with the UCIS connected.*

With the computer off, connect the UBEC as shown in Chap. 7. Now turn on the computer. No ICs should be inserted in UBEC or in IBPC when used. You should read +5 Vdc on your meter, if the computer is supplying power to the UBEC. If you read 0 V, recheck your connections. If correct, use the VOM to check for +5 V at the appropriate computer connector pins, on the CBAC, and at each cable connector. Find where you lose voltage, then establish a path for it to the UBEC.

If you have a computer other than what I covered in Chap. 7 and it doesn't supply +5 Vdc, then follow the procedure explained in Chap. 6 to add V1 to the UBEC and power the card with a separate 9-Vdc adapter.

☐ *Computer operation test.* Try running some of your regular computer programs with the UCIS connected. A malfunction at this stage indicates a problem with the CBAC or UBEC, most likely a solder bridge touching one of the address lines, data lines, or control lines.

Examine the ribbon cable headers, IC socket pins, and other closely spaced pads and traces for solder bridges. To help isolate the problem, repeat the operation test first with the IOMB disconnected, then the UBEC disconnected, and finally with the CBAC disconnected. If you've isolated the problem to a faulty card but can't see the problem, use your VOM to check each address, data, and control input to that card to find the offending line. Most likely a solder bridge will connect one of the input lines to another input, to ground, or to +5 Vdc. Trace until you find the solder bridge and remove it. Repeat until the computer operates properly with the CBAC, UBEC, and IOMB all connected.

☐ *UBEC DIP switch test.* With the computer turned on and the UCIS connected, connect your (−) meter lead to ground and move the (+) lead to touch the IC socket pinouts listed in Table 12–1. At each position turn the switch segment on and off to see if the corresponding IC socket point reads +5 Vdc when the switch is off, and 0 Vdc when the switch is on. If all readings are reversed, you have the DIP switch installed backward. If only some segments are incorrect, check around U5, U6, SW1, and R4–R11 for solder bridges or open circuits. Don't proceed until all eight switch segments work correctly.

☐ *IC installation.* Turn off your computer, then install ICs U1–U9 on the UBEC and, if you are using the IBPC CBAC, then install its U1 as well. Be sure you have the ICs as specified for your machine in Chap. 6 and that you put them in the right sockets with correct pin 1 orientation.

☐ *Set the UBEC DIP switch.* Now is the time to set SW1 to the UCIS starting address for your interface. Either use the setting for your computer shown

Table 12-1.
UBEC DIP Switch Tests

✓	UBEC		
	IC SOCKET	PIN	DIP SWITCH SEGMENT
	S5	12	A8
	S5	9	A9
	S6	12	A10
	S6	9	A11
	S5	2	A12
	S5	5	A13
	S6	2	A14
	S6	5	A15

With switch segment set to "ON"
IC socket pin should read 0V.
With switch segment set to "OFF"
IC socket pin should read +5VDC

in Chap. 7, or calculate the starting address for another machine or application following the guidelines established in Chap. 6. If your computer is one of the IBM PC family up through the Pentium then use the procedures covered in Chap. 5, being sure to set SW1 on the CBAC as well.

☐ *UBEC IC power test.* Set your VOM to measure +5 Vdc. With the computer on, follow Table 12–2 to check that each UBEC IC is receiving 5 Vdc and ground. If you find an IC not getting power and ground, work along the circuit path until you locate the open circuit and correct it.

☐ *Computer operation test.* Check your computer's operation by again running some programs. If it doesn't work properly, recheck the placement and orientation of all ICs in the UBEC. If necessary, remove ICs from their sockets one by one—turning the computer off and on again each time—to see if your problem is in the part or in the circuit itself. If you do trace a problem to an IC, replace it with a new part. It's handy to keep at least one spare of each IC type in your UCIS for debugging and replacement. If you still have a problem, it's most likely that the starting address and/or DIP switch setting are not correct for your computer configuration.

☐ *IOMB power test.* Turn off the computer and connect your power supply to IOMB's +5-Vdc and GND terminals. With the power supply off, turn on the computer and check its operation, then turn on your power supply and make sure the computer still operates correctly. Turning power on and off while a program is running should have no effect on the computer's operation. If this stops the program, it might be causing a line transient in the power outlet feeding your computer. Try plugging the computer and

Table 12-2. UBEC IC Power Tests

✓	IC	+ METER LEAD ON PIN NO.	–METER LEAD ON PIN NO.
	U1	20	10
	U2	20	10
	U3	20	10
	U4	20	10
	U5	14	7
	U6	14	7
	U7	14	7
	U8	14	7
	U9	14	7

Some applications don't use all ICs. You should read + 5VDC on all that are installed

power supply into outlets on different circuits and/or installing a power strip with power surge and noise-filtering protection to power your computer.

☐ *RRON test.* Connect a VOM lead to UBEC ground and its (+) lead to pin 4 of U9, then turn the 5-Vdc supply on and off. The meter should read 0 V when the supply is off, and about +4.5 V when it's on. If not, check along the circuit path for a bad solder joint, solder bridge, missing jumper, or a bad or incorrectly inserted U7 and/or Q1.

Warning: *Turn off the 5-Vdc power supply whenever you add or remove an I/O card on the motherboard. Turn off both the computer and the +5-Vdc supply when plugging in or unplugging the UBEC and/or CBAC. You need not turn off your computer or your 5-Vdc supply when making connections to or disconnections from the I/O cards, such as the Test Panel.*

Output Card Test Program

From this point forward, the testing of your UBEC-based system is nearly identical to that performed using IBEC. I'll simply point out a few of the differences in the remaining sections of this chapter. To conduct the rest of your tests, mainly follow Chap. 9. Each time you read IBEC, simply substitute UBEC.

The UBEC test setup is identical to that in Fig. 9–3, and if you are using UBEC with an IBM PC up through the Pentium or the IBM System/2 Model 30, the test program in Fig. 9–2 is also identical. If your computer is one of the other machines covered in Chap. 7, then you need to remove the DEF SEG statement and substitute the few statements defined in Table 12–3.

Table 12-3. Output Card LED-Driver Test Program Changes When Using UBEC

If using UBEC with IBM PC up through Pentium computers with memory-mapped I/O, then use the program defined in Fig. 9-2.
If using UBEC with port-mapped I/O then use the program defined in Fig. 9-2 incorporating the changes noted at the bottom of listing for port-mapped I/O.
If using UBEC with an adapter card other than IBPC and for memory-mapped I/O, then use the program defined in Fig. 9-2 with the following changes:

```
    70  **DELETE LINE**
    80    SA = insert your defined UCIS starting address per chapter 7
   130    POKE SA + 3, CO
   200    POKE SA + P, D
   250    POKE SA + P, 0
```

Without using DEF SEG you simply define the UCIS starting address as variable *SA,* and then incorporate *SA* in the appropriate POKE statements. Recommended addresses for use with different computers are defined in the corresponding figures in Chap. 7.

Conducting the test is identical to the presentation in Chap. 9. Success with the test program proves not only that the output card works correctly, but also that the address decoding and data-out portions of UBEC and CBAC are working along with the IOMB and your software. It's an effective visual test of the UCIS output. Total interface success is just a small step ahead.

In Case of Difficulty

If these instructions (when tailored for your own computer and UCIS address) don't control the LEDs on your test panel, then follow the troubleshooting procedures in Chap. 9. The UBEC has fewer function on it than does the IBEC, so your debugging should be easier. However the UBEC doesn't have the built-in address selected test LED, so you will need to skip over those sections. Alternatively, you might want to build that circuit on a piece of perforated circuit board and mount it on your UBEC.

The one area you do need to check separately is the correct placement of all the regular jumpers, plus the program jumpers on UBEC. For example, if you are using your UBEC with an IBM PC, *x*86, Pentium, or IBM System/2 Model 30, have you left J48 out altogether, installed J50 in the alternate position, and added J51? Also, for your particular computer, make sure you have the program jumpers installed per Fig. 6–6.

If you tailored your own CBAC and/or UBEC to adapt the UCIS to a computer not covered in this book, you might have a line or lines hooked up wrong. See Fig. 6–3 and check for improper logic inversions between the computer bus and the I/O cards.

Once you have successfully completed the output card test, you are ready to test the UCIS input operations.

Automated Wraparound Test Program

We could test the input card by connecting a DIP switch to the card inputs, using a test program to read the switch settings, and displaying the results on the computer screen. However, it's so much easier to use the computer to perform the whole test automatically. The test setup with UBEC is identical to that shown in Fig. 9–4, and if you are using the UBEC with an IBM PC up through the Pentium or the IBM System/2 Model 30, the test program in Fig. 9–5 is also identical. If your computer is one of the other machines covered in Chap. 7, then you need to remove the DEF SEG statement and substitute the few statements defined in Table 12–4.

Table 12-4. Automated Wraparound Test Program Changes When Using UBEC

If using UBEC with IBM PC up through Pentium computers with memory-mapped I/O, then use the program defined in Fig. 9-5.
If using UBEC with port-mapped I/O then use the program defined in Fig. 9-5 incorporating the changes noted at the bottom of listing for port-mapped I/O.
If using UBEC with an adapter card other than IBPC and for memory-mapped I/O, then use the program defined in Fig. 9-5 with the following changes: 80   SA = *insert your defined UCIS starting address per chapter 7* 150 POKE SA + 3, CO 160 POKE SA + 7, CI 180 POKE SA + P, D 200 C = PEEK(SA + 4 + P) 330 POKE SA + P, 0

Conducting the test is identical to the presentation in Chap. 9. Once you have a proven input card, you can use it to test the rest of your output cards automatically, just as your first proven output card can be used to test input cards. Keep your test programs handy for use if you ever suspect faulty operation in your UCIS.

At this point you can feel very proud. Your UCIS is fully operational, and you've written and executed two interface programs. In the next chapter we'll connect a variety of external hardware devices to your interface.

13
CHAPTER

Connecting External Hardware to the I/O Cards

Now it's time to start connecting your tested and operational UCIS to your external hardware. In this chapter I'll show you how to make output and input connections from your external hardware to the general purpose I/O cards. I'll cover how lamps, LEDs, relays, logic gates, optoisolators, panel switches, push buttons, solenoids, stepper motors, and other assorted devices are easily connected to your general purpose I/O cards. You'll also receive a peek preview at new cards being developed.

I'll also show how to make wiring simpler and easier still, with decoding and encoding circuits that control or monitor groups of devices over single USIC I/O lines. That can lower the cost of your interface by requiring fewer I/O cards, plus reducing the wiring requirements for a large interface system.

I give numerous presentations around the country on the application of the UCIS, and a most frequently asked question is simply "How are devices XYZ connected up to the I/O cards?"

Basically, the hookups are so straightforward that I think people have trouble comprehending them. Looking at the breadth of UCIS material and the somewhat overwhelming capabilities with multiple options available, we often fail to see the simplicity in the UCIS application.

The great majority of all interface connections are made directly to the connector pins on the output and input cards. These hookups are identical whether using the original-design or the new-design I/O cards.

Basic Output Connections

Remember that each output card can control 24 discrete signals to the external hardware using open-collector transistors as the output devices. The easiest way to understand the open-collector transistor circuit is to think of it as an electrical switch with one end connected to ground, as in Fig. 13–1. The computer, under software control as I'll explain in later chapters, simply turns the transistor on and off.

When the transistor is on, its collector, or C terminal, is in effect connected to ground, so the switch is closed. When the transistor is off, the collector is in effect an open circuit, and the switch is open. Software controls the switch.

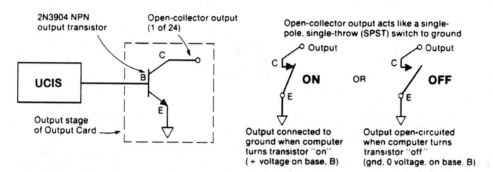

13-1 Open-collector output.

Any manner in which you connect such a regular SPST switch with one end grounded, you can also connect an output card line. The only restriction is that you do not exceed the capacity of the switch, meaning the transistor, which has a capacity of .3 A and 40 V.

Figure 13–2 shows circuits for several common items. Lamp operation, Fig. 13–2a, is straightforward: switch closed (to ground), lamp on; switch open, lamp off. The LED in Fig. 13–2b is the same, except that you need a current-limiting resistor in series with the diode.

With a relay, Fig. 13–2c, the power supply voltage must be compatible with the relay, and a diode should be added across the coil. The diode protects the output transistor by suppressing inductively induced voltage spikes when the relay is turned off. Diode polarity is important, and in this case the banded end of the diode should be toward the positive terminal of the power supply.

Figure 13–2d needs a bit more explanation. With the output transistor turned off (switch open), you have +5 Vdc into the TTL IC input, and with the transistor on (switch closed), the collector is grounded and you get 0 V into the TTL input. The 2.2-k Ω resistor pulls the TTL input up to +5 Vdc when the transistor is in the off (open circuit) state. That's why this is often called a pull-up resistor.

The collector voltage levels are not exactly 0 V and 5 V. On the high end, collector open, there is a small current draw from the TTL input, which causes a small voltage drop through the 2.2-k Ω resistor. The low end is not zero because of diode losses in the transistor junctions. The TTL gate input swing will actually be from a low of about 0.25 V to a high of about 4.75 V, but that's okay because TTL circuits are designed to analyze any input less than 0.8 V as low, and above 2.4 V as high.

The connections in Fig. 13–2e are identical to Fig. 13–2d, except that CMOS (complementary metal oxide semiconductor) circuits typically use a power supply in the range of 9 to 12 Vdc.

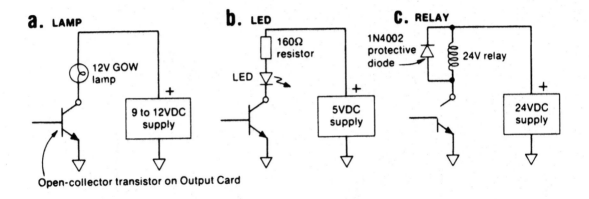

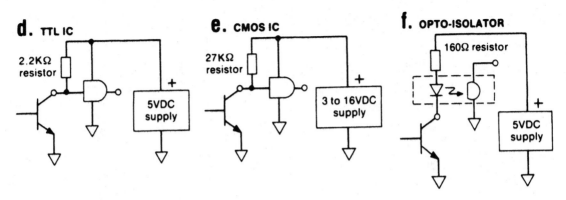

13-2 Connections for UCIS output.

If you have excessively noisy electrical conditions and/or some special output that must be electrically isolated from the UCIS, you can use an optoisolator as in Fig. 13–2*f*. This is an LED and a photo diode in a single package, and light is the only coupling between the two devices. When the output transistor is on it turns on the LED, and that in turn puts the photo diode into its low-resistance (on) state. When the transistor is off, the LED is off too, and the photo diode approaches an open circuit (off) state.

In all cases except TTL or CMOS ICs, supply voltages and loads can be varied as necessary, such as a 24-Vdc lamp with a 24-Vdc supply. The only limitations are that the load mustn't draw more than 0.3 A and the supply must be less than 40 Vdc, the rating of the output transistor. By substituting transistors as covered in Chap. 8, these limits can be increased to 1 A and 60 Vdc.

Supply voltage can be different for different output lines, but it's best to stay with +5 Vdc for as many applications as possible. That way it's easiest to combine loads on a single computer output line. Figure 13–3 shows one output line controlling four LEDs, a relay, a lamp, and two TTL logic gates.

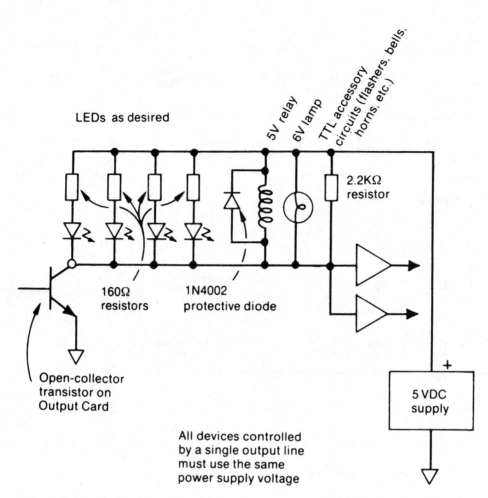

LEDs as desired

5V relay

6V lamp

TTL accessory circuits (flashers, bells, horns, etc.)

2.2KΩ resistor

160Ω resistors

1N4002 protective diode

Open-collector transistor on Output Card

5 VDC supply

All devices controlled by a single output line must use the same power supply voltage

13-3 Multiple loads on a single UCIS output line.

For cases where you have a definite requirement to drive different voltage loads from the same UCIS output line, the circuit shown in Fig. 13–4 could come in handy. When the output transistor is ON, its collector is effectively ground along with the input to the TTL circuits and 12 Vdc appears across the lamp and relay.

When the output transistor is off, its collector is pulled high, near 12 Vdc, thereby dropping the voltage across the lamp and relay to near zero while the 2.2-k Ω resistor and zener diode combination hold the input to the TTL circuits at 4.7 Vdc, which is near the zener voltage. For the circuit to work, the parallel combination of the 12-Vdc loads should be small compared to the 2.2-k Ω resistor. For best results from the 4.7-Vdc zener, be sure to use the higher-wattage 1N5230B rather than a 1N4732A.

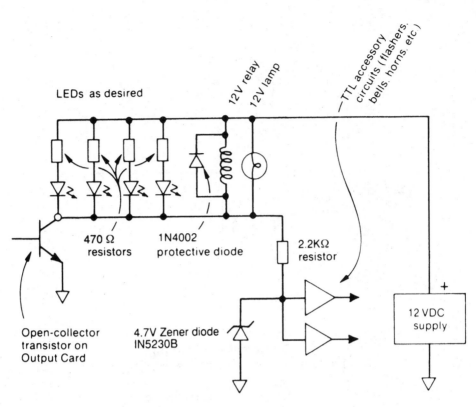

LEDs as desired

12V relay
12V lamp

TTL accessory
circuits (flashers,
bells, horns, etc.)

470 Ω
resistors

1N4002
protective diode

2.2KΩ
resistor

12 VDC
supply

Open-collector
transistor on
Output Card

4.7V Zener diode
IN5230B

13-4 Different voltage loads on the same line.

Driving Heavy Loads

For loads exceeding the output transistor ratings, 40 Vdc and 0.3 A for the 2N3904 and 60 Vdc and 1 A for the 2N6039, you need a booster circuit between the output card and the load. Figure 13–5 shows the one I use to drive surplus rotary relays powered by a 32-Vdc, 35-A supply. Cards for this circuit are available via JLC Enterprises on a special-order basis.

These relays are real current hoggers, and have a holding coil and a throwing coil. A built-in SPST contact cuts out the throwing coil once it throws, and the relay is then held in the energized position by the holding coil. When power goes off, a return spring restores the relay to its unenergized position. The throw current is around 4A, far exceeding the capacity of the output card's transistor.

The circuit in Fig. 13–5 handles a steady 10 A with surges to 15 A and supply voltages up to 60 V, all from your computer! Q2 is controlled by PNP transistor Q1. When the input terminal is brought low by the output card transistor turning on, Q1 turns on, and it turns on the Darlington power transistor Q2. With Q2 on, the output switches from open-circuit to ground, energizing the relay.

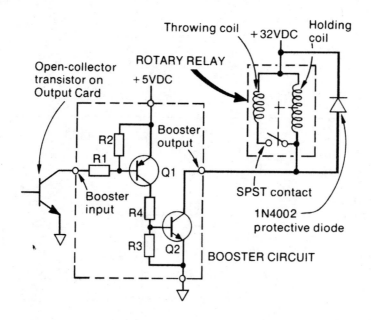

Qnty.	Description
R1	470Ω resistor [yellow-violet-brown]
R2,R3	1000Ω resistor[brown-black-red]
R4	47Ω, 1/2W resistor [yellow-violet-black]
Q1	2N2907 PNP small signal transistor (Mouser 592-PN2907)
Q2	2N6387 NPN power Darlington transistor (Mouser 511-2N6387)

Author's recommendations for suppliers given in parentheses above, with part numbers where applicable. Resistors are ¼W, 5 percent except as specified, and color codes are given in brackets. Equivalent parts may be substituted.

13-5 Booster circuit for heavy output loads.

Any time you drive an inductive load—one with a coil such as a relay or solenoid—add a diode across the coil with its banded end toward the positive power supply terminal.

Basic Input Connections

Now we'll look at connections to the input cards, so that your UCIS can read the status of important parameters from your external hardware. When the input card's 24 input lines are open-circuited, each is held high, at +5 Vdc or logic 1, by the card's built-in 2.2-kΩ pull-up resistor. All the external hardware needs to do to signal the computer of a change in input is to pull the connection, or line, to ground. Grounding an input line changes its state from logic 1 to logic 0.

Figure 13–6 shows UCIS input-line connections for various devices. The circuit in Fig. 13–6a lets the computer read the position of an SPST switch. If the switch is open the computer reads +5 Vdc, or logic 1, through the pull-up resistor. Closing the switch forces the input line to ground with the 5 Vdc dropped across the resistor, and the computer reads 0 V or logic 0. You can't get much simpler than that; with the switch in one position the computer reads logic 1, and in the other logic 0.

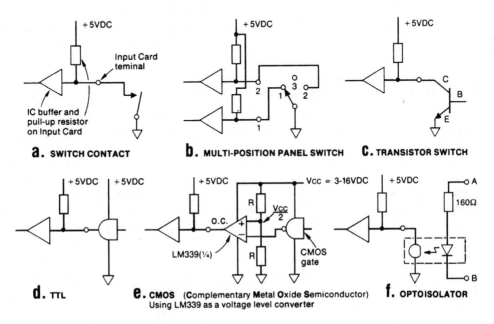

13-6 Connections for UCIS input.

To read a three-position switch, use Fig. 13–6b. If line 1 is logic 0 the switch is in position 1, if line 2 is logic 0 the switch is in position 2, and if neither line is logic 0 the switch is in position 3.

Input lines can also be driven by an open-collector transistor switch as in Fig. 13–6c. If the transistor is not conducting, the collector (C) is in effect an open circuit and the computer reads logic 1. When the transistor is on, the collector is tied to ground and the computer reads logic 0.

Figure 13–6d shows the direct hookup of a standard TTL circuit. The input card already has a built-in pull-up resistor on each line, so either regular TTL chips or those with open-collector outputs can be coupled directly to the input card.

Figure 13–6e shows one way to drive UCIS input lines with logic voltage levels other than +5 Vdc and 0. For example, a CMOS gate can operate with anywhere from +3 to +16 Vdc, with +10 V being a frequently used value. To make the conversion, this circuit uses part of an LM339 quad voltage comparator, a very handy IC to have in your repertoire.

Each of the four comparators in the LM339 has two inputs and an open-collector output. If the voltage applied at the positive input is greater than that at the negative input, the output is an open circuit. If the voltage at the positive input is less than at the negative, the output is pulled to ground.

The two resistors (R) are of equal value (10-kΩ values work well) and hold the positive input voltage constant at the midpoint of the logic switching levels out of the CMOS gate. As an example using a Vcc = 10 Vdc supply, the positive input would be held at +5 Vdc, which then is the switching level for the voltage applied to the negative input terminal.

Figure 13–6*f* shows how to connect an optoisolator. Current through inputs A and B lights the LED, which is seen by the photo diode, changing the signal to the input card from logic 1 to logic 0.

Simplified Wiring

The UCIS can perform monitoring and control functions with many fewer wires than an equivalent noncomputerized system, and its modular, plug-in circuit cards help keep your wiring neat and well-organized.

Yes, the UCIS really can simplify your wiring. Many UCIS I/O lines typically carry low currents, such as LED outputs and panel switch inputs, and for low-cost applications surplus multi-conductor telephone cable is a good solution. That helps keep wiring neat, and the prebundled, color-coded cables are easy to use. Talk to people renovating an old office or factory building, and you can often have what you need for the asking. Don't use telephone cable for driving heavy current loads; it's too small for that.

The smart way is to run no more wire than you need for a given job. With UCIS there are many things you can do to reduce interface wiring. For very large systems, the most effective is to use the distributed serial form of UCIS, with multiple USICs each driving separate IOMBs distributed throughout your external hardware; but making use of simple I/O decoding and encoding circuits can also make significant wiring reductions for either the serial or parallel form of UCIS. Let's look at decoding circuits first.

Decoding Output

Assume you have a conveyor stacking system with 16 levels, and you want your computer to light an LED on your panel to show which level is currently being accessed. Using the methods we've explained, you'd run 16 wires and tie up 16 lines of one output card; however, you can do the job with only four wires, as shown in Fig. 13–7.

Because you need only four lines to define up to 16 binary codes, you can use a 74154 IC (a 1-of-16 data distributor) to get a 1-low-out-of-16 output whose address is selected by the binary code on the four input address lines. Send out a decimal 4 (binary 0100) on the input lines, and the LED for level 4 will light. Send out a 9 (binary 1001) and level 9's LED would light.

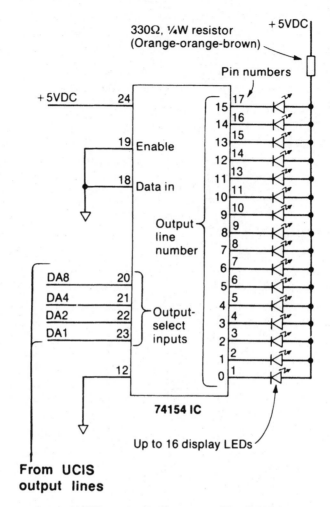

Four UCIS output lines and a 74154 IC drive up to 16 display LEDs

13-7 Decoding outputs.

The same circuit will work, of course, for fewer levels, and if necessary you could cascade 74154s to control up to 32 devices with five wires and up to 64 devices with only six. For 10 or fewer levels you can substitute a 7442 IC for the 74154.

Rather than have a separate LED for each level, you could use a single seven-segment numeric display as in Fig. 13–8. It uses a 7447 IC, a BCD-to-seven-segment decoder/driver, to decode the four input lines and light LED segments to display the number sent on the four input lines.

The same decoding-at-destination approach can control other devices, too. You can use the 7442 or the 74154, for example, to provide control logic for automatic rotary-table indexing alignment with one of up to 10 or 16 positions, respectively.

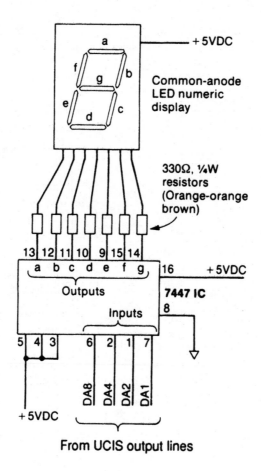

Common-anode
LED numeric
display

+5VDC

330Ω, ¼W
resistors
(Orange-orange
brown)

Four UCIS output lines and a 7447 IC
drive a seven-segment decimal display

13-8 Decoding for numeric display.

Other general or specialized decoder circuits can handle other cases, but you should evaluate each one on its own merit. Parallel wiring with multiconductor cable is often easier to connect and debug, it requires no special circuitry outside of your UCIS, and it can often be the more cost-effective solution.

For most of my applications I use some direct wiring, one line/one device, and some decoding at destinations. A balanced approach with some direct control and some decoding generally gives the optimum utilization of your UCIS.

Encoding Input

Encoding data at the source can also make significant wiring reductions. Say you have eight pushbuttons you want the computer to read. Rather than running eight wires and tying up eight input card lines, you can use just three wires as shown in Fig. 13–9.

The 74148 IC, an eight-bit priority encoder, puts out a three-line binary equivalent of the most significant (largest) input. All inputs and outputs are active low, and two chips can be ganged together to provide 16-line-to-4-line encoding, as also shown in Fig. 13–9. You could sense up to 16 positions of a rotary switch with only four input lines to your UCIS. If you have 9 or 10 lines to encode, you could use one 74147 IC, a 10-bit-to-4-line encoder, in place of the two 74148s.

Rather than build your own encoding circuits to connect to panel switches, you could use digital-type switches that have built-in encoding. For example, a 16-position hexadecimal rotary switch with positions 0–9 and A–F has only four outputs to carry the 16 binary codes for its 16 positions.

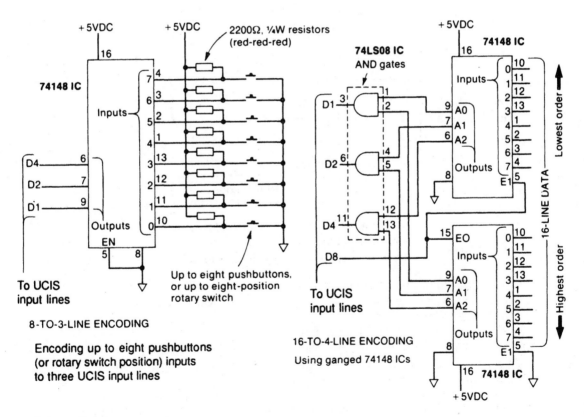

13-9 Encoding input.

Such a binary-coded hexadecimal or decimal switch can be connected directly to your UCIS input card as shown in Fig. 13–10, providing up to 16 or 10 inputs with only four wires. At the software end you simply read the decimal number of the switch position. You can't get things easier than that!

Where you already have a conventional rotary switch, consider using the encoding circuits shown in Fig. 13–11. Two lines can cover three positions (0, 1, and 2) as

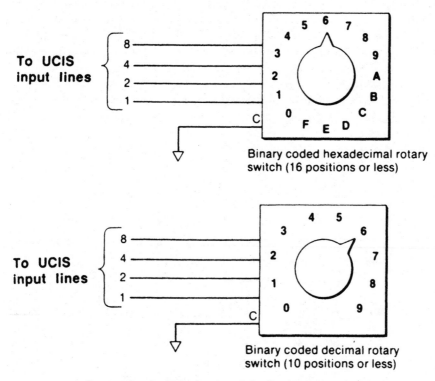

Binary coded hexadecimal rotary switch (16 positions or less)

Binary coded decimal rotary switch (10 positions or less)

Connecting hexadecimal and decimal rotary switches

13-10 Binary coded switch input.

shown in Fig. 13–11a. On the input card each input line goes through an inverter, so the software sees logic 1 when the input is grounded and logic 0 when it's open. That way the binary codes of the two lines correspond exactly to the switch position numbers: 00 = position 0 (or off), 01 = position 1, 10 = position 2, and 11 = position 3. Adding two diodes, two lines can encode a 4-position switch (positions 0 through 3) as shown in Fig. 13–11b. In switch position 3, the two diodes pull down both lines A and B to achieve the code 11, the binary equivalent for 3.

For as many as eight positions (0 through 7), three input lines are enough. Figure 13–11c shows a circuit using diodes. An alternative approach is to use a three-pole switch wired as shown in Fig. 13–11d.

You can also employ a few output lines to make significant reductions in the number of input lines. The output lines are used as address lines to select at the

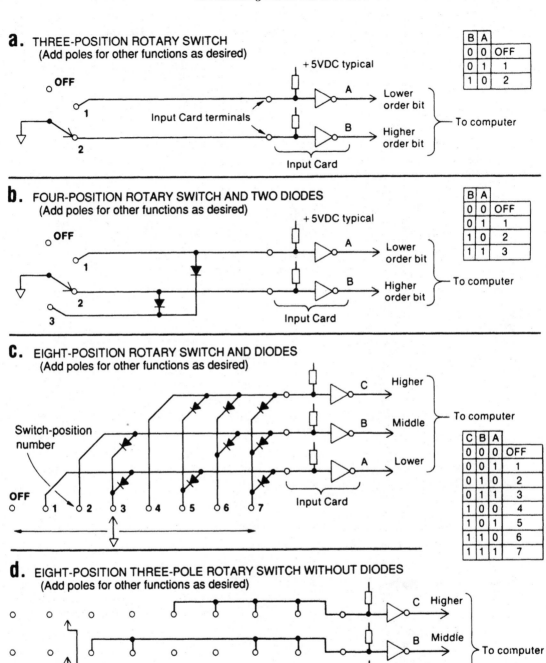

a. THREE-POSITION ROTARY SWITCH
(Add poles for other functions as desired)

B	A	
0	0	OFF
0	1	1
1	0	2

b. FOUR-POSITION ROTARY SWITCH AND TWO DIODES
(Add poles for other functions as desired)

B	A	
0	0	OFF
0	1	1
1	0	2
1	1	3

c. EIGHT-POSITION ROTARY SWITCH AND DIODES
(Add poles for other functions as desired)

C	B	A	
0	0	0	OFF
0	0	1	1
0	1	0	2
0	1	1	3
1	0	0	4
1	0	1	5
1	1	0	6
1	1	1	7

d. EIGHT-POSITION THREE-POLE ROTARY SWITCH WITHOUT DIODES
(Add poles for other functions as desired)

13-11 Encoding conventional rotary switch input.

source what particular input is to be read. The TTL chips that perform this selection for us are appropriately called data-selectors/multiplexers.

These ICs can be thought of as one-way logic versions of a toggle or rotary switch. The 74150 single 16-position selector works like an automatic 16-position rotary switch. The software sends a binary code on four address lines representing a decimal number from 0 through 15 to set the switch to a position for reading the desired 74150 input line. In addition to the 74150, you can get four two-position selectors per package (74157), two four-position selectors (74153), or one eight-position selector (74152).

As an example of how the data selector/multiplexers can cut down the number of I/O lines to your UCIS, suppose you have up to 16 remote points on a security system you are building and you want your computer to sample the points periodically to determine if an intruder has closed a switch. Instead of running all 16 switches directly to 16 inputs on an input card, you could use a 74150 IC to multiplex the 16 detector output lines onto a single UCIS input line (Fig. 13–12).

In operation, the software would send binary code N on the four address lines to read the state of switch N from the single input line. For example, to sample switch 5 it would send 0101 (binary for the decimal number 5) on the four address lines and read the input line.

Because the UCIS always reads one port at a time and a port is eight bits, you could group eight 74150s together, driving them all from the same four address lines, but connecting each of the selector outputs to a different line on the same UCIS input port. This way your software could increment through sending 16 different addresses (0–15) and sequentially read back the equivalent of 16 different input ports, but all on a single port! That's the same as 128 (16 × 8) discrete lines. You can use the other data-selector/multiplexer ICs I listed above to set up multiplexing schemes of your own.

There are similar one-way switches you could use to distribute UCIS output. Called data-distributors, these ICs include such chips as the 74154, which takes one input line and routes it to one of 16 output lines. These output lines are not latched, so data isn't retained when you switch addresses. Without adding more circuitry, these devices can't help much to expand the output capability of the UCIS.

Sample Decoder Card

For some applications it can be very cost-effective to design special decoder circuits to handle repetitive applications. As an example, consider a situation on my model railroad where I had a requirement to drive sets of up to 20 separate LEDs at 14 different remote locations.

Figure 13–13 shows the required truth table logic that must be satisfied to control the 20 LEDs properly. The resulting signal decoder card circuit schematic is shown in Fig. 13–14. The circuit drives track-side and panel signals, either two- or three-color, for a track section around a powered turnout (called an OS section) or any pair of adjacent block signals. Figure 13–15 shows the track and control panel

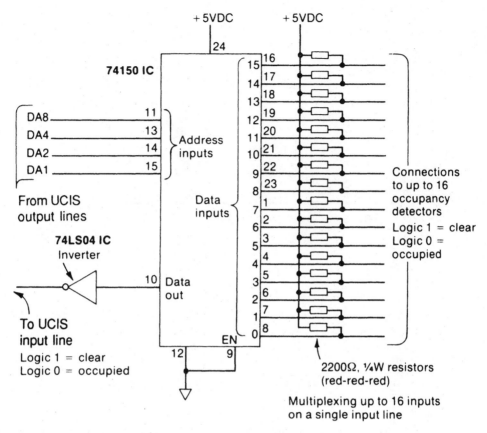

13-12 Multiplexing input.

schematics. The specific application is not the important point; however, this example that shows what can be accomplished with simple decoding is important.

The 20 LEDs are controlled using only three lines from the UCIS (SF*, ST*, and S2*) and one line from the local turnout switch motor (SR*) as shown in Fig. 13–14 and shown as inputs for the required truth table.

The benefits of remote decoding using the SDC (or a similar circuit set up to meet your specific application requirements) are easy to calculate. For example, I have a need for 14 of these circuits, and at 20 LEDs per circuit that's 280 LEDs. With SDC cards I need only 42 UCIS output lines to control these LEDs, versus 280 lines with direct wiring. Even with the cost of the SDCs, that still gives almost a 3:1 cost reduction and nearly a 4:1 reduction in the amount of wiring.

Figure 13–16 shows the parts layout and Fig. 13–17 is a photo of the completed SDC. Ready-to-assemble SDC cards are available from JLC Enterprises and fully assembled and tested cards from ECO Works, EASEE Interfaces, and AIA. See Appendix A for details. Table 13–1 gives the SDC parts list. Resistors for R1–R20 can be deleted for any LED outputs not required. Here's how to assemble a complete SDC:

| LED LOGIC TABLE | | | | | | | | | | | | | | | | | | | | | | | | |
| INPUTS | | | | LED OUTPUTS | | | | | | | | | | | | | | | | | | | | |
SF*	ST*	SR*	S2*	L1	L2	L3	L4	L5	L6	L7	L8	L9	L10	L11	L12	L13	L14	L15	L16	L17	L18	L19	L20
I	I	I	I			•	•									•	•	•		•			
I	I	I	O			•	•									•	•	•		•			
I	I	O	I			•		•								•	•	•		•			
I	I	O	O			•		•								•	•	•		•			
O	I	I	I	•			•				•				•	•	•	•		•			
O	I	I	O	•			•				•					•	•	•		•			
O	I	O	I	•				•					•	•		•	•			•			
O	I	O	O	•				•					•	•		•	•						
I	O	I	I		•			•								•	•	•			•		
I	O	I	O		•			•	•							•	•	•	•				
I	O	O	I		•			•	•							•	•			•		•	
I	O	O	O		•			•								•	•			•			•

13-13 Sample decoder truth table.

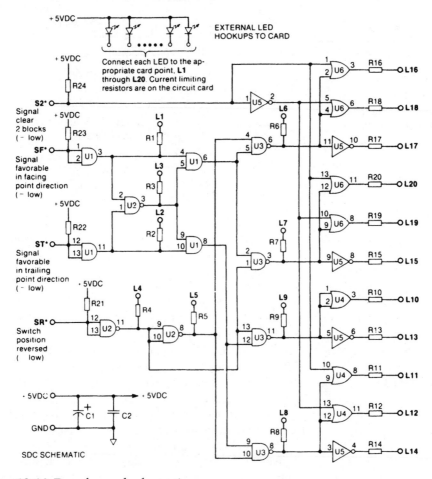

13-14 Decoder card schematic.

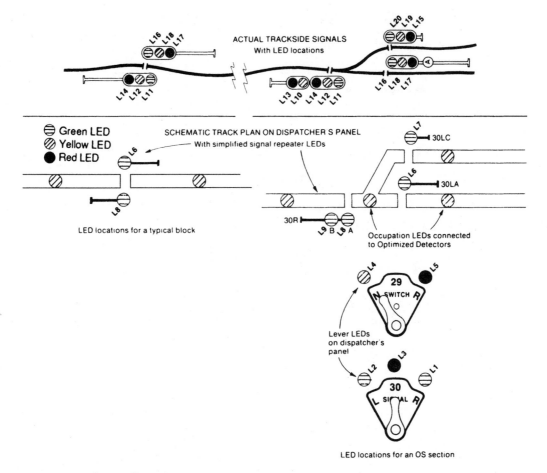

ACTUAL TRACKSIDE SIGNALS
With LED locations

⊖ Green LED
⊘ Yellow LED
● Red LED

SCHEMATIC TRACK PLAN ON DISPATCHER S PANEL
— With simplified signal repeater LEDs

30LC

30LA

30R

LED locations for a typical block

Occupation LEDs connected
to Optimized Detectors

29 N SWITCH R

Lever LEDs
on dispatcher's
panel

30 L SIGNAL R

LED locations for an OS section

13-15 Decoder application.

☐ *Card inspection.*
☐ *Terminal screws.* Insert four 4-40 screws from the top side, tighten four
nuts on the bottom, and solder the nuts to the circuit pads.
☐ *J1–J17.*
☐ *R1–R24.*
☐ *S1–S6[+].*
☐ *C1[+].*
☐ *C2.*
☐ *U1–U6[+].*
☐ *Cleanup and inspection.*
☐ *Card test.* Temporarily solder a test lead to the GND hole and another to
the +5-Vdc hole. Connect these to the +5-V supply and turn on the power.
Make an LED test probe as shown in Fig. 13–18, attach the clip to the +5-
Vdc output from your supply, and sequentially touch the LED end to each of

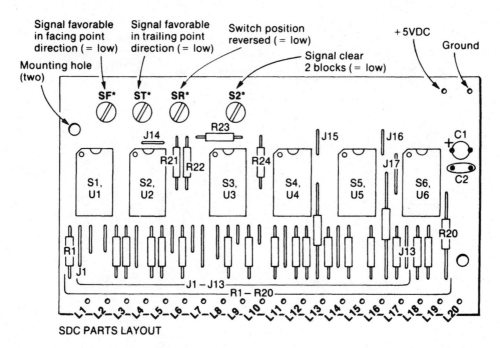

13-16 Decoder card parts layout.

the 20 LED output pads. The test LED should light only for positions L3, L4, L13, L14, L15, and L17 as indicated by the top row in the logic table. Arrange clip leads to ground the input screws according to each of the remaining rows in the logic table, attaching a lead where the table has a 0 (logic 0 = ground), and removing any lead where the table has a 1. For each case check all 20 pads, or all with parts installed, and make sure only those with a dot cause the test LED to light. Check the truth table (Fig. 13–13) as you pass each test. If you have problems, look for a faulty solder joint, a part left out, or a solder bridge. Correct and continue until you pass all tests.

Any serial/multiplexing scheme reduces the number of individual wires compared with an all-parallel approach, but you add complexity and in some cases lose speed both in the hardware and software. These are trade-offs you have to make on your own, but when in doubt, I suggest that you lean toward using one wire per device. It's simpler to use and easier to understand.

Stepper Motor Control

Figure 13–19 shows a circuit schematic for connecting a stepper motor to a UCIS standard output card, either DOUT or COUT24. Two motors can be connected to each output port for a total of six motors controlled from each card.

The Sprague UDN 5703A, a quad two-input NAND driver, provides the drive capability for most laboratory-level and light-industrial 12-Vdc stepper motors. With

13-17 Decoder card complete.

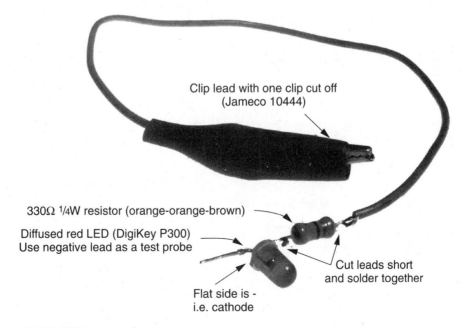

Clip lead with one clip cut off
(Jameco 10444)

330Ω ¼W resistor (orange-orange-brown)

Diffused red LED (DigiKey P300)
Use negative lead as a test probe

Cut leads short
and solder together

Flat side is -
i.e. cathode

13-18 LED test probe.

Table 13-1. Decoder Card Parts List

Qnty.	Symbol	Description
4	—	4-40 x ¼" Pan-head machine screws (Digi-Key H142)
4	—	Brass nuts
17	J1-J17	Jumpers, make from no. 24 uninsulated bus wire (Belden no. 8022)
20	R1-R20	220Ω resistors [red-red-brown]
4	R21-R24	2.2KΩ resistors [red-red-red]
6	S1-S6	14-pin DIP sockets (Digi-Key A9314)
1	C1	2.2μF, 35V tantalum capacitor (Jameco 33734)
1	C2	100000pF, 50v ceramic disk capacitor (Digi-Key P4164)
1	U1	7408 quad 2-input AND gate (Jameco 49146)
1	U2,U3	7400 quad 2-input NAND gates (Jameco 48979)
2	U4,U6	7432 quad 2-input OR gates (Jameco 50235)
1	U5	7404 hex inverter (Jameco 49040)

Author's recommendations for suppliers given in parentheses above with part numbers where applicable. Equivalent parts may be substituted. Resistors are ¼W, 5 percent and color codes are given in brackets.

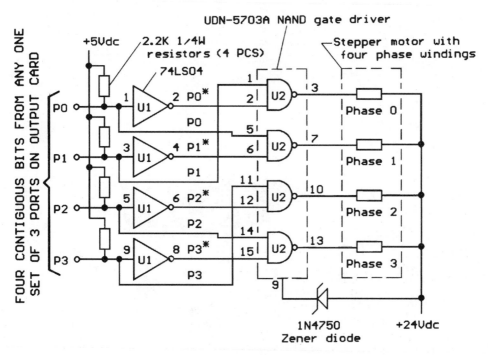

13-19 Stepper motor control schematic.

the inputs connected as shown with the 74LS04 inverters, the circuit prevents more than two of the motor windings from being on simultaneously, regardless of the output value from the computer or the state of the 8255 outputs during the power-up sequence. This prevents harm to the motors and to the 12-Vdc power supply.

To step the motor, a sequence of binary phase patterns is output to the desired motor, one pattern per step as defined by Table 13–2. For clockwise rotation, the patterns are output sequentially from top to bottom. When the end of the table is reached, the software simply wraps around to the other end of the table and continues sequentially. To change motor direction, simply reverse the direction for sending out the entries.

Table 13-2. Stepper Motor Rotation Phase Patterns

	Phase Winding Values				Decimal Equivalent	
	P3	P2	P1	P0		
Clockwise (↓)	0	1	0	1	5	Counter Clockwise (↑)
	0	1	0	0	4	
	0	1	1	0	6	
	0	0	1	0	2	
	1	0	1	0	10	
	1	0	0	0	8	
	1	0	0	1	9	
	0	0	0	1	1	

The particular entries shown generate a sequence known as *half stepping,* where the steps are half the size specified by the motor manufacturer. Compared to full stepping, half stepping produces smoother slow-speed motions, doubles the resolution, and reduces the power required. For example, using a motor where the manufacturer specifies 48 steps per revolution, half stepping would provide 96 steps per revolution.

A simple QuickBASIC program to accomplish the movement of any one of six stepper motors connected to a single DOUT or COUT24 card is presented in Fig. 13–20 and included on the enclosed disk.

Preview of Additional Circuits under Development

Several new UCIS cards are under development for release by JLC Enterprises. These include a control card to drive two-lead and three-lead tricolor LEDs. The card is set up to drive multiple LEDs via special circuit layout enabling cutting the card to different lengths for distributing control of a varied number of LEDs at each location. Two control line inputs are used per LED, with binary 00 = dark, 01 = red,

```
REM**PROGRAM TO CONTROL STEPPER MOTORS ON RTI ROBOT
REM**Programmed by Bruce Chubb on Jan 16, 1990
REM**DEFINE GENERAL VARIABLES AND CONSTANTS**
    DEFINT A-Z
    DIM DD$(2), SM(8)
    'MR = ROBOT MOTOR NUMBER TO BE CONTROLLED (1 TO 6)
    'DR = DIRECTION OF DESIRED MOVE
    'SP = NUMBER OF STEPS TO BE MOVED
    DD$(0) = "CLOCKWISE": DD$(1) = "COUNTER CLOCKWISE"
    CW = 0  'CLOCKWISE
    CC = 1  'COUNTER CLOCKWISE
REM**DEFINE MOTOR ROTATION PHASE SEQUENCE
    SM(1) = 5:  SM(2) = 4: SM(3) = 6: SM(4) = 2
    SM(5) = 10: SM(6) = 8: SM(7) = 9: SM(8) = 1
REM**DEFINE I/O FOR INTERFACE ASSUMING IBM PC AND
    '1ST CARD IS OUTPUT CARD ATTACHED TO THE 6
    'STEPPER MOTORS TO BE CONTROLLED.
    DEF SEG = 40960 'UNUSED MEMORY SEGMENT AT 640KB
    POKE 3, 128    'INITIALIZE 8255 ON CARD TO OUTPUT
REM**READ KEYBOARD INPUT AND TRANSMIT ROBOT COMMAND...
BEGIN: CLS        '...SET UP IN A CONTINUOUS LOOP
    INPUT MR, SP, DR
    PRINT "COMMANDING MOTOR "; MR; "MOVE "; SP; "STEPS "; DD$(DR)
    GOSUB MOVE
    GOTO BEGIN


REM**SUBROUTINE TO MOVE STEPPER MOTOR
REM**SET UP DIRECTION FOR SEQUENCE OF PHASES
MOVE: IF DR = CW THEN
    N1 = 1: N2 = 1: NS = -1
    ELSE N1 = 8: N2 = 1: NS = -1
    END IF
REM**DEFINE PORT AND PACKING OFFSETS CORRESPONDING...
    IF MR = 1 THEN PO = 0: BO = 1 '..TO MOTOR TO BE MOVED
    IF MR = 2 THEN PO = 0: BO = 8
    IF MR = 3 THEN PO = 1: BO = 1
    IF MR = 4 THEN PO = 1: BO = 8
    IF MR = 5 THEN PO = 2: BO = 1
    IF MR = 6 THEN PO = 2: BO = 8
REM**TRANSMIT SEQUENCE OF 8 PHASES TO MOVE COMMANDED MOTOR...
    FOR JJ = 1 TO SP        '...SP NUMBER OF STEPS
    FOR II = N1 TO N2 STEP NS
      OB = SM(II) * BO
      POKE PO, OB
    NEXT II
    NEXT JJ
    RETURN
```

13-20 Stepper motor control program.

10 = green, and 11 = yellow. Each LED has its own trim potentiometer for optimizing the yellow indication.

Also under development is a UCIS-controlled 110-Vac, 15–20-A dimmer circuit designed mainly for incandescent lighting brightness control. Optical isolation of the UCIS low-level dc control circuitry from the high-voltage ac is one of the circuit's main attributes.

If your UCIS application might have a need for either one of these new circuits, contract JLC Enterprises for the latest technical and availability details.

Using Current-Sourcing Output Cards

Up to this point, the output control circuits presented in this chapter have connected to the standard current-sinking version of the output card. This is the preferred method of driving external hardware from digital circuits. Hopefully it should handle most all your interfacing needs. Sometimes, though, you need to connect UCIS to external hardware that is all pre-wired with a common ground. What's needed then, short of rewiring the external hardware, is to be able to use UCIS to switch the hot side of the line.

An obvious solution is to insert a relay contact in the line's hot side and drive the relay coil with an output line from a standard-configuration UCIS output card, one that is configured for current-sinking. This way the UCIS activates the relay by switching one side of the coil to ground, and that in turn activates the load. The relay itself can be either mechanical or solid-state. This is the best way for controlling high-voltage and/or high-current ac or dc loads such as large industrial motors, heating coils for mixing tanks, and so on.

Figure 13–21*a* shows an alternative solution for less demanding common-ground loads. The hot side of each load is connected directly to a DOUT card assembled in its current-sourcing configuration. The loads can be a set of lamps, relay coils, LEDs with appropriate current-limiting resistors, or any other devices connected via common ground. Additional output lines can drive additional loads, all connected with the same common ground. The more conventional hookup, using the DOUT card in its standard current-sinking configuration, is illustrated in Fig. 13–21*b*.

Figure 13–22 illustrates another common application for the modified DOUT card. Two output lines are used to drive a tricolor three-lead common-cathode LED. A typical example for hundreds of these hookups is driving searchlight-type trackside signals on a model railroad system. With two control lines per signal, you achieve four separate signal indications defined as binary 11 = yellow, 01 = red, 10 = green, and 00 = dark. Flashing indications of each color are also easily achievable via UCIS software. Adjusting one of the current-limiting resistors sets the quality of the yellow indication.

Driving Matrixed Outputs

Another neat interfacing approach is to combine the features of both output card configurations into a matrix hookup, as illustrated in Fig. 13–23. This example

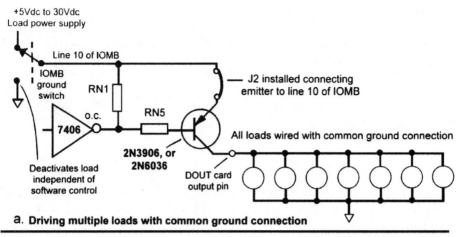

a. Driving multiple loads with common ground connection

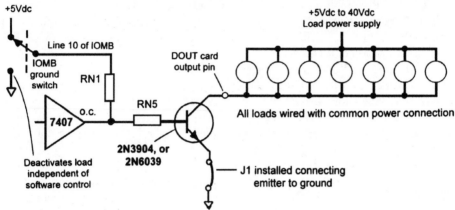

b. **Driving multiple loads with common power connection**

13-21 Driving common power and common ground loads.

uses an A-port from a DOUT card configured for current-sinking and a B-port from a second DOUT card configured for current-sourcing. Each of the 64 loads is connected with one lead connecting to an A-port line and the other to a B-port line. When all lines are in their normal open-circuit state, no loads are activated. Any one of the 64 different loads can be activated by software simply setting the corresponding B-port line high and the A-port line low. This enables 16 lines to separately control 64 devices.

Expanding to a full card of each type, yielding 48 lines total, would control 24×24 or 576 devices. A real-world example application of matrix outputs is the automatic controlling of fireworks displays. One system uses three cards of each type with each matrix crossing of lines tied to an igniter, called a fuse. This way UCIS software can control the music, the lighting, and also with 142 lines (72 of each type) separately control the ignition of 72×72 or 5184 fireworks elements. The result is a very impressive demonstration of UCIS capability!

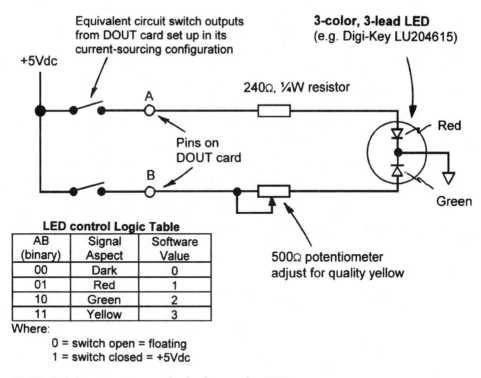

13-22 Driving common-cathode three-color LEDs.

Your application may not be as lively as a large fireworks display, but the use of matrixed outputs can benefit many applications. For example, each load in Fig. 13–23 could be one of many LEDs in a master control panel, for example a rail-road dispatcher's panel showing the location of trains, track switch and signal status, and so forth, or a control panel in a power plant or a manufacturing facility. By loop-ing though the display at a fast rate and each time through setting different combi-nations of lines off and on, it is possible to have any combination of lights appear on at any point in time.

The system works because the software display update rate is much faster than the eye can detect, so even when each LED is on for only a portion of the cycles, the viewer's perception is that they are on throughout all the cycles. The payoff is that if you're using two cards of each DOUT type, 96 lines total, you could control a panel with 48 × 48 or 2304 separate LEDs. Each LED can be independently controlled so it appears on or off.

Good Ground Wiring

Any electrical hardware system should have good ground wiring, but when you in-troduce electronics as part of the system, good ground wiring is particularly important. With TTL or CMOS circuits, a voltage shift of only a volt or two can change a logic state.

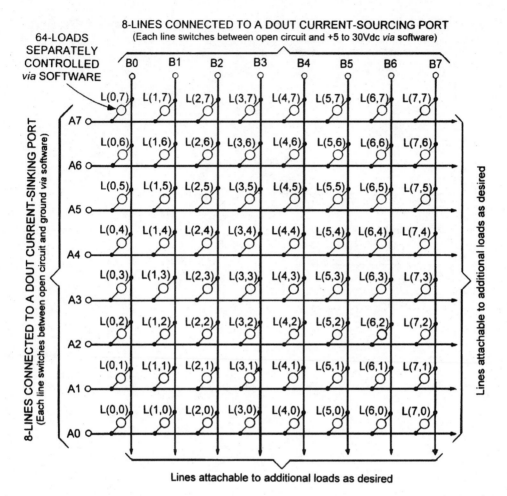

13-23 Driving matrixed outputs.

Typical IC response times are in the range of a few nanoseconds, while heavy electrical equipment systems are full of large current surges. For example, when a solenoid is energized in a piece of machinery, a heater coil is turned on, or a pump motor started, it's not unusual to get large-step current changes of 10–20 A or more.

A 10-A current in a 50-ft ground line of #18 copper wire (0.64 Ω per 100 ft) will cause an end-to-end voltage differential of 3.2 V, enough to give any TTL circuit fits. Throw in a few resistive solder joints and some dirty relay or switch contacts, and the loss can be big even for short lengths of wire. Smaller wire makes the problem even worse. You don't have to understand the theory of voltage loss, or worry about it either, as long as you follow three basic practices for good ground wiring.

☐ Use heavy wire for ground lines. You might get by with #18 wire on very small low-power interfaces, but #16 is better. For larger interface systems the investment in #14 or #12 would be worthwhile, with #10 or larger best for the main feeder grounds on very large interface systems. (Note that in American wire gauge or AWG sizes, lower numbers mean bigger wire.)

☐ Use separate ground wires for high-current circuits (such as relay control and power feeds) and for logic and computer circuits. In electronics, these two types of grounds are often called *power ground* and *signal ground.*

☐ Tie all of your various ground lines together at only one central point, as close as possible to your power supplies.

These practices should eliminate significant voltage shifts in signal grounds and reduce the electrical noise level at IC inputs.

Good ground wiring practices go hand-in-hand with the good signal line practices we've been following in assembling the UCIS, like always using either an open-collector transistor output (as in the output card and Optimized Detector, which I'll discuss later) or a special driver IC (like the 74LS244 in the IBEC, UBEC, and USIC) when driving signals long distances.

When your external hardware exhibits large current surges, large (for example, 48,000-μF) electrolytic capacitors liberally distributed throughout your external hardware wiring system will also enhance its immunity to electrical noise. While a large electrolytic capacitor is an excellent ripple filter for a power supply, it can't provide much high-frequency bypassing. In fact, at high frequencies an electrolytic capacitor usually acts more like a noise source than a noise filter.

The large electrolytic capacitors need 10-μF tantalum and 0.1-μF ceramic-disk capacitors parallel with them for good noise suppression. The big capacitors can be almost anywhere, but the small ones should be as close as possible to potential sources of high-frequency noise, such as the IC chips themselves and motor commutators.

TTL and CMOS circuits make a lot of electrical noise on their own. When their outputs change state at nanosecond speeds, ICs draw heavy current spikes from their supply lines. These must be kept from propagating through the supply system, where they could ring the lines with oscillating transients and interfere with other circuits.

Local despiking capacitors momentarily provide energy to stabilize supply lines during output state transitions. Use small monolithic or ceramic-disk capacitors in the 0.01–0.1-μF range, with leads cut as short as possible. They should be distributed around any TTL or CMOS ICs, with at least one for every four 14- to 16-pin DIP packages and a separate capacitor for each larger MSI (medium-scale integration) package such as the 40-pin 8255. It's also good practice to include a 1–10-μF tantalum capacitor on each circuit card where the +5-Vdc supply line enters the card.

These smaller capacitors are already included in the UCIS circuits. For example, the IBEC, UBEC, DIN, and DOUT cards all have a .1-μF despiking capacitor for each IC! For reliability, use these same solid design practices in any circuits you add to your UCIS.

This completes the general discussion of setting up to handle discrete I/O. In the next chapter I'll show you how to expand your UCIS capabilities to handle analog I/O.

14
CHAPTER

Adding Analog Interface Cards to a UCIS

Up to now we've been working with digital devices using on-off binary codes: either +5 or 0 Vdc, but nothing in between. In this chapter I'll explain how the UCIS can monitor and control analog devices, those with a continuous range of input or output voltages.

A dc permanent magnet motor is an example of an analog device; its speed range is continuous from stop to full speed. The motor speed is controlled with an analog signal, a continuously variable voltage. With the digital-to-analog converter output card, or DAC card, UCIS software can generate analog voltages for such features as motor speed control and light dimming. At the end of this chapter I'll cover a special DAC-type circuit useful for controlling dc motors.

I'll also cover analog-to-digital converter input cards (ADC cards) to let the UCIS receive analog voltage inputs. With this device your computer can read such inputs as variable voltages and potentiometer settings.

Digital-to-Analog Converter (DAC) Chips

A DAC converts digital signals into analog voltage or current. The output of the DAC is a stairstep approximation of continuous voltage. An eight-bit DAC gives steps of $1/255$ of the full analog output. For example, for an eight-bit DAC set for a range of 0 to +12 Vdc, the smallest voltage step (the resolution) would be 12 $1/255$, or about 0.047 V.

A digital 0 command to the DAC would give an output of 0 V, a digital 1 would give +.047 Vdc, a 2 would give +.094 Vdc, and so on up to 255 for the full-scale +12 Vdc. This would give you smooth speed control of a 12-Vdc instrument motor; note, however, that a DAC's resolution is a function of the number of bits that carry the digital information. There are 10-, 12-, and 16-bit converters, but I'll stick to eight-bit (one byte) converters. Eight-bit converters have ample resolution for most UCIS applications.

DAC ICs are flexible devices that, with different external circuit components, can give a variety of outputs. To keep costs down I'll cover only two inexpensive chips: the DAC0807LCN, the lowest-priced DAC IC in the Jameco catalog; and the DAC0800LCN, a slightly higher-priced but more flexible chip. Both come in 16-pin DIPs, and I've designed the UCIS DAC output card to work with either chip.

Looking at the pinouts in Fig. 14–1, the eight-bit digital input to both DACs is via pins 5 through 12. The *MSB* (most significant bit) is B1, pin 5, and the *LSB* (least significant bit) is B8, pin 12. Both chips have multiplying (MDAC) capability, but for our use we'll keep the differential analog reference input on pins 14 and 15 constant, so the analog output is simply proportional to the digital input.

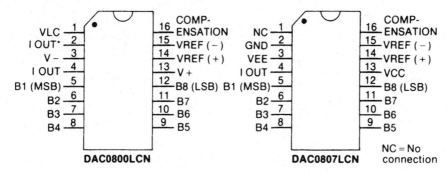

14-1 DAC pinouts.

The 0800's output is a differential current between pins 4 and 2, while the 0807's output current is single-ended on pin 4 with pin 2 as ground. Pin 1 of the 0800 sets the logic-switching levels of the digital input. For the UCIS, pin 1 is grounded for logic levels compatible with standard TTL logic (+5 Vdc and 0 V).

Figure 14–2 shows a typical application circuit for the more elegant of the two ICs. In each application circuit I've attached an off-loading *op amp* (operational amplifier) to the DAC outputs. The op amp buffers and amplifies the weak signal current coming from the DAC to help maintain conversion accuracy and to drive heavier loads than could the DAC alone. Varying the decimal value of the digital input from 0 to 255 (by either a POKE A, N or an OUT A, N instruction in the software, where A is the DAC's 8255 port address and N is the desired value from 0 to 255) will vary the analog output *VOUT* between –9.92 Vdc and +9.92 Vdc, as shown in Table 14–1. The full-scale value, here 9.92 V, can be made adjustable by inserting the optional trim potentiometer in the input reference circuit.

Figure 14–3 is a simplification of Fig. 14–2, for positive analog output only. Either chip can be used, but the 0807 is usually less expensive. The equation defining the output voltage as a function of how the input bits stack up in a binary powers-of-two fashion is defined as:

EQUATION FOR OUTPUT VOLTAGE – **VOUT**

$$\mathbf{VOUT} = \mathrm{VFS}\left[\frac{B1}{2} + \frac{B2}{4} + \frac{B3}{8} + \frac{B4}{16} + \frac{B5}{32} + \frac{B6}{64} + \frac{B7}{128} + \frac{B8}{256}\right]$$

$$\text{Full-scale output voltage} = \frac{255}{256} \times \frac{\mathrm{VREF}}{\mathrm{RREF}} \times \mathrm{RF} = .0019921\ \mathrm{RF}$$

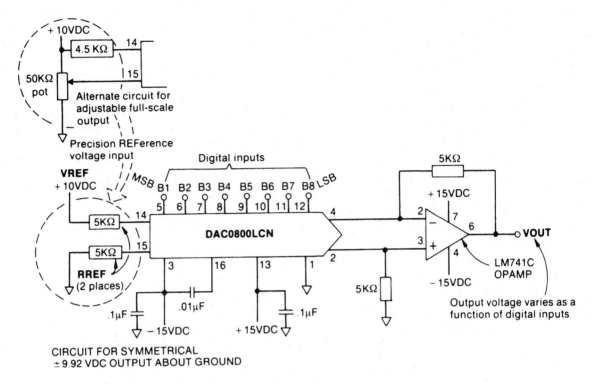

14-2 DAC application circuit.

Table 14-1. DAC Truth Table

	B1	B2	B3	B4	B5	B6	B7	B8	EO
Pos. Full-scale	1	1	1	1	1	1	1	1	+ 9.920
Pos. Full-scale-LSB	1	1	1	1	1	1	1	0	+ 9.840
(+) Zero scale	1	0	0	0	0	0	0	0	+ 0.040
(−) Zero scale	0	1	1	1	1	1	1	1	− 0.040
Neg. Full-scale + LSB	0	0	0	0	0	0	0	1	− 9.840
Neg. Full-scale	0	0	0	0	0	0	0	0	− 9.920

Each B value is a 1 if the software has that bit position on, and 0 if off.

The full-scale voltage *VFS* is a function of the *VREF* and *RREF* values and the op amp feedback resistor *RF*. By using standard values of *VREF* = +10 Vdc and *RREF* = 5 kΩ, then selecting *RF* from a range of 1–6 kΩ, you can achieve full-scale outputs between 2 and 12 V. The relationship of *VOUT*'s full-scale value to *RF* is shown by the graph in Fig. 14–4; it is linear up to the point where the op amp saturates as VFS approaches the +15-Vdc supply voltage.

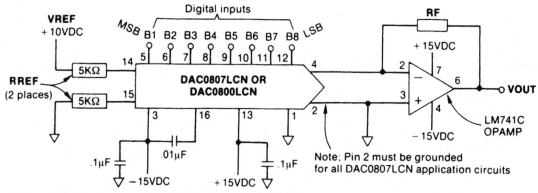

CIRCUIT FOR POSITIVE OUTPUT OPERATION

14-3 Alternate DAC application circuit.

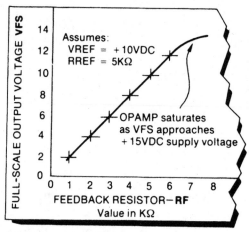

14-4 Op amp performance curve.

DAC Output Card

Figure 14–5 shows a completed DAC output card. It's designed to plug into the motherboard just like the general-purpose output card. The DAC card uses three DAC ICs for three independent analog outputs. Your external hardware connects to the 10-pin right angle header at the top; its lines include ±15-Vdc power for the DACs and their op amps, three analog outputs, and five ground connections. The DAC card uses the same DIP-switch circuit for address decoding as the output card, and the same 8255 chip for switching data bytes to the appropriate DAC IC.

The three independent DAC circuits let you install different parts and jumpers in each to yield symmetrical operation about ground with the 0800, or positive-only output voltages with either the 0800 or the 0807. Also, you can install fixed resistors for fixed full-scale settings or trim pots for adjustable full-scale values.

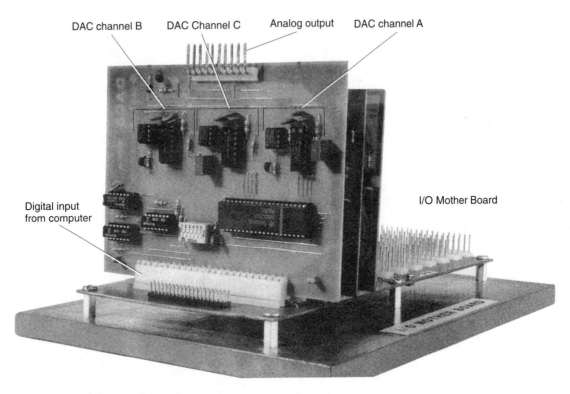

14-5 DAC output card.

The example illustrates the DAC card's flexibility. I've configured the analog channel A circuit, corresponding to Port A of the 8255, to use the 0800 IC for symmetrical output about ground and an adjustable full-scale value. The channel B circuit is set up to use the 0807 IC for positive-only output with a fixed full-scale value of +12 Vdc. Channel C is the same as B, except that the full-scale value is adjustable.

Figure 14–6 is the DAC card schematic. Its left side, including the 8255, is identical to that of the COUT24 output card presented in Chap. 16. The 8255's eight-bit ports A, B, and C connect directly to the digital inputs of the three DACs. One +10-Vdc zener diode, Z1, is the +VREF source for all three DACs.

The four resistors around each DAC/op amp combination vary in different configurations. For example, you'd use 5-kΩ resistors as R12–R14 for symmetrical output with the 0800, but replace them with jumpers (0 Ω) for positive-only output. Note that R12–R14 must always be jumpers when using the 0807. The op amp feedback resistors R15–R17 can be fixed resistors selected for one full-scale value, or they can be replaced by potentiometers for adjustable output. Likewise, reference input resistors R18–R20 can be fixed or adjustable resistors.

Feedback capacitors C14–C16 are in parallel with the op amp feedback resistors to reduce the amplifier's high frequency gain, which improves circuit stability and reduces output noise.

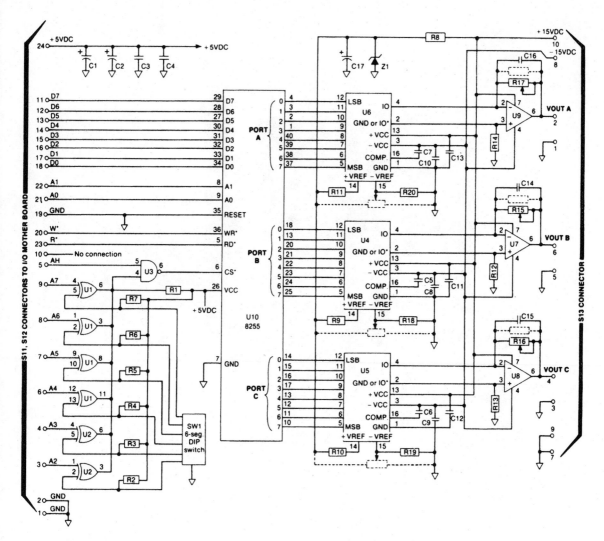

14-6 DAC card schematic.

Building the DAC Card

Figure 14–7 includes the parts layout and Table 14–2 the parts list for the DAC card. Ready-to-assemble circuit cards are available from JLC Enterprises, and fully assembled and tested cards are available from ECO Works, EASEE Interfaces, or AIA.

In the parts list, the parts that can vary depending on how you want to configure each DAC circuit are marked with asterisks; select those for your applications as described above. Potentiometers are more expensive than fixed resistors, but they provide easy adjustment.

For R9–R20 I've called out the nearest standard ¼-W, 5 percent resistor values, for example 5100 Ω in place of 5000 Ω for *RREF* as in Fig. 14–2. That's fine for most

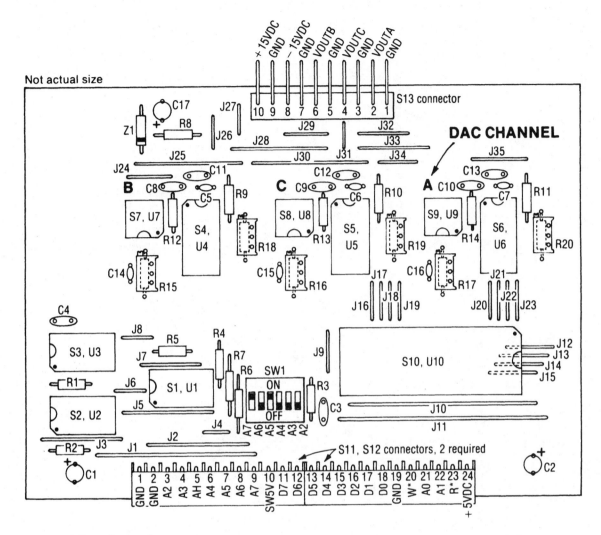

14-7 DAC card parts layout.

applications, but for greater accuracy and less shifting of output voltage with temperature variation (for example, during warmup), substitute metal film resistors with a 1 percent tolerance. These are listed in the Digi-Key catalog, and their price is about double that of 5 percent resistors. They come in finer increments, so you can get for example a 4.99-kΩ resistor for RREF. To assemble the DAC card, refer to Fig. 14–7 and Table 14–2 and follow these steps:

□ *J1–J35*. Jumpers J12–J15 will lie partially underneath S10.
□ *R1–R20*. Where parts are marked by asterisks, be sure to install the right component for the type of DAC output you want, whether they are fixed resistors, jumpers (R12–R14 positions), or potentiometers (R15–R20 positions). There are five holes at each set of R15–R20 locations; use them as shown in Fig. 14–8.

Table 14-2. DAC Card Parts List

Qnty.	Symbol	Description
35	J1-J35	Jumpers, make from no. 24 uninsulated bus wire (Belden no. 8022)
1	R1	1.0KΩ resistor [brown-black-red]
6	R2-R7	2.2KΩ resistors [red-red-red]
1	R8	150Ω ½W resistor [brown-black-brown]
3*	R9-R11	5.1KΩ resistors [green-brown-red] or substitute 4.7KΩ resistors [yellow-violet-red] if using 50KΩ trim pot for R18-R20
3*	R12-R14	Use jumpers (R=0Ω) of no. 24 uninsulated bus wire (Belden no. 8022) for use with DAC0807LCN and also with DAC0800LCN for single-ended output. Use 5.1KΩ resistors [green-brown-red] with DAC 0800LCN for +/- symmetrical output
3*	R15-R17	Use 5.1KΩ resistors [green-brown-red] with DAC0800LCN if +/- symmetrical output is desired, otherwise select fixed resistor values per graph or equation for desired full-scale output or install 10KΩ trim potentiometers (Digi-Key CT9X103) for adjustable full-scale output voltage.
3*	R18-R20	Use 5.1KΩ resistors [green-brown-red] or with DAC0800LCN install 50KΩ potentiometer (Digi-Key CT9X503) for adjustable full-scale output voltage.
1	Z1	10V, 1W zener diode (Digi-Key 1N4740A)
3	S1-S3	14-pin DIP sockets (Digi-Key A9314)
3	S4-S6	16-pin DIP sockets (Digi-Key A9316)
3	S7-S9	8-pin DIP sockets (Digi-Key A9308)
1	S10	40-pin DIP socket (Digi-Key A9340)
2	S11,S12	12-pin Waldom side entry connectors (Digi-Key WM3309)
1	S13	10-pin Waldom right angle header (cut from Digi-Key WM4510)
1	SW1	6-segment DIP switch (Digi-Key CT2066)
2	C1,C2	2.2µF, 35V tantalum capacitors (Jameco 33734)
2	C3,C4	100000pF, 50V ceramic disk capacitors (Digi-Key P4164)
3	C5-C7	10000pF, 50V ceramic disk capacitors (Digi-Key P4158)
6	C8-C13	100000pF, 50V ceramic disk capacitors (Digi-Key P4164)
3	C14-C16	1000pF, 50V ceramic disk capacitors (Digi-Key P4431)
1	C17	10µF, 25V tantalum capacitor (Jameco 94078)
2	U1,U2	74LS136 quad EXCLUSIVE-OR gates with open collector (Jameco 46586) or (JDR 74LS136)
1	U3	74LS00 quad 2-input NAND gate (Jameco 46252)
3	U4,U6	DAC0807LCN 8-bit D/A converters (Jameco 14939) or for special high-accuracy applications (see text) use DAC0800LCN (Jameco 14904)
3	U7-U9	LM741CN operational amplifiers (Jameco 24539)
1	U10	8255 or 8255A-5 programmable peripheral interface (Jameco 52742 or 52732) or (JDR 8255-5)

Author's recommendations for suppliers given in parentheses above with part numbers where applicable. Equivalent parts may be substituted. Resistors are ¼W, 5 percent and color codes are given in brackets.

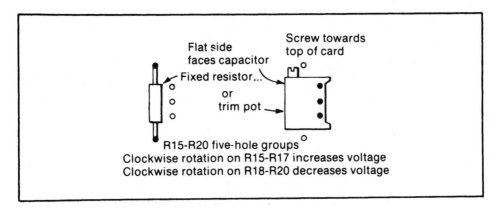

14-8 Mounting resistors on the DAC card.

☐ *Z1[+]*.
☐ *S1–S10[+]*. Be sure S10 is tight against the card, not held away by J12–J15.
☐ *S11–S13*.
☐ *SW1[+]*. Install this component so that the DIP switch contacts are closed when thrown toward S13.
☐ *C1, C2[+]*.
☐ *C3–C16*.
☐ *C17[+]*.
☐ *U1–U10[+]*.
☐ *Cleanup and inspection*.
☐ *IC power test*. Set SW1 to the desired card slot address. Plug the DAC card into the motherboard, and connect pins 7–10 of S13 to a ±15-Vdc supply. If you don't have such a supply handy, you can use Fig. 11–2 to build one, but change the T1 transformer to 25.2 VCT, 450 mA (Radio Shack 273-1281), and the regulators V1 to 7815T and V2 to 7915T (both available from Jameco). Turn on both the ±15-Vdc and the +5-Vdc motherboard supplies, then use your VOM and Table 14–3 to see that power and ground reach each IC. If not, look for a missing or shorted jumper, an IC reversed, or bad soldering. Keep checking until you pass all power tests.
☐ *Reference voltage test*. Touch your VOM's positive lead to the banded end of Z1 and the negative lead to U4 pin 1. You should read between +9.5 and +10.5 Vdc. If not, then Z1 is backward or defective, your soldering is faulty, or one or more of ICs U4–U6 is defective. Locate and correct the problem before proceeding.
☐ *Channel A output voltage test*. Attach the negative lead of your VOM to Channel A S13 pin 1 (ground), and the positive lead to pin 2 (VOUTA). Plug the card into a vacant IOMB and set SW1 to all segments off. Then copy the VOUT test program from the disk enclosed and listed in Table 14–4. This

assumes a PC through Pentium with memory-mapped I/O at 640K. Modify the program as required for your setup. If you are using a USIC, modify your setup per Chap. 10. Executing the software should initialize the VOM to 0 Vdc, then each program loop should increment the output voltage by one step, reaching full scale when $N = 255$. The voltage of each step should equal $N \times$ (full-scale value)/255.

Table 14-3. DAC Card IC Power Tests

IC POWER TEST

✔	IC	+ METER LEAD ON PIN NO.	– METER LEAD ON PIN NO.	VOLTAGE READING
	U1	14	7	+5VDC
	U2	14	7	+5VDC
	U3	14	7	+5VDC
	U4	13	1	+15VDC
	U4	1	3	+15VDC
	U5	13	1	+15VDC
	U5	1	3	+15VDC
	U6	13	1	+15VDC
	U6	1	3	+15VDC
	U7	7	*	+15VDC
	U8	7	*	+15VDC
	U9	7	*	+15VDC
	U7	*	4	+15VDC
	U8	*	4	+15VDC
	U9	*	4	+15VDC
	U10	26	7	+5VDC

* = U4 pin number 1

If your card has potentiometers, you can stop the program at $N = 255$ and adjust the potentiometers to set the full-scale value. If you see no voltage when the program runs, you have at least one of the following: an error in the software, an address error in your UBEC/USIC or DAC card DIP switches, or a fault in the DAC card itself.

☐ *Channel B and C tests.* Use the same procedure to test the other two DAC channels, moving the VOM leads to appropriate S13 pins, and changing program line 90 to *PO* = 1 for channel B and *PO* = 2 for C. This completes the DAC card.

Table 14-4. VOUT Test Program

```
10    PRINT "VOUT TEST PROGRAM FOR DAC3 CARD"
20 REM DEFINE UCIS STARTING ADDRESS
30 REM THIS STEP IS APPLICATION DEPENDENT
40 ASSUMES IBM PC THRU PENTIUM WITH
50    DEF SEG = 40960   'MEMORY MAPPED I/0 WITH SA = 640KB
60 REM INITIALIZE THE 8255 ON DAC3 CARD
70    POKE 3, 128
80 REM DEFINE THE CHANNEL YOU WANT TO TEST
90    CH = 0 ' use 0 for channel A, 1 for channel B, and 2 for channel C
100 REM LOOP THROUGH FULL RANGE OF DIGITAL VALUES
110    FOR N = 0 TO 255
120    PRINT N
130 REM TRANSMIT DIGITAL VALUE TO DAC3 CARD
140    POKE CH, N
150 REM WAIT TO CHECK METER READING VS. DISPLAY
160    INPUT "CHECK METER VS. DISPLAY, PRESS ENTER TO CONTINUE",CR
170 REM INCREMENT TO OUTPUT NEXT VOLTAGE STEP
180    NEXT N
190    GOTO 10   'REPEAT TEST UNTIL HALT PROGRAM
```

Analog-to-Digital Converter (ADC) Chips

An ADC converts analog voltage into digital data, like a DAC in reverse. Again the precision of conversion is a function of the number of data bits. ADCs are necessarily more expensive than DACs, and for the UCIS I'll stick to one eight-bit chip, the ADC0804LCN. It comes in a 20-pin DIP, and is very flexible even though it's the lowest priced ADC in Jameco's catalog.

Figure 14–9 summarizes the features of the 0804. Let's look at the pinouts first in Fig. 14–9*a*. Pins 6 and 7 can receive differential analog input, or, if the pin 7 VIN(–) input is grounded, pin 6 can receive single-ended input. Differential input reduces common-mode noise, as in RS-422 I/O. Pin 7 can also be used to subtract a fixed voltage value from the input reading automatically. For maximum accuracy and noise immunity the AGND (analog ground) on pin 8 is separate from the DGND (digital ground) on pin 10.

The 0804 ADC generates an eight-bit digital output on pins 11–18 in proportion to the analog input voltage measured across pins 6 and 7. The MSB is DB7, pin 11, and the LSB is DB0, pin 18. The eight output lines are all logic zero when the differential voltage across pins 6 and 7 is zero (off), and all ones (full scale) when the voltage across pins 6 and 7 equals reference voltage *VREF*.

Unless the 0804 receives an input on pin 9, *VREF* equals supply voltage *VCC* from pin 20. With +5 Vdc on pin 20, the digital output on pins 11–18 will be all ones when the differential analog input across pins 6 and 7 reaches +5 Vdc.

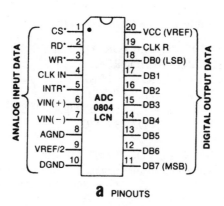

a PINOUTS

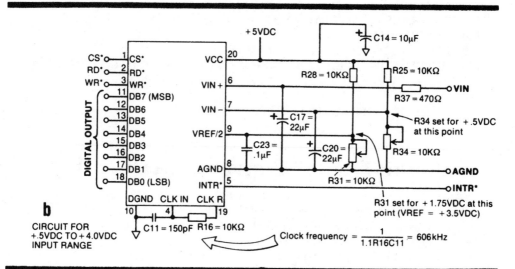

b

CIRCUIT FOR
+.5VDC TO +4.0VDC
INPUT RANGE

Clock frequency = $\dfrac{1}{1.1R16C11}$ = 606kHz

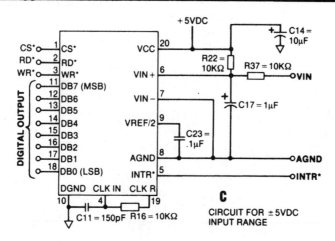

c

CIRCUIT FOR ±5VDC
INPUT RANGE

14-9 Basic ADC features.

For greater flexibility, pin 9's *VREF*/2 input can adjust the span or range of the analog input voltage. For example, suppose you want to handle a dynamic input range of +0.5 to +4 Vdc, a span of 3.5 V. Apply +0.5 Vdc to the VIN( ) pin to absorb the +0.5 Vdc zero offset, and make the *VREF*/2 input equal to $\frac{3.5}{2}$, or +1.75 Vdc. Now the ADC encodes the input on pin 6 so that +0.5 Vdc gives a digital output byte of all zeros, and +4 Vdc gives a full-scale output byte of all ones. That way you get full, eight-bit resolution even with a reduced input range.

Figure 14–9*b* shows an application circuit for the example above. Resistor R25 and pot R34 are selected and set so that VIN(-), pin 7, is held at the zero-offset value of +0.5 Vdc. Likewise R28 and R31 hold *VREF*/2, pin 9, at +1.75Vdc. Capacitors C17, C20, and C23 suppress electrical noise into the 0804; such noise could reduce the conversion accuracy. The 10-μF tantalum capacitor C14 minimizes transients on the supply/reference line.

The conversion process is triggered by simultaneous lows on the chip select (CS*) and write (WR*) control lines; the 0804 is designed to follow changes on a computer bus and has its own CS*, RD*, and WR* inputs just like the 8255. The ADC makes its conversion in a sequential manner using successive approximation logic, a test to see which bits must be turned on to match the input voltage best.

The 0804 tests the highest-order bit first, and after eight comparisons (64 clock cycles) the eight-bit code is set on the output pins. C11 and R16 together determine the chip's internal clock frequency, in this case 606 kHz. Total conversion time, from request until the digital byte can be read, is under 100 μs.

When the output byte is ready, the 0804 sets pin 5 low. A low pin 5 is an interrupt or INTR* signal to tell the outside world that the ADC data can be read. To read the data, lines RD* and CS* are both brought low, enabling the chip's tri-state output buffer.

Figure 14–9*c* shows the configuration to read an analog input range of ±5 Vdc. Its operation is easy to understand if you start with the endpoints of the range. When VIN is –5 Vdc, and with R22 equal to R37, pin 6 is midway between +5 and –5 Vdc, 0 V, giving a digital output of all zeros. At the other end of the range, with *VIN* equal to +5 Vdc, pin 6 is +5 Vdc (the full-scale *VREF* value) so the digital output is all ones. Any input between the end points yields a proportional in-between digital output.

Change R22 to 5 kΩ, and add an R19 of 10 kΩ between pin 6 and AGND pin 8, and the effective input range changes to ±10 Vdc. Keeping R19 at 10 kΩ, changing R37 to 14 kΩ, and deleting R22 shifts the input range to 0–12 Vdc.

ADC Input Card

Figure 14–10 shows a completed ADC input card, which plugs into the motherboard just like a general-purpose input card. The ADC card uses three 0804 ICs for three independent analog inputs. Your external hardware connects to S11, the 14-pin right angle header at the top, and the address decoding/DIP switch circuit is like the input card's.

Each of the three ADC circuits is designed so you can use different parts and jumpers to tailor it for reading inputs with different zero offsets and dynamic ranges. You can install fixed resistors for preset ranges or potentiometers for adjustment.

14-10 ADC input card.

The example illustrates the ADC card's flexibility. I've configured the analog A circuit with fixed resistors to handle an input range of 0 to +5 Vdc. The analog B circuit is set up for a range of 0 to some positive full-scale voltage set by a potentiometer. The analog C circuit is adapted for a range of +0.5 to +4 Vdc.

Figure 14–11 is the ADC card schematic. The bottom left side is identical to the DAC card. The U5 1-of-10 decoder/driver decodes the A0 and A1 inputs, along with the card-selected signal from U3's pin 8, to pull the proper CS* line low to activate the ADC being addressed.

The U5 outputs have open collectors, so when W* goes low along with pin 8 of U3, U4's pin 1 output goes low. Through diodes D1–D3, that pulls all ADC CS* inputs low to trigger data conversions simultaneously. When the R* line goes low to read the digital data, only the addressed CS* line goes low, enabling only one ADC output buffer at a time.

In a serial, USIC-based interface, the ADC card is defined as an input card in the *CT( )* array. You can't write to an input card through a USIC, so I've included an alternate conversion-start circuit on the ADC card using leftover gates from U2 and U3.

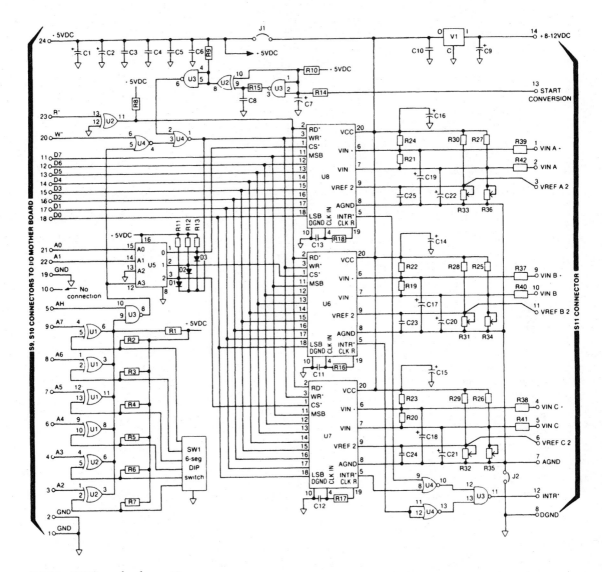

14-11 ADC card schematic.

The conversion start circuit attached to pin 13 of S11 operates to generate a short, negative-going, narrow pulse out of U2 pin 8, the 74LS136 exclusive-OR IC, every time the start conversion input on pin 13 of S11 changes state. For explanation I've labeled the input to U3, pin 2 as X, and its inverted output on U3, pin 3 as X*. The 74LS136 output goes low only when its two inputs are identical (00 or 11). Because X* and X can't be identical, by definition, it might seem that the 74LS136 would always be a logic high.

R15 and C8 cause a small delay in the path to X* every time X changes state, so the X* signal arrives at the 74LS136 a bit later than the X. That gives a 1-μs, logic

0 trigger pulse out of U2, pin 8, which in turn triggers the three ADCs via U3 and U4 to start converting their analog inputs to digital outputs.

Each transition of the conversion-start line (high-to-low or low-to-high) triggers the ADCs to start their conversions. You can drive any number of ADC cards with a single line from an output card that changes state each time through the real-time loop.

Each time your program writes to the output cards it toggles the conversion-start line and triggers the ADC conversions. Because the 100 μs the ADCs take to make conversions is quick compared to most real-time loop processing, by the time your program reads back the input cards, the ADCs are updated to reflect last-iteration data.

For the greatest accuracy, use a separately regulated supply for ADC reference voltage and separate grounds AGND and DGND. Connect a +8–12-Vdc supply through pin 14 of S11 to +5-V regulator V1. This serves as a stable local reference-voltage supply for the ADCs.

If accuracy isn't that important in your application, V1, C9, and C10 can be omitted, and jumper J1 used to connect the ADC reference line to the +5-Vdc motherboard power. If you use V1, do not install J1. Noise-filter capacitors C17 through C25 are all tied to AGND. For applications where separate grounds aren't necessary, use jumper J2 to tie AGND into the common card digital ground on the motherboard.

If you want $VREF = VCC = +5$ Vdc, do not install resistors R28–R33. If you want lower $VREF$ values for reduced input range, use 10-kΩ fixed resistors for R28–R30 and select R31–R32 for the desired pin 9, $VREF/2$ voltage. If you want a zero-offset voltage, install R25–R27 as 10-kΩ fixed resistors and select R34–R36 for the desired pin 7 offset voltage.

When you want neither differential input nor an offset zero, replace resistors R34–R36 with jumpers for 0 Ω resistance. Also delete C20–C22, R25–R27, and R40–R42 if jumpers are substituted for R34–R36. If the range of VIN is to include negative values, install R22–R24. Similarly, if the range of VIN is to go above +5 Vdc, install R19–R21.

Resistors R37–R42 should never be less than 470 Ω, to help protect the ADC if you accidentally connect the VIN pins of S11 to high voltages. If you want the effective input range to include voltages above +5 Vdc, use 10-kΩ resistors for R19–R21 and select R37–R39 to provide a level of attenuation such that full-scale input corresponds to +5 Vdc at pin 6.

To minimize the chance of a noise spike input causing an output error, it's good practice to make capacitors C17–C19 (as well as C20–C22 when they are used) as large as the board area and update rate permit. Larger values give more filtering, but values can be too large, causing sampling errors in which the ADC inputs lag behind inputs at S11 by a period longer than the time between conversion updates. For maximum accuracy, keep the capacitor values on the high side, but not greater than the time between conversions (in seconds) divided by the product of five times the value (in ohms) of R37–R42.

For example, in Fig. 14–9 I've assumed a real-time-loop update period of 0.06 s, so the largest value recommended for C17 with R37 at 470 Ω is $^{0.06}/_{(5\times470)}$, or 26 μF. I picked the nearest standard value of 22 μF. In Fig. 14–9c R37 is 10 kΩ, so for the same update period C17 is decreased to 1 μF. Table 14–5 lists recommended standard capacitors to fit the card.

Building the ADC Card

Figure 14–12 is the parts layout and Table 14–6 the parts list for the ADC card. Ready-to-assemble circuit cards are available from JLC Enterprises, and fully assembled and tested cards are available from ECO Works, EASEE Interfaces, or AIA. As with the DAC, asterisks in the parts list mark parts to be substituted for the particular ADC circuit configurations you want. For R16–R42 you might want to substitute 1 percent tolerance precision resistors. To assemble the ADC card, refer to Fig. 14–12 and Table 14–6 and follow these steps:

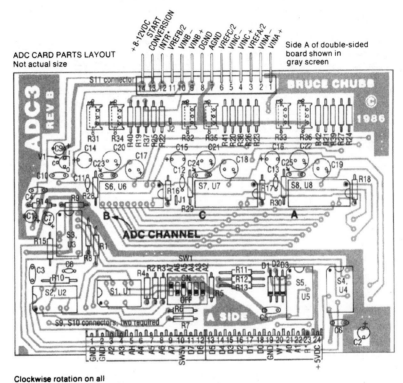

Clockwise rotation on all
potentiometers increases voltage

14-12 ADC card parts layout.

☐ *J1, J2.* Omit J1 to use V1, and omit J2 to separate AGND from DGND.

☐ *R1–R42.* For parts marked by asterisks, be sure to install components for the configuration(s) you want. The options are no part (open circuit), a fixed resistor, a trim pot, or a jumper (0Ω). Where you use a potentiometer, its adjustment screw should face S11.

☐ *D1–D3[+].*

☐ *S1–S8[+].*

☐ *S9–S11.*

☐ *SW1[+].* Install this part so that the DIP switch contacts are closed when thrown toward S11.

☐ *C1, C2[+].*
☐ *C3–C6.*
☐ *C7[+].*
☐ *C8, C11–C13.*
☐ *C14–C16[+].*
☐ *C17–C19[+].* Check Table 14–5 for values.

Table 14-5. Capacitor Size Selection for C17–C22

Capacitor	Source	Lead Spacing
1µF, 35V tantalum	Jameco 33662	.098"
2.2µF, 35V tantalum	Jameco 33734	.098"
4.7µF, 100V electrolytic	Digi-Key P6292	.079"
10µF, 100V electrolytic	Digi-Key P6293	.098"
22µF, 63V electrolytic	Digi-Key P6280	.098"
47µF, 50V electrolytic	Digi-Key P6267	.098"
100µF, 25V electrolytic	Digi-Key P6239	.098"
220µF, 10V electrolytic	Digi-Key P6215	.098"

☐ *C20–C22[+].* Omit these if you are substituting jumpers for resistors R34–R36.
☐ *C23–C25.*
☐ *U1–U8[+].*

Skip to cleanup and inspection unless you're using alternate supply V1.

☐ *V1[+].*
☐ *C9[+].*
☐ *C10.*
☐ *Cleanup and inspection.*
☐ *IC power test.* Plug the ADC card into your motherboard. If you are using V1, connect a separate +8–12-Vdc input to pin 14 of S11, and the corresponding ground to pin 8. Turn on the IOMB supply, and the V1 supply if it is used; use your VOM and Table 14–7 to see that power reaches each IC. Correct any problems before proceeding.
☐ *Input voltage test.* Plug your card into a vacant IOMB and set SW1 to all segments off. Connect an external potentiometer as in Fig. 14–13 to provide a variable input voltage to pin 1 of S11, the ADC channel A input. Set SW1 to all segments off. Copy the VIN test program from the disk enclosed and listed in Table 14–8. This assumes memory-mapped I/O at 640K. Modify this arrangement as required to match your computer setup. Then execute the program to display the decimal equivalent, between 0 and 255, of the input voltage. The screen readout should vary as you turn the pot. Once channel A passes, repeat the test for the other channels. That completes the ADC card.

Table 14-6. ADC Card Parts List

Qnty.	Symbol	Description
2	J1-J2	Jumpers, make from no. 24 uninsulated bus wire (Belden no. 8022). Omit J1 if using alternate power supply with V1. Omit J2 to separate analog and digital grounds.
1	R1	1.0KΩ resistor [brown-black-red]
12	R2-R13	2.2KΩ resistors [red-red-red]
2	R14,R15	47Ω resistor [yellow-violet-black]
3	R16-R18	10KΩ resistors [brown-black-orange]
3*	R19-R21	Open circuit for VIN+ full scale less than +5Vdc or install 10KΩ resistors [brown-black-orange] if VIN+ full scale to be greater than +5 Vdc
3*	R22-R24	Open circuit for VIN+ only positive values or select as fixed per text to handle negative VIN+ voltage values
3*	R25-R27	Open circuit for single-ended inputs with VIN- grounded or install 10KΩ resistors [brown-black-orange] to set zero-offset via R34-R36
3*	R28-R30	Open circuit for VREF = VCC or install 10KΩ resistors [brown-black-orange] to set VREF/2 via R31-R33
3*	R31-R33	Open circuit for VREF = VCC or install fixed resistor or 10KΩ potentiometer (Digi-Key CT9X03) to set VREF/2
3*	R34-R36	Jumper for single-ended inputs with VN- grounded or select as either fixed resistor or 10KΩ potentiometer (Digi-Key CT9X03) for setting non-zero offset (along with R25-R27)
3*	R37-R39	470Ω resistors [yellow-violet-brown] for VIN+ full scale less than +5Vdc or install greater fixed resistor to attenuate (along with R19-R21) VIN+ values greater than +5Vdc
3*	R40-R42	Open circuit for single-ended inputs with VIN- grounded, otherwise install 470Ω resistor [yellow-violet-brown]
3	D1-D3	1N4148 switching diodes (Digi-Key 1N4148CT)
4	S1-S4	14-pin DIP sockets (Digi-Key A9314)
1	S5	16-pin DIP socket (Digi-Key A9316)
3	S6-S8	20-pin DIP sockets (Digi-Key A9320)
2	S9,S10	12-pin Waldom side entry connectors (Digi-Key WM3309)
1	S11	14-pin Waldom right angle header (cut from two Digi-Key WM4510)
1	SW1	6-segment DIP switch (Digi-Key CT2066)
2	C1,C2	2.2μF, 35V tantalum capacitors (Digi-Key P2061)
4	C3,C6	100000pF, 50V ceramic disk capacitors (Digi-Key P4164)
1	C7	1μF, 35V tantalum capacitor (Digi-Key P2059)
1	C8	.022μF, 50V ceramic disk capacitors (Digi-Key P4164)
3	C11-C13	150pF, 50V ceramic disk capacitors (Digi-Key P4431)
3	C14-C16	10μF, 25V tantalum capacitor (Digi-Key P2049)
3*	C17-C19	Select using equation in text and pick nearest standard part in table between 1μF and 220μF
3*	C20-C22	Open circuit when R34-R36 is jumper or install 1μF, 35V tantalum capacitor (Digi-Key P2059) with non-zero zero-offset or for differential input of VIN- select using equation in text and pick nearest standard part in table between 1μF and 220μF
3	C23-C25	100000pF, 50V ceramic disk capacitors (Digi-Key P4164)
2	U1,U2	74LS136 quad EXCLUSIVE-OR gates with open collector (Jameco 46586) or (JDR 74LS136)
1	U3	74LS00 quad 2-input NAND gate (Jameco 46252)
1	U4	74LS02 quad 2-input NOR gate (Jameco 46287)
1	U5	74LS145 BCD to decimal decoder with o.c. outputs (Jameco 46666)
3	U6-U8	ADC0804LCN 8-bit A/D converters (Jameco10153)

Table 14-6 *continued*

Optional parts used for separate ADC power supply

1	V1	78L05 5V, 100mA positive regulator (Jameco 51182)
1	C9	.33μF, 35V tantalum capacitor (Digi-Key P2056)
1	C10	100000pF, 50V ceramic disk capacitor (Digi-Key P4164)

Author's recommendations for suppliers given in parentheses above with part numbers where applicable. Equivalent parts may be substituted. Resistors are ¼W, 5 percent and color codes are given in brackets.

Table 14-7. ADC Card IC Power Tests

✔	IC	+ METER LEAD ON PIN NO.	− METER LEAD ON PIN NO.	VOLTAGE READING
	U1	14	7	+ 5VDC
	U2	14	7	+ 5VDC
	U3	14	7	+ 5VDC
	U4	14	7	+ 5VDC
	U5	16	8	+ 5VDC
	U6	20	10	+ 5VDC
	U7	20	10	+ 5VDC
	U8	20	10	+ 5VDC

Motor Speed Control

For applications where you don't need the precision of the DAC chip but you do need more power output, you might want to consider building your own conversion circuit. As an example, I'll show you a circuit that I have found very handy for driving dc permanent-magnet motors.

Figure 14–14 shows the circuit. I call it a computer-controlled throttle or CCT because I use it to control motor speed. The resistor network in front of op amp U1 serves as a four-bit DAC. A second stage of U1 is used to make the logic inputs active low, so the throttle is off if the inputs are not connected. The UCIS can drive the CCT directly with five lines from an output card. Lines B0*–B3* control motor speed, and RV* controls direction by activating reversing relay RLY.

The four speed-control lines give 16 discrete levels of motor voltage. Regulator V1 provides a constant 15 Vdc for the op amp, and zener diode Z1 and R20 set the 6.8-V reference for the ladder network.

Regulator V2 controls the motor voltage. Its output can only be controlled down to about 2 Vdc, but the drop across series diodes D3–D6 cancels the effect of the

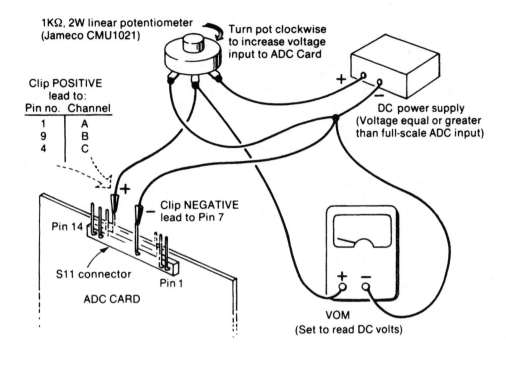

1KΩ, 2W linear potentiometer
(Jameco CMU1021)

Turn pot clockwise
to increase voltage
input to ADC Card

Clip POSITIVE
lead to:

Pin no.	Channel
1	A
9	B
4	C

DC power supply
(Voltage equal or greater
than full-scale ADC input)

Clip NEGATIVE
lead to Pin 7

Pin 14

S11 connector

Pin 1

ADC CARD

VOM
(Set to read DC volts)

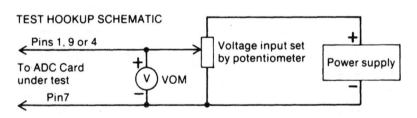

TEST HOOKUP SCHEMATIC

Pins 1, 9 or 4

To ADC Card
under test

Pin7

VOM

Voltage input set
by potentiometer

Power supply

14-13 VIN test schematic.

minimum *VOUT,* and drops minimum motor voltage nearly to zero. Depending upon how much voltage it takes to make your motor just start to rotate, you might get by with only one, two, or three of the diodes, replacing those omitted with jumpers. Each diode gives about a 0.7-V drop in motor voltage.

The two motor leads are connected using S3 (pins 1 and 2) if you don't want the reverse feature, and pins 3 and 4 if the computer-controlled, reversing feature is desired. The full-scale motor voltage can be adjusted with software, by selecting the voltage rating of zener diode Z1, or by a combination of both. Table 14–9 gives several choices for Z1 ratings; the change in the full-scale voltage will be twice the change in the rated zener voltage. Resistor R19 gives the correct voltage across the relay coil when RV* is pulled low by the UCIS. Diodes D1 and D2 protect V1 and the output card, respectively.

Table 14-8. VIN Test Program

```
10   PRINT "VIN TEST PROGRAM FOR ADC3 CARD"
20   REM DEFINE UCIS STARTING ADDRESS
30   REM THIS STEP IS APPLICATION DEPENDENT
40   REM ASSUMES IBM PC THRU PENTIUM WITH
50   DEF SEG = 40960  'MEMORY MAPPED I/0 WITH SA = 640KB
60   REM DEFINE THE CHANNEL YOU WANT TO TEST
70   CH = 0  ' use 0 for channel A, 1 for channel B, and 2 for channel C
80   REM WRITE TO START CONVERSION PROCESS FOR ADC
90   POKE 3, 128
100  REM WAIT FOR ADC CONVERSION TO COMPLETE
110  FOR IJ = 1 TO 50000: NEXT IJ  'VARY AS REQUIRED
120  REM READ INPUT FROM ADC AND PRINT VALUE ON MONITOR
130  N = PEEK(CH)
140  PRINT N
150  REM WAIT TO CHECK METER READING VS. DISPLAY
160  INPUT "CHECK METER VS. DISPLAY, PRESS ENTER TO CONTINUE",CR
170  REM INCREMENT TO READ IN NEXT INPUT VOLTAGE
180  NEXT N
190  GOTO 10   'REPEAT TEST UNTIL HALT PROGRAM
```

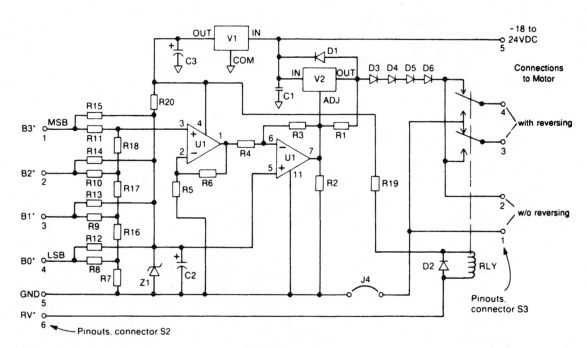

14-14 Computer-controlled throttle (CCT) schematic.

Table 14-9. Zener Diode Options

Zener diode		Approx. max. track voltage*
Part no.	Reference voltage	
1N4736	6.8V	12
1N4735	6.2V	10.8
1N4734	5.6V	9.6
1N4733	5.1V	8.6
1N4732	4.7V	7.8

*Based upon 3 diode drops D3-D6

Power is 18–24 Vdc applied to terminal 5 of S3, with ground on terminal 5 of S2. You typically need just one set of power connections per card, and multiple cards may be powered in parallel. The throttle power supply must be filtered but need not be regulated. It can be built following Fig. 8–12, with V1, C2, C3, R1, and L1 omitted and T1 selected to have a 14–18 V secondary. The transformer and bridge rectifier, B1, must have current capacity greater than the expected maximum simultaneous current requirement from each group of throttles. C1's value should be increased or other capacitors added in parallel so the total capacitance equals about 5000 µF per amp of supply capacity. The bridge and the filter capacitors should be rated for at least 100 V.

Building the CCT Card

Figure 14–15 is a photo of the CCT4 card, which includes four identical throttles to control speed and direction of up to four motors independently, all from your computer. The parts layout and assembly details for the CCT4 card are shown as Figs. 14–16 and

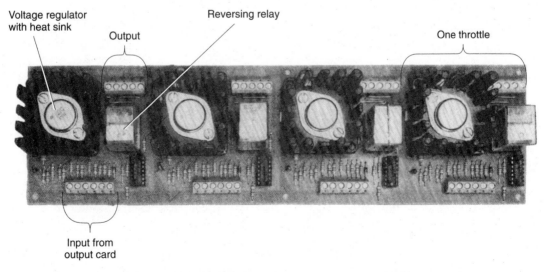

14-15 Computer-controlled card (CCT4).

14–17, and the parts list is Table 14–10. Ready-to-assemble circuit cards are available from JLC Enterprises and fully assembled and tested cards from ECO Works, EASEE Interfaces, and AIA.

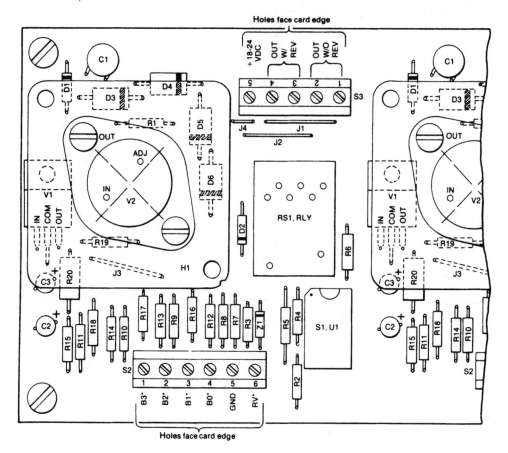

14-16 CCT parts layout.

You can build up to four CCTs on each card, but as all are identical I've repeated the parts nomenclature from throttle to throttle. The parts list quantities are for a single throttle. You may cut the CCT4 card along the dashed lines on the trace side to make four individual CCTs.

I've listed three V2 regulators for different maximum currents. This is a function of the maximum, steady-state current you expect to need for each motor. The higher the regulator current rating and the higher the input voltage to the card, the more power needs to be dissipated by the regulator when you have a stalled motor or a short circuit at the throttle output. Thus it's best to use the lowest-current regulator that you get by with. The regulators will handle surge currents greater than their continuous ratings, and if overloaded they automatically go into thermal shutdown without damage. When the short is removed, V2 will quickly cool and automatically resume operation.

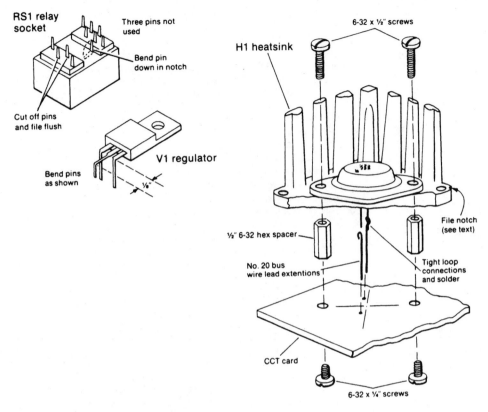

RS1 relay socket

Three pins not used

Bend pin down in notch

Cut off pins and file flush

V1 regulator

Bend pins as shown

⅛"

H1 heatsink

6-32 x ½" screws

File notch (see text)

½" 6-32 hex spacer

No. 20 bus wire lead extentions

Tight loop connections and solder

CCT card

6-32 x ¼" screws

14-17 CCT construction details.

If you need the three-amp regulator, you might want to add metal to the heat sink, and it's best to mount the five-amp regulator off the card on a large heat sink like Digi-Key HS118. To assemble a CCT card refer to Figs. 14–14 to 14–17 and Table 14–10 and follow these steps:

☐ *J1–J4.*
☐ *R1–R20.*
☐ *D1, D2[+].*
☐ *D3–D6[+].* Use needle-nose pliers to bend the heavy leads of these power diodes at right angles so they drop into the holes. The banded ends must face in the direction indicated in Fig. 14–16. Leaving a ¹⁄₁₆-in space between the diode and the card helps ventilate the diodes and protects the card. If you omit any diodes, use #20 or larger jumpers in their place.
☐ *Z1[+].*
☐ *C1.* Once it is soldered, bend this part flat against the card as in the parts layout.
☐ *C2, C3[+].*
☐ *S1–S3[+].*

Table 14-10. CCT Parts List

Qnty.	Symbol	Description
4*	J1-J4	Jumpers, make from no. 20 uninsulated bus wire (Belden no. 8020 to carry track current). Omit J4 if using CCT as separate throttle per block for Computer Block control (CBC) with Optimized Detector
1	R1	240Ω resistor [red-yellow-brown] except for V2 equal LM338K use 120Ω resistor [brown-red-brown]
1	R2	470Ω resistor resistors [yellow-violet-brown]
3	R3-R5	10KΩ resistors [brown-black-orange]]
1	R6	12KΩ resistor [brown-red-orange]
5	R7-R11	36KΩ resistors [orange-blue-orange]
4	R12-R15	3.0KΩ resistors [orange-black-red]
3	R16-R18	18KΩ resistors [brown-gray-orange]
1	R19	47Ω resistor [yellow-violet-black]
1	R20	160Ω ½W resistor [brown-blue-brown]
2	D1,D2	1A, 100V diodes (Digi-Key 1N4002GICT)
4*	D3-D6	3A, 100V diodes (Digi-Key 1N5401GICT) may replace 1 and 2 diodes with jumpers(s) made from no. 20 uninsulated bus wire (Belden no. 8020, to carry track current), if such a low throttle-off output voltage isn't needed. For V2 equal LM338K substitute 6A diodes (Digi-Key GI751CT)
1	Z1	6.8V, 1W zener diode (Digi-Key 1N4736ACT)
1	C1	100000pF ceramic disk capacitor (Digi-Key P4164))
2	C2,C3	1µF, 35V tantalum capacitors (Jameco 33662)
1	S1	14-pin DIP socket (Digi-Key A9314)
1	S2	6-position terminal block (Mouser 153-2506)
1	S3	5-position terminal block (Mouser153-2505)
1	V1	15V,1.5A positive regulator (Mouser 511-L7815CV)
1	V2	LM317K 1.5A positive adjustable voltage regulator (Jameco 23544) LM350K 3A positive adjustable voltage regulator (Jameco 23931) LM338K 5A positive adjustable voltage regulator (Jameco 23835)
1	H1	25W heat sink for TO-3 case (Digi-Key HS103-1.25) For LM338K may need to substitute larger off-card heat sink such as Digi-Key HS118
2	—	Hex 6-32 threaded ½"-long spacer (Digi-Key J178)
2	—	6-32 x ½" pan-head machine screws (Digi-Key H160)
2	—	6-32 x ¼" pan-head machine screws (Digi-Key H154)
1	RS1	Relay socket (Potter & Brumfield 27E128 or Digi-Key PB134)
1	RLY	2PDT, 12Vdc relay (Potter & Brumfield R10E1Y2V185 or Digi-Key PB117)
1	—	Relay hold down clip (Potter & Brumfield 20C249 or Digi-Key PB136)
1	U1	LM324 low power quad operational amplifier (Jameco 23683)

Author's recommendations for suppliers given in parentheses above with part numbers where applicable. Equivalent parts may be substituted. Resistors are ¼W, 5 percent and color codes are given in brackets.

☐ *V1[+]*. Bend the leads as shown in Fig. 14–17; install the component flat against the card.

☐ *V2[+]*. Mount V2 to the heat sink as in Fig. 14–17. Then make and solder tight hook joints between two 2-in lengths of #20 jumper wire and the V2 leads. These leads are only slightly off-center; to avoid blowing the regulator by installing it backwards, file a notch in the heat sink as shown in Fig. 14–17.

Feed the jumpers through the IN and ADJ holes in the card, making sure the notch filed in the heat sink is near D2 and the leads are in the correct holes, and secure the assembly with 6–32 × ¼-in screws. After soldering, check that the jumpers are tight and not touching, and use your VOM to make sure the leads are not shorted to the case of V2.

☐ *RS1, RLY[+].* Modify the pins of relay socket RS1 as shown in Fig. 14–17, and insert relay RLY into the socket. Snap the hold-down clip into the socket and position it over the top of the relay. Hold RS1 (with RLY in place) firmly against the card as you solder, so the pins extend as far as possible on the trace side.

☐ *U1[+].*

☐ *Power-up test.* Leaving the motor connection terminals open, connect +18–24-Vdc power to the card. Touch the negative lead of your VOM to U1 pin 11; you should read about +15 V with the positive lead on U1 pin 4, and +6.8 V with the positive lead on U1 pin 5.

☐ *Relay test.* Use a clip lead to ground the RV* terminal of S2 temporarily. The relay should throw, and the voltage drop across the relay coil should be close to 12 Vdc. If off more than a few volts, vary R19 until you get the correct value.

☐ *Operational voltage test.* Use clip leads to attach a 50-Ω 10-W resistor between S2, terminal 5, and S3, terminal 2. Attach the negative lead of your VOM to S2, terminal 5, and touch the positive lead to the test points in Table 14–11. Check the boxes next in the inputs' Open Circuited column as you pass each test. A reading that is off by much more than a volt indicates

Table 14-11. CCT Voltage Tests

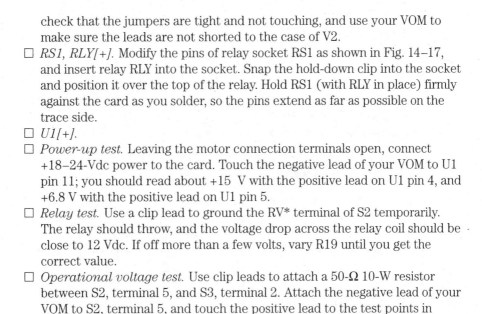

INPUTS B0*-B3* SET AT	MINIMUM OUTPUT VOLTAGE TESTS OPEN CIRCUITED		MAXIMUM OUTPUT VOLTAGE TESTS GROUNDED	
+ meter lead on *	✔	Voltage reading	✔	Voltage reading
U1 pin 1		13.9		.71
U1 pin 2		6.3		.34
U1 pin 3		6.3		0.00
U1 pin 4		15.3		15.3
U1 pin 5		6.9		6.9
U1 pin 6		7.4		6.8
U1 pin 7		.94		12.9
Case V2		2.3		14.2
D3 cathode•		1.6		13.4
D4 cathode•		1.1		12.7
D5 cathode•		.60		12.0
D6 cathode•		.06		11.3

* – meter lead on S2 terminal 5

• banded end

Above readings taken with load of 50Ω, 10W resistor
(Radio Shack 271-133) connected between S3 term 2 and S2 term 5

a soldering problem, an incorrect resistor, or a faulty U1. Use clip leads to ground all four inputs, B0*–B3*, and recheck the previous test points, comparing voltages read to the inputs' Grounded column. Again, a reading that is off by much more than 1 V indicates a problem.

☐ *Variable voltage test.* Use clip leads to ground temporarily various combinations of the B0* to B3* terminals of S2. Grounding all four terminals should give a maximum output near 12 V. Each other grounding combination should give voltage proportional to the decimal equivalent of the binary code set on B0*–B3*, according to Table 14–12.

Table 14-12. CCT Input/Output Tests

Decimal number in software

N	B3*	B2*	B1*	B0*	VOUT
0					0
1				●	.8
2			●		1.6
3			●	●	2.4
4		●			3.2
5		●		●	4.0
6		●	●		4.8
7		●	●	●	5.6
8	●				6.4
9	●			●	7.2
10	●		●		8.0
11	●		●	●	8.8
12	●	●			9.6
13	●	●		●	10.4
14	●	●	●		11.2
15	●	●	●	●	12.0

● Indicated line pulled low by software (or manually with clip lead to ground)

☐ *Software test.* Wire the S2 terminal to a UCIS output card as in Fig. 14–18 and connect a ground line between your CCT and your 5-Vdc supply. Then copy the CCT test program from the disk enclosed and listed in Table 14–13. Modify this program as required. In executing the program, the measured voltage should vary as shown in Table 14–12 with each pass through the real-time loop. You've already tested the CCT card, so any problems here are in your software, card addresses, or output-card-to-CCT-card wiring.

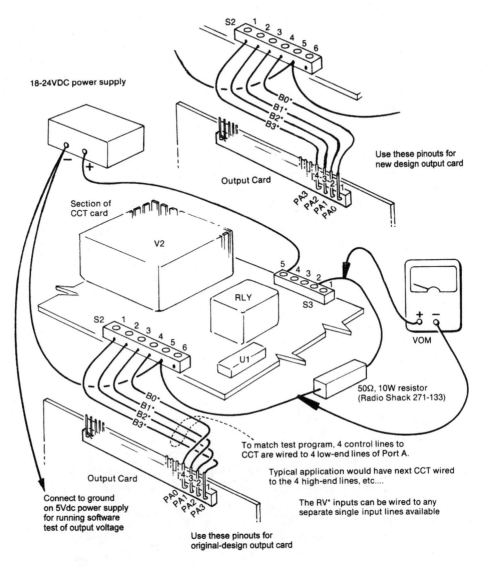

18-24VDC power supply

S2

B0*
B1*
B2*
B3*

Output Card

Use these pinouts for
new design output card

PA3 PA2 PA1 PA0

Section of
CCT card

V2

RLY

S3

U1

VOM

50Ω, 10W resistor
(Radio Shack 271-133)

S2

B0*
B1*
B2*
B3*

To match test program, 4 control lines to
CCT are wired to 4 low-end lines of Port A.

Typical application would have next CCT wired
to the 4 high-end lines, etc....

Output Card

PA0 PA1 PA2 PA3

Connect to ground
on 5Vdc power supply
for running software
test of output voltage

The RV* inputs can be wired to any
separate single input lines available

Use these pinouts for
original-design output card

14-18 CCT test schematic.

Pulse Power Control

The CCT circuit works great as is for many applications; however, when you desire to achieve extremely smooth slow-speed motor operation, the performance can be improved by adding *pulse power control* (PPC) to the CCT circuit. Figure 14–19 shows a photo of the PPC card. Its schematic and parts layout are shown as Figs. 14–20 and 14–21. The parts list is given in Table 14–14.

Table 14-13. CCT Test Program

```
10    PRINT "COMPUTERIZED THROTTLE TEST PROGRAM"
20  REM DEFINE UCIS STARTING ADDRESS
30  REM THIS STEP IS APPLICATION DEPENDENT
40  ASSUMES IBM PC THRU PENTIUM WITH
50     DEF SEG = 40960   'MEMORY MAPPED I/0 WITH SA = 640KB
60  REM INITIALIZE THE 8255 ON OUTPUT CARD
70     POKE 3, 128
80  REM DEFINE VOLTAGE STEP SIZE
90     VA = .8    'calculated as full scale at 12Vdc divided by 15 steps
100 REM ASSUME CCT4 CONNECTED TO PORT A (bits A0-A3)
110 REM LOOP THROUGH ALL VOLTAGE STEPS TO CCT4
120    FOR N = 0 TO 15
130    VO = VS * N  'CALCULATE OUTPUT VOLTAGE
140    POKE 0, N  'SEND BIT PATTERN FOR VOLTAGE TO CCT4
150    PRINT N, VO  'DISPLAY DIGITAL COMMAND AND VOLTAGE
160 REM WAIT TO CHECK METER READING VS. DISPLAY
170    INPUT "CHECK METER VS. DISPLAY, PRESS ENTER TO CONTINUE",CR
180 REM INCREMENT TO OUTPUT NEXT VOLTAGE STEP
190    NEXT N
200    GOTO 10   'REPEAT TEST UNTIL HALT PROGRAM
```

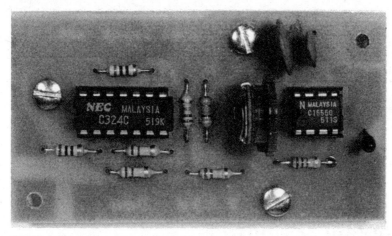

14-19 Pulse power control card.

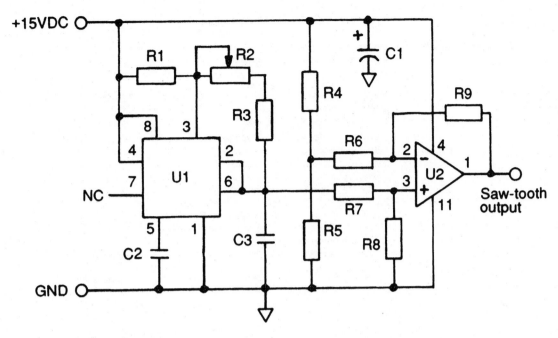

14-20 PPC schematic.

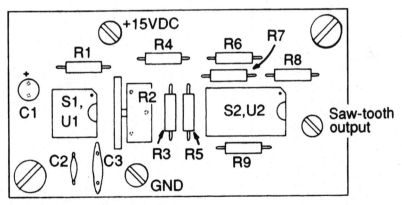

14-21 PPC parts layout.

Table 14-14. PPC Parts List

Qnty.	Symbol	Description
4	—	4-40 x ¼" Pan-head machine screws (Digi-Key H142)
4	—	Brass nuts
1	R1	1.0KΩ resistor [brown-black-red]
1	R2	100KΩ trim potentiometer (Jameco 94730 or Mouser 320-1510-100K)
1	R3	47KΩ resistor [yellow-violet-orange]
1	R4	22KΩ resistor [red-red-orange]
1	R5	10KΩ resistor [brown-black-orange]
2	R6,R7	820KΩ resistors [gray-red-yellow]
2	R8,R9	2MΩ resistors [red-black-green]
1	S1	8-pin DIP socket (Digi-Key A9308)
1	S2	14-pin DIP socket (Digi-Key A9314)
1	C1	2.2µF, 35V tantalum capacitor (Jameco 33734)
1	C2	10000pF, 50v ceramic disk capacitor (Digi-Key P4161)
1	C3	100000pF, 50v ceramic disk capacitor (Digi-Key P4164)
1	U1	LM555CN timer (Jameco 27422)
1	U2	LM324N low power quad operational amplifier (Jameco 23683)

Author's recommendations for suppliers given in parentheses above with part numbers where applicable.   Equivalent parts may be substituted.   Resistors are ¼W, 5 percent and color codes are given in brackets.

Ready-to-assemble PPC circuit cards are available from JLC Enterprises, and fully assembled and tested cards can be ordered from ECO Works, EASEE Interfaces, and AIA. To assemble a PPC card, follow these steps:

☐ *Terminal screws.*
☐ *R1–R9.*
☐ *C1[+].*
☐ *C2, C3.*
☐ *S1, S2[+].*
☐ *U1, U2[+].*

To connect the PPC card to each CCT, you need to modify each CCT a bit as follows:

☐ *Remove R3.* Simply clip both ends of the resistor on each CCT if you have them already installed, or don't install them to begin with.
☐ Cut the circuit trace midway between Z1 and pin 5 of S1.
☐ Solder one end of a wire to pin 5 of S1 and connect the other end to the sawtooth output of the PPC Card.
☐ Connect the GND terminal to PPC to the GND terminal of one of the CCTs.
☐ Connect the +15-Vdc input to PPC to the output terminal pad of one of the CCT V1 regulators.

That completes the pulse power control. In the next chapter I'll cover generating I/O handling software.

15
CHAPTER

Generating I/O Handling Software

In this chapter I'll cover the basics of setting up I/O interfacing software for use with your UCIS. This will include basic steps to follow for setting up your UCIS application, the programming format for your real-time loop, recommended I/O worksheet tables to organize your specific UCIS input and output requirements, making I/O addressing easy, reading and unpacking input bytes, packing and writing output bytes, and writing a small I/O handling program. Apple BASIC as supplied with the Apple II series of computers requires some special care in the unpacking and packing operations, so I'll cover this as well.

Basic UCIS Application Steps

Although it's impossible to cover every possible application of a UCIS even in generality, the five basic application steps I recommend you take are:

- ☐ Draw a functional schematic of your planned external hardware that shows the devices you want your computer to control and/or receive input from, such as switches, LEDs, solenoids, lamps, relays, pushbuttons, and motors. Assign unique, meaningful, two-character alphabetic labels to each device, such as SW for switches, LD for LEDs, and so on. Where you have more than one of a given device, number them consecutively from one through the maximum for each device and distinguish them with sequential identifiers; for example, SW1 for switch number one, SW2 for switch number two, and so on.
- ☐ Set up I/O tables as I'll explain shortly, grouping your inputs and outputs to keep contiguous bits together for each variable, and to fill most of your ports for efficient use of your I/O cards.
- ☐ Program your software to read inputs from the input cards, perform calculations on the inputs to generate outputs, and then write the outputs

to the output cards. As you do the programming, you will probably go back to change the names or number of variables, and to rearrange the I/O tables, all to simplify the programming. This is normal procedure in engineering design. I'll be explaining the details of application programming in Chap. 18 by leading you through a sample application program.

☐ Connect devices to your I/O cards as defined by your I/O tables. As you make connections to output cards you can write short test programs to turn the devices on and off, to see that everything works as you go. Test your inputs by writing a program to operate in a loop to read an input port for, say, switch positions; separate the bits that define individual switches—SW0, SW1, and so forth—and print them on the monitor. Throw switches and your display will be updated, making it easy to check your connections.

☐ Test and debug the complete system. You tested your UCIS in Chaps. 9, 11, or 12, and if you do a good job of testing I/O connections as you go, most of your problems will be in the software. To help, I'll provide a short example in Chap. 18 of an effective simulation subroutine that uses keyboard input to let you test your program independently of actual UCIS hardware inputs.

Incremental Addressing to I/O Cards

Figure 15–1 illustrates a collection of I/O cards attached to a computer via IBEC. Let's use it to look at how the computer sends a message to an example UCIS starting address, say at 640K, and then increments this address to reach all the other cards. Once you understand this you'll have a working knowledge of how the UCIS communicates with all your external hardware.

The UCIS starting address always corresponds to Port A of the first I/O card attached to your external hardware. Following this example, to transmit a message to the external hardware via Port A, the computer places the required 640K address (655,360, which is 640 × 1024) on the address bus. Every device connected to the address bus tries to decode the 640K to see if it is the location addressed, but only the I/O card at 640K gets the proper decoding to make it receptive to the data lines. While decoding is taking place the computer places eight bits of data, the message, on the eight data lines.

In less than 1 μs the lines will have stabilized and the address will have been decoded all the way up to the third level of decode on the selected card. Then the computer triggers the W* (write, active low) line for another fraction of a microsecond. This completes the enabling of the level 3 decode to connect Port A's eight data lines to the data bus. This manner of operation is what makes the computer's transmitted data available to turn devices connected to the interface on or off as desired.

Our first I/O card responds any time the computer addresses 655360, 655361, 655362, or 655363. This is because for these four addresses lines A2 and greater are identical, and only A0 and A1 become binary 00, 01, 10, and 11 respectively.

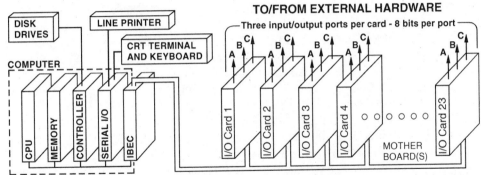

For each additional card addresses increment by four

Memory-mapped I/O above 640KB

UCIS starting address	Card 1	Card 2	Card 3	Card 4	Card 23	Individually Addressed Port on Card
640KB = 655360 A0000h DEF SEG = 40960	0 1 2 3	4 5 6 7	8 9 10 11	12 13 14 15	83 84 85 86	Port A Port B Port C Control Byte
684KB = 700416 AB000h DEF SEG = 43776	0 1 2 3	4 5 6 7	8 9 10 11	12 13 14 15	83 84 85 86	Port A Port B Port C Control Byte
884KB = 864256 D3000h DEF SEG = 54016	0 1 2 3	4 5 6 7	8 9 10 11	12 13 14 15	83 84 85 86	Port A Port B Port C Control Byte

Where address sent = (DEFSEG x 16) + (offset address noted in above table))

Port-mapped or memory-mapped I/O below 64KB

UCIS starting address	Card 1	Card 2	Card 3	Card 4	Card 23	Individually Addressed Port on Card
1KB = 1024 400h SA = 1024	SA SA + 1 SA + 2 SA + 3	SA + 4 SA + 5 SA + 6 SA + 7	SA + 8 SA + 9 SA + 10 SA + 11	SA + 12 SA + 13 SA + 14 SA + 15	SA + 83 SA + 84 SA + 85 SA + 86	Port A Port B Port C Control Byte
48KB = 49152 C000h SA = 49152	SA SA + 1 SA + 2 SA + 3	SA + 4 SA + 5 SA + 6 SA + 7	SA + 8 SA + 9 SA + 10 SA + 11	SA + 12 SA + 13 SA + 14 SA + 15	SA + 83 SA + 84 SA + 85 SA + 86	Port A Port B Port C Control Byte
60KB = 61440 F000h SA = 61440	SA SA + 1 SA + 2 SA + 3	SA + 4 SA + 5 SA + 6 SA + 7	SA + 8 SA + 9 SA + 10 SA + 11	SA + 12 SA + 13 SA + 14 SA + 15	SA + 83 SA + 84 SA + 85 SA + 86	Port A Port B Port C Control Byte

Where address sent = SA + (offset address noted in above table)

15-1 Addressing increments for I/O cards.

Thus we'd program the computer to transmit address 655360 when we want Card 1 Port A, 655361 for Card 1 Port B, 655362 for Card 1 Port C. The fourth address, 655363, is required to set a control byte within the original-design I/O cards (and the DAC card), and is explained in Chap. 16. To address the next card we simply add four to the previous addresses, and for the next add four again.

The lower portions of Fig. 15–1 illustrate this phenomenon whereby we add an incremental offset address to the UCIS starting address to move from one port to the next, then to the next, and so on until we've worked our way through every I/O card attached to the interface. Three examples are shown for memory-mapped I/O, starting with our 640K example, and three for port-mapped I/O. The offset addresses, meaning 1, 2, 3, 4, 5, and so forth, are totally independent of the I/O mode and totally independent of the starting address. This makes it easy to switch between port-mapped and memory-mapped I/O, and even easier to change starting addresses. The memory-mapped examples above 640K assume that you are operating on an IBM PC or compatible machine using BASIC with its DEF SEG statement.

The value defined in the DEF SEG = statement is always calculated as the UCIS starting address divided by 16. For example, with our starting address set at 640K, which is 655,360, the corresponding DEF SEG is set at 655360 divided by 16, or 40,960. Once you have defined the DEF SEG, all subsequent PEEK or POKE instructions calculate the actual card-port address as the DEF SEG value times 16 plus the incremental offset address defined in the PEEK or POKE and listed in Fig. 15–1.

For example, a POKE 9, 255 with the segment defined as 40960 places the decimal value 255 into address location (40960 × 16) + 9 = 655360 + 9, or 655369. Looking only at the offset 9, according to Fig. 15–1, we see this location corresponds to port B of the card 3. Changing the instruction to POKE 14, 0 we would deactivate all external devices attached to port C of card 4. An instruction $XY =$ PEEK(4) would read the external hardware devices on port A of card 2 and store their values in variable XY.

If you are operating with port-mapped I/O on an IBM PC up through a Pentium, or with memory-mapped I/O with one of the other computers covered in Chap. 7 where the maximum address space is 64K, you simply drop the use of DEF SEG and add the SA + to each incremental offset entry. This situation is illustrated in the lower portion of Fig. 15–1.

For example, using memory-mapped I/O with the CBAC plugged in slot 4 of an Apple II (see Fig. 7–8), we would use the statements:

```
SA = 50176 `Defines the UCIS starting address per Fig. 7-8
POKE SA + 9, 255 `Activates all external devices on port B card 3
POKE SA + 14, 0 `Deactivates all external devices on port C card 4
XY = PEEK(SA + 4) `Reads all external devices on port A of card 2 and stores them in variable XY
```

If using port-mapped I/O with an IBM AT/286 up through the Pentium, the above statements are simply replaced with:

```
SA = 50176 `Defines the UCIS starting address per Fig. 5–6 output
OUT SA + 9, 255 `Activates all external devices on port B card 3
OUT SA + 14, 0 `Deactivates all external devices on port C card 4
XY = INP(SA + 4) `Reads all external devices on port A of card 2 and stores them in variable XY
```

To resummarize for the memory-mapped cases above 640K, say for a Pentium with a UCIS starting address set at 844K (decimal address 864256, or 844 × 1024), we would use:

```
DEF SEG = 54016 `Defined as 864256 divided by 16
POKE 9, 255 `Activates all external devices on port B card 3
POKE 14, 0 `Deactivates all external devices on port C card 4
XY = PEEK(4) `Reads all external devices on port A of card 2 and stores them in variable XY
```

Let's put this information to work by preparing an example I/O program.

Functional I/O Example

Following the steps outlined at the beginning of this chapter, Fig. 15–2 shows an interface schematic as suggested by step 1. The example employs nine SPST switches and one four-position rotary switch connected to the single input card in slot 1. Connected to the single output card in slot 2 we have an LED display digit (defined in Fig. 13–8), a relay used to control a bank of ac-driven lamps, and an annunciator horn.

The requirement is for us to read the inputs and display via the LED display digit the lowest switch number *SW( )* that is turned on. If all switches are open, meaning that they are turned off, then a zero is displayed and no further action is taken. If one or more of the switches are turned on, then the specific additional action to be taken is dependent upon the position of the rotary switch RS. If RS is in position 0, or off, then no additional action is to be taken. If RS is in position 1, the lowest-order switch number turned on is to be displayed on the CRT. If RS is in position 2, the lamp relay is to be energized, and if in position 3 the annunciator horn is to be turned on.

Recommended I/O Tables

Table 15–1 provides a blank copy of the general-purpose I/O worksheet, which I like to use for organizing the I/O for UCIS applications. Make multiple photocopies for your use, or better yet set up this format as a special form on your personal computer. The three leftmost columns list ports, bits, and card pinouts. Because the new and the original-design I/O cards have different pinout arrangements, I've listed both under Pinouts. The left subcolumn is for the original-design cards, and the right subcolumn is for the new cards. Cross out the subcolumn representing the card you are not using. In the blank area on the right I write in the function of each I/O line, including the variable name to be used in the software as derived from step one.

The line locations are mostly arbitrary, but it's handy to group related or similar lines. When it takes more than one bit to represent a variable, the bits should be contiguous on the same port.

Table 15–2 shows how I filled out the tables for our I/O example. Assignments are quite arbitrary, except that the two lines required for the rotary switch should be contiguous bits on the same port with the low-order position connected to the lower-order bit. Likewise, on the display digit the four lines should be contiguous on the

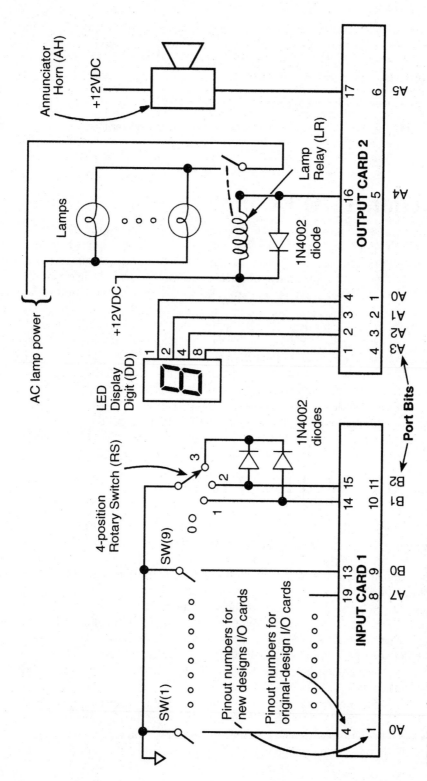

15-2 Interface schematic for the I/O example.

Table 15-1. Blank Digital I/O Worksheets for Master

UCIS I/O WORKSHEET

CARD No.____ INPUT OR OUTPUT

PORT	BIT	PIN		DESCRIPTION OF FUNCTION PERFORMED
A	0	4	1	
	1	3	2	
	2	2	3	
	3	1	4	
	4	16	5	
	5	17	6	
	6	18	7	
	7	19	8	
B	0	13	9	
	1	14	10	
	2	15	11	
	3	24	12	
	4	23	13	
	5	22	14	
	6	21	15	
	7	20	16	
C	0	9	17	
	1	10	18	
	2	11	19	
	3	12	20	
	4	8	21	
	5	7	22	
	6	6	23	
	7	5	24	

NOTE: Cross out the left column under PIN when using the new-design I/O cards, and cross out the right column under PIN when using the original-design I/O cards.

same port and the 1, 2, 4, and 8 inputs for the display digit arranged in ascending bit order on the output card.

To help when interface wiring is hooked up in step 4, I have gone back to take the line numbers generated from the I/O tables (both for the new and the original-design I/O cards) and placed them on the schematic derived in step 1.

Programming Format for a Real-Time Loop

The top-level format for most every UCIS application program will follow the flow diagram illustrated in Fig. 15–3. Once initialized, the computer loops through its set of instructions over and over again for as long as you want to have the computer interacting with your external hardware. Such an operation is called a tight-infinite or real-time loop.

The program initializes the UCIS by defining variables and constants and sending command bytes to the 8255s. Once into the real-time loop, the program first reads the computer keyboard inputs (if any are used) and then the input cards to receive the status of the external hardware.

From these inputs the program determines what, if any, action needs to be taken and, using the inputs, it calculates the outputs. Once the outputs are calculated, the program updates the computer monitor display (if it is used) and then writes the outputs to the external hardware via the output cards. Then it branches back to reread the inputs and begin the loop over again.

The loop typically repeats itself very rapidly, and the time it takes per loop is called the *loop time*. Its reciprocal is called the *iteration rate,* and represents the number of loops per second. For UCIS applications performing data acquisition, the term *sample rate* is frequently used. The loop time varies greatly from application to application and can range from a few microseconds up to several seconds. Numbers are highly dependent on the software language used, how it's programmed, the amount of calculations required, the number of I/O lines, and so on. As examples, I have a QuickBASIC data acquisition program on my Pentium that loops every 5 μs, in other words taking 200,000 samples per second. Larger systems with more I/O cards, a slower computer, and more calculations would more typically be in the 100-μs to, say, the 500-ms range.

Independent of the loop time, the process is the same. As inputs change, new outputs are calculated from the inputs, thereby generating the desired monitoring/control interaction between the computer and the external hardware. In the next few sections I'll cover a few fundamentals essential to developing application programs, and then we'll be all set to program our little I/O handling example.

Making I/O Addressing Easy

Table 15–3 is a handy reference for setting your I/O card DIP switches and establishing the corresponding incremental software address to be used in your

Table 15-2. Completed I/O Worksheets

UCIS I/O WORKSHEET

CARD No. 2 — ~~INPUT~~ OR OUTPUT

PORT	BIT	PIN		DESCRIPTION OF FUNCTION PERFORMED
A	0	4	1	1 ⎫
	1	3	2	2 ⎬ LED DISPLAY DIGIT
	2	2	3	4 ⎬ DD
	3	1	4	8 ⎭
	4	16	5	LAMP RELAY LR
	5	17	6	ANNUNCIATOR HORN AH
	6	18	7	
	7	19	8	
B	0	13	9	
	1	14	10	
	2	15	11	
	3	24	12	
	4	23	13	
	5	22	14	
	6	21	15	
	7	20	16	
C	0	9	17	
	1	10	18	
	2	11	19	
	3	12	20	
	4	8	21	
	5	7	22	
	6	6	23	
	7	5	24	

UCIS I/O WORKSHEET

CARD No. 1 — INPUT ~~OR OUTPUT~~

PORT	BIT	PIN		DESCRIPTION OF FUNCTION PERFORMED
A	0	4	1	INPUT SWITCH SW(1)
	1	3	2	INPUT SWITCH SW(2)
	2	2	3	INPUT SWITCH SW(3)
	3	1	4	INPUT SWITCH SW(4)
	4	16	5	INPUT SWITCH SW(5)
	5	17	6	INPUT SWITCH SW(6)
	6	18	7	INPUT SWITCH SW(7)
	7	19	8	INPUT SWITCH SW(8)
B	0	13	9	INPUT SWITCH SW(9)
	1	14	10	Pos 1 ⎫ ROTARY SWITCH RS
	2	15	11	Pos 2 ⎭
	3	24	12	
	4	23	13	
	5	22	14	
	6	21	15	
	7	20	16	
C	0	9	17	
	1	10	18	
	2	11	19	
	3	12	20	
	4	8	21	
	5	7	22	
	6	6	23	
	7	5	24	

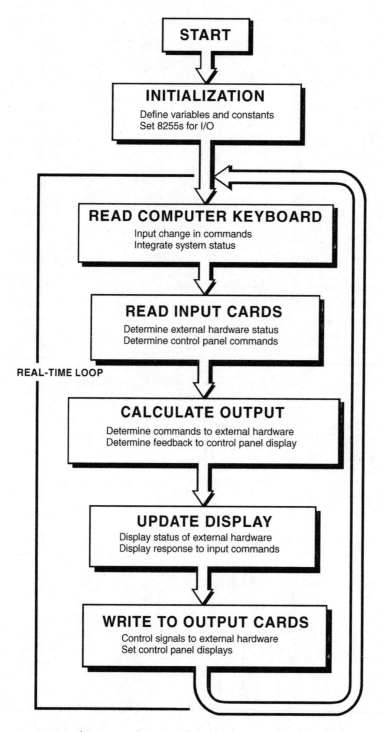

15-3 Software application flow diagram.

computer program. For example, your I/O card placed in the first slot (second slot if you are using a USIC) will have each of its DIP switch segments set to OFF with corresponding port address definitions given by: port A = *SA*, port B = *SA* + 1, port C = *SA* + 2, and the 8255 control byte = *SA* + 3. Corresponding DIP switch settings and software addresses are shown for progressively higher card slots up to the 64th card, which would have all DIP switch segments set to ON and addresses *SA* + 252, *SA* + 253, *SA* + 254, and *SA* + 255. When using DEF SEG you use the same table but simply drop the *SA* + and only use the remaining numerals 1, 2, 3, 4, 5, and so forth to increment from port to port.

Reading and Unpacking Input Bytes

We always talk to the external hardware one byte at a time using PEEK and POKE or INP and OUT instructions. Each I/O bit (each separate line) can be connected to a separate external hardware device as, for example, with input switches SW1–SW9 in Fig. 15–2; or groups of bits can go to single devices, as with the LED display digit. For input the software must unpack or separate the eight bits received in each input byte to assign their values to appropriate variables. For output the software packs, or combines, the values of different variables into bytes for transmission to the external hardware.

Special constants will be set up as part of the initialization section of each application program to help us unpack and pack the input and output bytes received and transmitted over the UCIS. This lets us make the most effective use of our I/O hardware by packing lots of information into each byte.

The pattern for each unpacking operation is the same, starting with a PEEK operation (or INP if using port I/O) to read in the input byte *IB*. The variable to be defined is calculated by dividing the input byte by a bit-position constant *B0* through *B7*, depending on the position of the variable's least significant bit (LSB). This effectively shifts the byte to the right by the required number of bits so that the LSB is in the *B0* position.

That result is ANDed with a second constant, *W1* through *W7*, depending on the variable's width in bits. The AND operation *zeros out* or *masks* all but the bits defining the variable in question.

I'll give you the values for the bit position constants *B0* through *B7* and width constants *W1* through *W7* shortly. These constants remain identical for every application program.

Unpacking is easier to do than to explain. For example, *SW(1)* in our example is in bit position zero and has width of one bit. Thus, the value of *SW(1)* is derived using the BASIC statement SW(1) = IB/B0 AND W1. Likewise, with *SW(2)* wired to bit position one and also one bit wide, the statement is SW(2) = IB/B1 AND W1. Once all the variables from one port are defined, you move on to the next with another PEEK (or INP) to read the next input byte.

The rotary switch, used in our example, has a width of two bits with its LSB at bit position one; thus the value of the *RS* variable is given by RS = IB/B1 AND W2. That's all there is to reading external hardware inputs: a PEEK (or INP) to bring in

Table 15-3. I/O Card DIP Switch Settings and Addresses

CARD SLOT ADDRESS	DIP SWITCH SETTINGS	SOFTWARE ADDRESS			COMMAND BYTE
		PORT			
		A	B	C	
1		SA	SA+1	SA+2	SA+3
2		SA+4	SA+5	SA+6	SA+7
3		SA+8	SA+9	SA+10	SA+11
4		SA+12	SA+13	SA+14	SA+15
5		SA+16	SA+17	SA+18	SA+19
6		SA+20	SA+21	SA+22	SA+23
7		SA+24	SA+25	SA+26	SA+27
8		SA+28	SA+29	SA+30	SA+31
9		SA+32	SA+33	SA+34	SA+35
10		SA+36	SA+37	SA+38	SA+39
11		SA+40	SA+41	SA+42	SA+43
12		SA+44	SA+45	SA+46	SA+47
13		SA+48	SA+49	SA+50	SA+51
14		SA+52	SA+53	SA+54	SA+55
⋮	⋮	⋮	⋮	⋮	⋮
64		SA+252	SA+253	SA+254	SA+255

the port data byte, followed by repeated executions of the same basic unpacking instruction format to separate the input bits defining the respective variables. This unpacking process will become pretty automatic once you have a couple applications under your belt.

In Chap. 18 I'll show you various types of shortcuts to look for whereby you can skip some of the input operations. Learning such tricks of the trade is important to speed up your programs.

Packing and Writing Output Bytes

The last part of each application program is a section to send the calculated variables to the external hardware as output. As for inputs, we'll use repetitions of a standard procedure to pack variables into bytes and send output to the external hardware a byte at a time.

For handling each port, the first step is to initialize the output byte *OB* to zero. This is followed with a bit-packing instruction, used once for each variable to go into the byte. The bit-packing instruction format is *OB* = the name of the variable to be packed multiplied by the position constant of the variable's LSB, then ORed with the previous value of *OB*. The multiply operation effectively shifts the variable to the left by the required number of bits so that the LSB is in the desired position in the output byte, and the OR operation folds it into the output byte. This instruction format is repeated once for each variable to go into the byte.

Following that format, the instruction for the LED display would be OB = DD*B0 OR OB. The corresponding instruction for the lamp relay would be OB = LR*B4 OR OB.

Once all the bits are packed, a POKE (or OUT) instruction sends the output byte to the appropriate port of an output card. The remainder of the output lines follow the same procedure. Only the variable to be output is changed along with the bit position and card/port address, all following the I/O tables.

Sample I/O Program Using the IBEC or UBEC

Table 15–4 shows a listing for the example I/O program using MicroSoft BASIC. This program plus equivalent programs in Pascal and C are included on the disk provided with this second edition. I've avoided exotic codes, so most of the statements should work on any computer with few changes. The program assumes memory-mapped I/O on an IBM PC up through a Pentium with the UCIS starting address set at 640K. If you are using one of the other computers in Chap. 7, specifically one with an address range of 64K, then drop the DEF SEG in line 120, insert your SA = statement, and insert the SA + into each of your PEEKs and POKEs. If you are using port I/O you need to change PEEK and POKE to INP and OUT. Later I'll explain some special coding for the Apple II series and show the program modified to operate with USIC instead of IBEC and UBEC.

Because this is our first application-type UCIS program, I'll walk you through it. For easy understanding, the program is grouped in blocks, with a REM line at the

Table 15-4. Example I/O Program Using an IBEC or UBEC

```
10  PRINT "EXAMPLE I/O PROGRAM USING IBEC OR UBEC"
20  REM**Define variable types and array sizes**
30  DEFINT A-Z
40  DIM SW(9)
50  REM**Define constants for packing and unpacking bits**
60  B0=1: B1=2: B2=4: B3=8: B4=16: B5=32: B6=64: B7=128
70  W1=1: W2=3: W3=7: W4=15: W5=31: W6=63: W7=127
80  REM**Define I/O interface constants**
90  CI=155          '8255 Control byte for input
100 CO=128          '8255 Control byte for output
110 REM**define UCIS starting address
120 DEF SEG = 40960  'Assumes Memory-Mapped I/O at 640KB
130 REM**Transmit control bytes to set up I/O ports**
140 POKE 3, CI          'Card 1 input
150 POKE 7, CO          'Card 2 output
160 REM
170 REM**Begin real-time loop**
180 REM**Read inputs and separate out bit areas**
190 IB=PEEK(0)          'Card 1 port A
200 SW(1)=IB AND W1
210 SW(2)=IB/B1 AND W1
220 SW(3)=IB/B2 AND W1
230 SW(4)=IB/B3 AND W1
240 SW(5)=IB/B4 AND W1
250 SW(6)=IB/B5 AND W1
260 SW(7)=IB/B6 AND W1
270 SW(8)=IB/B7 AND W1
280 IB=PEEK(1)          'Card 1 port B
290 SW(9)=IB AND W1
300 RS=IB/B1 AND W2
310 REM**Initialize outputs to off
320 DD= 0       'Set Display Digit off
330 LR= 0       'Set Lamp Relay off
340 AH= 0       'Set Horn Annunciator off
350 REM**Determine lowest switch turned on**
360 FOR I = 1 TO 9
370 IF SW(I) = 1 THEN DD=I: GOTO 410
380 NEXT I
390 GOTO 460
400 REM**Found switch on - take appropriate action**
410 IF RS=0 THEN GOTO 460   'Ignore
420 IF RS=1 THEN PRINT "SWITCH",DD," IS TURNED ON": GOTO 460
430 IF RS=2 THEN LR=1: GOTO 460
440 AH = 1
450 REM Write outputs to external hardware
460 OB = 0                'Card 2 port A
470 OB = DD*B0 OR OB
480 OB = LR*B4 OR OB
490 OB = AH*B5 OR OB
500 POKE 4, OB
510 REM**Return to begining of loop**
520 GOTO 190
```

beginning of each block to describe the code lines that follow. Anything that follows an apostrophe (') in a code line is also explanatory and will not be executed.

Line 10 prints the program title. Line 30 is a MicroSoft BASIC statement defining program variables starting with letters A through Z as integers, meaning they can have only whole-number values: 0, 1, 2, 3, and so forth. UCIS programs typically don't need real numbers (numbers with decimals), and integer variables use less memory, thereby speeding up your program's execution. If it doesn't apply to your computer or application, line 30 can be omitted.

Line 40 defines the size of the switch array variable *SW( )* to have nine elements, while lines 60 and 70 set up the special constants to unpack and pack the input and output bytes received and transmitted over the UCIS. These latter two statements will be an integral part of nearly every interface application program.

Lines 90 and 100 define the input and output control bytes for the original-design I/O cards. Line 120 defines the UCIS starting address via the DEF SEG statement. Lines 140 and 150 transmit the control bytes to set the 8255 on the first card for input and on the second card for output. These lines can be deleted, along with lines 90 and 100, if none of your I/O cards make use of the 8255 chip.

The real-time loop starts at line 190 with the reading of the input byte from Port A of card 1 and stores it at location *IB*. After each PEEK, we unpack the bits once for each variable read from that port.

As defined in Table 15–2, card 1, Port A includes eight variables: *SW(1)* through *SW(8)*. All take one bit each; therefore, the subsequent unpacking statements all employ the identical W1 width constant. The bit position constant selected is different for each unpacking statement to properly reflect the bit line (or wire) position for each input switch position.

Once the eight switch settings are unpacked from port A, you move to line 280 with another PEEK to read port B for the next input byte. Line 290 unpacks bit 0 to define the *SW(9)* variable and line 300 determines the variable *RS*, which represents the rotary switch position.

That's all there is to reading external hardware inputs. A PEEK brings in the port data, followed by repeated executions of the same basic unpacking instruction format to separate the input bits defining the respective variables.

Lines 320 through 340 initialize the output variables for the display digit, lamp relay, and annunciator horn to off. Lines 360 through 380 form a loop incrementing the index variable *I* from 1 up through 9 looking for the first switch that is turned on, meaning set to 1. When found, the display digit DD is set equal to *I*, the switch number, and the program branches to 410 to determine what other action needs to be taken. If no switch is found to be on, the program drops through the loop to line 390 and branches to line 460 to pack and transmit the outputs with *DD, LR,* and *AH* all remaining at off.

Assuming a switch is found to be on, line 370 causes a branch to line 410 whereby if the rotary switch is set to 0 position, meaning that *RS* = 0, then a branch is made to line 460 to ignore taking any further action. In like manner, line 420 checks if the rotary switch is in the 1 position, and if so then prints the required message on the monitor and branches to pack and transmit outputs. Line 430 does a similar check to turn on the lamp relay variable if the switch is in position 2. If the switch is not in position 0, 1, or 2, it must be in position 3; in that case the program reaches line 440, where the annunciator horn variable is turned on.

Lines 460 through 490 perform the packing operation. Note the jump in the use of the bit position constant from *B0* in line 470 to *B4* in line 480. This is because DD requires four lines spanning from *B0* through *B3,* and is also a direct result of the lamp relay LR being wired to the fourth bit position, all per the output table.

That's all there is to transmit external hardware outputs. Setting *OB* = 0 initializes the output byte for each port, followed by repeated executions of the same basic packing instruction format to combine the bits defining the respective variables, and then a POKE (or OUT) instruction to actually transmit the output.

Line 520 completes the real-time loop with a GOTO branch back to line 190, to repeat the loop.

Backslash vs Forward Slash

I've used the backslash in Table 15–4 to achieve the integer divide embedded in the unpacking instructions, lines 219 through 270. This is a requirement when using QuickBASIC. However, many regular interpretative BASIC languages let you use the more typical forward slash for achieving the integer divide.

If you have trouble with unpacking and packing of I/O bytes, I've included three different methods for bit unpacking and packing as test programs on the disk enclosed with this second edition. Each of the three programs simply simulates looping to read an input byte (with values running from 0 to 9) and unpacks the input byte into its separate bit positions 4, 3, 2, and 1 and then packs the bit positions back up to form an output byte. For each iteration, the programs print out the input byte, the unpacked bit positions, and the repacked byte. If the unpacking and packing operations are correctly handled, the output byte exactly matches the input byte.

Once you have run all three programs on your machine with your version of BASIC, you are free to use whichever program techniques produce the correct results.

Apple I/O

I've used AND and OR logic in packing and unpacking I/O bytes for simplicity and speed. I've tested this software on every computer covered in Chap. 7 plus on the Macintosh, and it works for all except the Apple II series of machines when used with either Applesoft BASIC or Apple Integer BASIC. Both these languages unfortunately use simplified functions rather than true boolean logic for the AND and OR operations.

The simplified logic gives the wrong answers from AND and OR operations with the Apple BASICs. You can test your machine by entering two BASIC commands:

```
A = 5 AND 2
PRINT A
```

If you get $A = 0$, the packing and unpacking operations explained above will work. If you get $A = 1$ or TRUE, replace lines 190 through 300 in Table 15–4 with the code as listed in Table 15–5, and change every OR to + in lines 470 through 490. This version of the packing logic takes a few more lines of code, but it does let you use the Apple II BASICs.

Table 15-5. Unpacking Code for Apple BASICS

```
190 IB=PEEK(SA)        'Card 1 port A
195 FOR I = 1 TO 9: SW(I)=0: NEXT I
200 IF IB>W7 THEN SW(8)=1: IB=IB-B7
210 IF IB>W6 THEN SW(7)=1: IB=IB-B6
220 IF IB>W5 THEN SW(6)=1: IB=IB-B5
230 IF IB>W4 THEN SW(5)=1: IB=IB-B4
240 IF IB>W3 THEN SW(4)=1: IB=IB-B3
250 IF IB>W2 THEN SW(3)=1: IB=IB-B2
260 IF IB>W1 THEN SW(2)=1: IB=IB-B1
270 SW(1)=IB
280 IB=PEEK(SA+1)      'Card 1 port A
285 RS=0: DD=0
290 IF IB>W7 THEN IB=IB-B7
291 IF IB>W6 THEN IB=IB-B6
292 IF IB>W5 THEN IB=IB-B5
293 IF IB>W4 THEN IB=IB-B4
294 IF IB>W3 THEN IB=IB-B3
295 IF IB>W2 THEN RS=RS+2: IB=IB-B2
296 IF IB>W1 THEN RS=RS+1: IB=IB-B1
300 SW(9)=IB
```

Sample I/O Program Using the USIC

Table 15–6 shows the program in Table 15–4 modified to operate with USIC following the guidelines explained in Chap. 10. The body of the program stays essentially the same, while the I/O portions change per the format presented in Fig. 10–8.

That completes our technology baseline by establishing all that's needed to build up a universal computer interface system and to get started on its application to meet your particular interfacing desires. In Chaps. 17 and 18 I'll provide additional assistance by presenting an example application of the UCIS, attaching it to model railroad hardware.

This detailed application example will serve to demonstrate the dynamic capabilities of the UCIS with its attachment to many types of hardware devices and the generation of real-time application-dependent software to help you better suit your particular UCIS application requirement.

Table 15-6. Example I/O Program Using a USIC

```
10  PRINT "EXAMPLE I/O PROGRAM USING USIC"
20  REM**Define variable types and array sizes**
30  DEFINT A-Z
40  DIM SW(9), CT(4), OB(20), IB(20), TB(40)
50 GOTO 3000    'Branch to user program


3000 REM************************************************
3010 REM**BEGIN USER I/O DEMONSTRATION PROGRAM**
3020 REM************************************************
3030 REM**Define general constants**
3040 PA = 760    'Assumes IBM compatible COM2 port
3070 UA = 0      'USIC ADDRESS
3080 DL = 4000 'USIC delay - adjust toward zero once operating
3090 NS = 1      'Number of card sets
3100 NI = 3      'Number of input card ports
3110 NO = 3      'Number of output card ports
3120 CT(1) = 9   'Card type definition
3130 GOSUB 1810        'Invoke initialization subroutine
3140 REM**Define constants for packing and unpacking bits**
3150 B0=1: B1=2: B2=4: B3=8: B4=16: B5=32: B6=64: B7=128
3160 W1=1: W2=3: W3=7: W4=15: W5=31: W6=63: W7=127
3170 REM
3180 REM**Begin real-time loop**
3190 REM**Read inputs
3200 GOSUB 1010        'Invoke read subroutine
3210 REM**Separate out bit areas**
3220 IB=IB(1)           'Card 1 port A
3225 SW(1)=IB AND W1
3230 SW(2)=IB/B1 AND W1
3240 SW(3)=IB/B2 AND W1
3250 SW(4)=IB/B3 AND W1
3260 SW(5)=IB/B4 AND W1
3270 SW(6)=IB/B5 AND W1
3280 SW(7)=IB/B6 AND W1
3290 SW(8)=IB/B7 AND W1
3300 IB=IB(2)           'Card 1 port B
3310 SW(9)=IB AND W1
3320 RS=IB/B1 AND W2
3330 REM**Initialize outputs to off
3340 DD= 0      'Set Display Digit off
3350 LR= 0      'Set Lamp Relay off
3360 AH= 0      'Set Horn Annunciator off
3370 REM**Determine lowest switch turned on**
3380 FOR I = 1 TO 9
3390 IF SW(I) = 1 THEN DD=I: GOTO 3430
```

```
3400 NEXT I
3410 GOTO 3480
3420 REM**Found switch on - take appropriate action**
3430 IF RS=0 THEN GOTO 3480    'Ignore
3440 IF RS=1 THEN PRINT "SWITCH",DD," IS TURNED ON": GOTO 3480
3450 IF RS=2 THEN LR=1: GOTO 3480
3460 AH = 1
3470 REM**Pack outputs**
3480 OB = 0               'Card 2 port A
3490 OB = DD*B0 OR OB
3500 OB = LR*B4 OR OB
3510 OB = AH*B5 OR OB
3520 OB(1) = OB
3530 REM**Transmit outputs**
3540 GOSUB 810           'Invoke transmit subroutine
3550 REM**Return to beginning of loop**
3560 GOTO 3200
```

16
CHAPTER

Original I/O Circuits and Test Card

In this chapter we'll cover how to build the original-design input and output cards. These cards still have a lot of merit. Being single-sided, with wide and well separated traces, these cards are especially suited for readers not versed in circuit card assembly and for those desiring to purchase the lower-cost cards, or to fabricate their own boards from scratch. The artwork for etching your own boards is included in Appendix A. For others, ready-to-assemble cards are available from JLC Enterprises and completely assembled and tested cards are available from ECO Works, EASEE Interfaces, and AIA.

Both the original-design and the new cards are designed to plug into the same motherboard and function with the IBEC, USIC, or UBEC. Operationally, the only difference is the ordering of the I/O lines. To minimize the number of jumpers required with the single-sided boards, the card outputs are arranged in an unusual pattern. This doesn't matter once you have the cards all connected to your external hardware, but tends to be a bit frustrating during initial system setup. Being double-sided, the new boards make it easier to group the lines for each port and assign pinouts in bit-position order.

It is totally OK to mix old and new cards in the same motherboard; but if you interchange a new and an original-design card within the same slot, then you must either change the cable wiring, or change the software to reflect the different ordering of the I/O connections. If you already have a number of the original-design cards, or if you need them for the special features they provide, you can alter them with added cuts and jumpers so that their I/O connections are identical to the new cards. This gives you 100 percent compatibility.

This chapter assumes that you have already read Chap. 8 on digital I/O circuits and the test card, so we'll keep this chapter short by avoiding duplication. For example, Chap. 8's functional description of address decoding is applicable to both designs. If you desire refreshing in this area, you might want to go back and review parts of Chap. 8.

The I/O Chip

The heart of the original I/O cards is a very neat IC, the 8255. It's a large, 40-pin DIP that goes by the complex-sounding name of Programmable Peripheral Interface. This chip is a general-purpose, programmable I/O device designed as an external interface for microprocessors. It gives the lowest cost per bit of any parallel I/O circuit on the market, and it helps to make the UCIS affordable and easily expandable. In the new output card design we need six separate ICs to accomplish the functions performed by the 8255! It's easy to see why I really like the 8255.

I didn't include the 8255 in the new designs because Intel, one of the chip's many manufacturers, is phasing it out of production. Even so, I'm sure versions of the 8255 (including the CMOS version 82C55) will be available on the market for many years to come. As long as you can find the 8255s, the original I/O cards provide a very attractive option for building your UCIS.

The 8255 can operate in many different modes, but its most basic, mode 0, meets the needs of the UCIS. Figure 16-1 summarizes our use of the 8255. Imagine that it's a high-speed, eight-pole, four-position switch, switching the eight data lines from the IBEC (or UBEC or USIC) to port A, B, or C, or to an open-circuit state with no port connected.

The four control-line inputs, RD*, WR*, RESET, and CS*, and the two address lines A0 and A1, determine which port is selected and the direction of data flow. The port switch is open-circuited any time that RESET is high or CS* (chip select active low) is high. Which port is connected at a given instant depends on the address lines A0 and A1. At the same time, the status of RD* and WR* determine if the data transfer is a read or write operation.

In this most basic mode we need to initialize the 8255s placed in output cards as output devices, and those placed in input cards as input devices. This is accomplished by software writing to the I/O card, with both the A0 and A1 address lines high, the control byte shown in the bottom right of Fig. 16–1; namely 128 for all ports output, and 155 for all ports input. We already have been doing this for each of our parallel IBEC and UBEC software applications, and for serial applications the USIC automatically takes care of this step.

Before we start building up one of the output cards, let's take a brief look at its schematic.

Output Card Schematic

Figure 16–2 is the schematic for the original output card. The input card, which I'll cover later, is nearly identical. As we've seen, the 8255 chip does most of the I/O work for us, but a few other parts are needed too.

From the motherboard, data lines D0–D7, address lines A0–A1, the grounded RESET line, and the two IBEC (or UBEC or USIC) control lines W* and R* are connected directly to the 8255. Address lines A2 through A7 are decoded by two 74LS136 ICs, U5 and U6, just as the higher address lines were decoded on the UBEC. However, for the I/O cards only six bits are decoded, so a six-segment DIP switch sets the card address.

IN FUNCTION, THE 8255 IS LIKE A HIGH-
SPEED, 8-POLE, 4-POSITION SWITCH

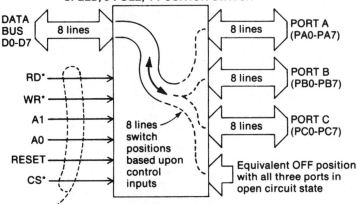

These control inputs determine which port is
selected, if any, and the direction of data flow

Switch is in OFF (i. e., all 3 ports = open circuit state) if RESET
= 1 or chip select (CS*) = 1. Data transfer can take place any
time that RESET = 0 and CS* = 0, according to the table below:

	A1	A0	RD*	WR*	DATA TRANSFERRED FROM	TO
INPUT	0	0	0	1	Port A	Data bus
	0	1	0	1	Port B	Data bus
	1	0	0	1	Port C	Data bus
OUTPUT	0	0	1	0	Data bus	Port A
	0	1	1	0	Data bus	Port B
	1	0	1	0	Data bus	Port C
	1	1	1	0	Data bus	Control byte

Control bytes	Binary	Hex	Decimal
All ports output	10000000	80h	128
All ports input	10011011	9Bh	155

8255 PIN CONFIGURATION TOP VIEW

PA3	1	40	PA4
PA2	2	39	PA5
PA1	3	38	PA6
PA0	4	37	PA7
RD*	5	36	WR*
CS*	6	35	RESET
GND	7	34	D0
A1	8	33	D1
A0	9	32	D2
PC7	10	31	D3
PC6	11	30	D4
PC5	12	29	D5
PC4	13	28	D6
PC0	14	27	D7
PC1	15	26	Vcc = +5 VDC
PC2	16	25	PB7
PC3	17	24	PB6
PB0	18	23	PB5
PB1	19	22	PB4
PB2	20	21	PB3

8255 (or 8255A)

PIN NAMES

D0-D7	Data bus (bidirectional)
RESET	Reset input
CS*	Chip select (active low)
RD*	Read input (active low)
WR*	Write input (active low)
A0-A1	Port address
PA0-PA7	Port A (bit)
PB0-PB7	Port B (bit)
PC0-PC7	Port C (bit)
Vcc	+5VDC
GND	0V

16-1 Basic features of the 8255 provide for easy I/O.

The address decode circuit is designed so that pin 6 of U7 is high except when the CPU is addressing this specific card; that is, when AH is high and lines A2–A7 correspond to SW1. When this is the case, pin 6 of U7 goes low, sending a low signal (logic 0) to the CS* input of the 8255, pin 6 of U8, and enabling data transfer across the I/O chip. For a write operation, the addressed port—A, B, or C—is latched by the 8255. The outputs stay constant until the CPU sends new data to the 8255.

Each port output line from the 8255 is connected through a 7407 hex noninverting buffer (U1, U2, U3, or U4) to a 2N3904 NPN output transistor. The open-collector transistor provides drive capability for most loads, including LEDs, lamps, and relays, as well as other logic circuits. You can make use of the IOMB's optional ground switch to eliminate random states on power-up, just like with the new cards.

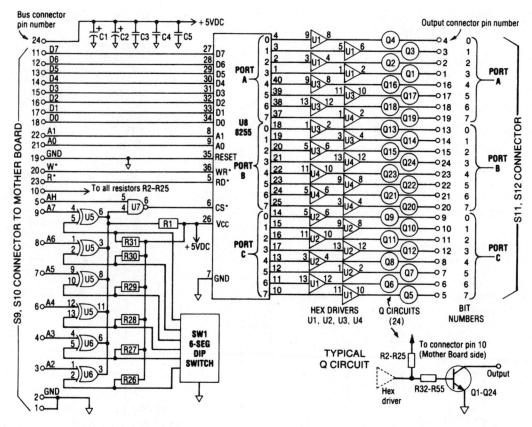

16-2 Original-design output card schematic.

Building the Output Card

Figure 16–3 shows a photo of the output card plugged into the I/O motherboard. The cable at the upper right connects the external hardware to the output card. Figure 16–4 is the parts layout and Table 16–1 gives the parts list.

Note that U8, the 8255, is static-sensitive. You won't be installing this part until later, but you ought to know that before handling it you should ground your hands by touching them to a large metal object. This helps discharge any static charge on your body that could damage the chip. You'll need one output card for every 24 lines of output desired from your interface. Here's how to assemble one:

☐ *Card test.*
☐ *J1–J23.*
☐ *R1–R55.*

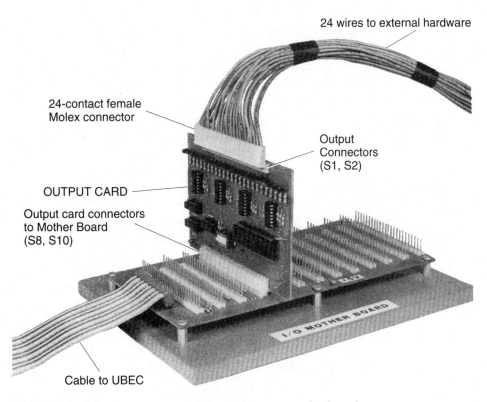

24 wires to external hardware

24-contact female
Molex connector

Output
Connectors
(S1, S2)

OUTPUT CARD

Output card connectors
to Mother Board
(S8, S10)

Cable to UBEC

16-3 Original-design output card plugged into a motherboard.

- ☐ *S1–S8[+]*. Note that the ICs face different directions, so be particularly careful that you align the sockets with the correct pin 1 orientation before soldering them in place.
- ☐ *SW1[+]*. Use your VOM to make sure you install this DIP switch so its contacts are closed when thrown toward S3. This direction will be on, and away from S3 will be off.
- ☐ *S9, S10*.
- ☐ *S11, S12*.
- ☐ *Q1–Q24[+]*. Slightly bend the leads of each transistor to fit its holes, and orient each with its flat side facing the R32–R55 resistors, with its center (base) lead in the hole nearest the resistors. You may have covered some of these holes with solder when installing the resistors. Either use desoldering braid (Radio Shack 64-2090) to remove the excess solder, or redrill the hole from the component side with a no. 72 bit (0.025 in). For a neat installation with Q1–Q24 all lined up in a row, I put a spacer of ⅛-in² stripwood under the curved back sides of the transistors, then press each one down until it just rests on the wood. Once all 24 transistors are soldered in place, I remove the wood and trim the leads.
- ☐ *C1, C2[+]*. Install these capacitors with their plus (+) leads in the plus holes as indicated in Fig. 16–4. Reversed polarity will damage the capacitors.

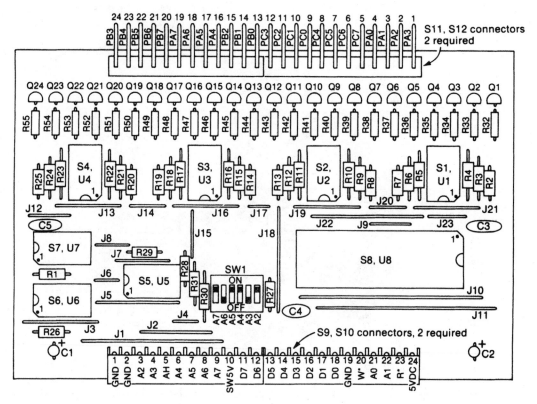

16-4 Original-design output card parts layout.

☐ *C3–C5*. Insert, solder, and trim leads.

☐ *U1–U8*. Note that the ICs face different directions, so you need to be particularly careful to make sure that you put them in the right sockets with correct pin 1 orientation. Remember, U8 is static-sensitive, so follow the special handling procedures noted above.

☐ *Cleanup and inspection*.

☐ *Card IC power test*. Turn off the +5-Vdc supply and plug the output card into any motherboard slot. Turn the power back on and follow Table 16–2 to see that power and ground reach each IC. If the voltage is way low, your output card probably has a short circuit. Look for and correct reversed polarity-sensitive capacitors, ICs backward, and/or solder bridges. If necessary, remove ICs one at a time, turning power off and on each time, to help isolate the problem.

☐ *DIP switch test*. Connect your (–) meter lead to ground and move the (+) lead to touch the IC pins listed in Table 16–3. At each position move the corresponding switch segment to off and you should read +5 Vdc; at

Table 16-1. Original-Design Output Card Parts List

Qnty.	Symbol	Description
23	J1-J23	Jumpers, make from no. 24 uninsulated bus wire (Belden no. 8022)
25	R1-R25	1.0KΩ resistors [brown-black-red]
6	R26-R31	2.2KΩ resistors [red-red-red]
24	R32-R55	120Ω resistors [brown-red-brown]
7	S1-S7	14-pin DIP sockets (Digi-Key A9314)
1	S8	40-pin DIP socket (Digi-Key A9340)
1	SW1	6-segment DIP switch (Digi-Key CT2066)
2	S9,S10	12-pin Waldom side entry connectors (Digi-Key WM3309)
2	S11,S12	12-pin Waldom right angle headers (Digi-Key WM4510)
11	C1,C2	2.2µF, 35V tantalum capacitors (Jameco 33734)
3	C3-C5	100000pF, 50v ceramic disk capacitors (Digi-Key P4164)
24	Q1-Q24	2N3904 NPN small signal transistors (Jameco 38359)
4	U1-U4	7407 hex buffer/drivers w/open collector output (Jameco 49120)
2	U5,U6	74LS136 QUAD EXCLUSIVE-OR gates with open collector (Jameco 46586) or (JDR 74LS136)
1	U7	74LS00 quad 2-input NAND gate (Jameco 46252)
1	U8	8255 or 8255A-5 programmable peripheral interface (Jameco 52724 or 52732) or (JDR 8255-5)

Mating connector for cable

2	—	12-pin Waldom terminal housing (Digi-Key WM2110)
24	—	Crimp terminals (Digi-Key WM2301 for wire sizes 18-24 or WM2301 for wire sizes 22-26)

Author's recommendations for suppliers given in parentheses above with part numbers where applicable. Equivalent parts may be substituted. Resistors are ¼W, 5 percent and color codes are given in brackets.

Table 16-2. Original-Design I/O Card IC Power Tests

✓	IC	+ METER LEAD ON PIN NO.	- METER LEAD ON PIN NO.
	U1	14	7
	U2	14	7
	U3	14	7
	U4	14	7
	U5	14	7
	U6	14	7
	U7	14	7
	U8	26	7

Each line should read +5Vdc

ON, you should measure 0 Vdc. If all readings are reversed you have the switch in backward. If only some segments are incorrect, check around for poor soldering.

Table 16-3. Original-Design I/O Card DIP Switch Tests

✓	IC	PIN NO.	DIP SWITCH SEGMENT
	U6	2	A2
	U6	5	A3
	U5	13	A4
	U5	10	A5
	U5	2	A6
	U5	5	A7

With switch segment set to "ON" IC pin should read 0Vdc.
With switch segment set to "OFF" IC pin should read +5Vdc.

That completes the output card. I'll discuss the input card's schematic next, but if you wish you may skip ahead to Building the Input Card.

Input Card Schematic

Figure 16–5 is the input card schematic. It's like the output card except for the area at the right, between the A, B, and C ports of the 8255 and the interface input connections. The input lines could be wired from the external hardware socket directly into the 8255, but I've found it best to include a buffer in each input line via U1–U4.

I use these four 74LS04 hex inverters not so much for logic inversion (either kind of input logic could be handled with software) but to protect the 8255. If you couple the 8255 directly to the external hardware and accidentally connect an input line to voltages outside the 0 to +5-Vdc range, you ruin the 8255 and lose about $3. It's better to ruin a 25-cent 74LS04. Since installing my first input card this precaution has saved me numerous 8255s. No latching is provided nor required when the 8255 is used as an input device. Once the computer has performed the READ operation, the data read is stored in memory.

Building the Input Card

You'll need one input card for every 24 lines of desired input from your external hardware. Figure 16–6 shows an input card plugged into the I/O motherboard. Figure 16–7 gives its parts layout and Table 16–4 the parts list. Here's how to build one:

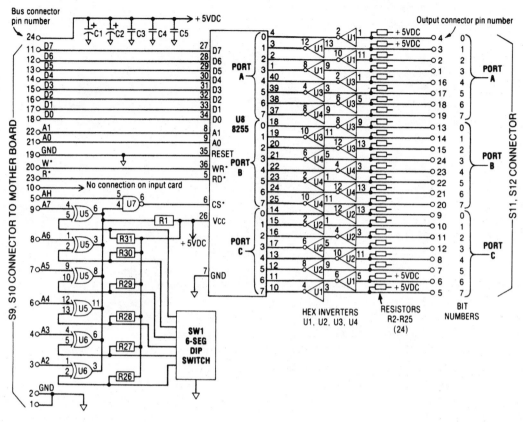

16-5 Original-design input card schematic.

☐ *Card test.*
☐ *J1–J17.*
☐ *R1–R31.*
☐ *S1–S8[+].* Note that the ICs U1–U4 face the opposite direction from those on the output card, so be particularly careful that you align the sockets with the correct pin 1 orientation before soldering them in place.
☐ *SW1[+].* Use your VOM to make sure you install this DIP switch so its contacts are closed when thrown toward S3. This direction will be on, and away from S3 will be off.
☐ *S9, S10.*
☐ *S11, S12.*
☐ *C1, C2[+].*
☐ *C3–C5.*

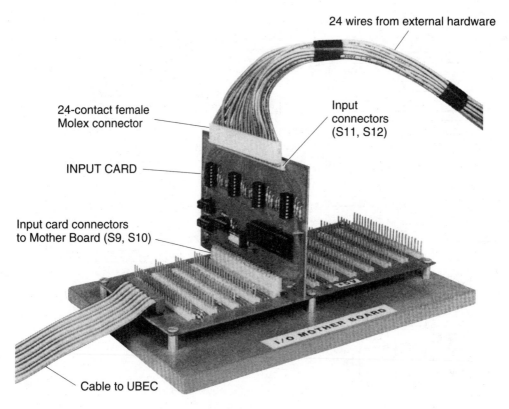

24 wires from external hardware

Input
connectors
(S11, S12)

24-contact female
Molex connector

INPUT CARD

Input card connectors
to Mother Board (S9, S10)

I/O MOTHER BOARD

Cable to UBEC

16-6 Original design input card plugged into a motherboard.

- [ ] *U1–U8[+].* Note that the ICs face different directions, so you need to be particularly careful to make sure that you put them in the right sockets with correct pin 1 orientation. Remember, U8 is static-sensitive, so follow the special handling procedures noted above.
- [ ] *Cleanup and inspection.*
- [ ] *Card IC power test.* Turn off the +5-Vdc supply and plug the output card into any motherboard slot. Turn the power back on and follow Table 16–2 to see that power and ground reach each IC. Correct any problems as you would with the output card.
- [ ] *DIP switch test.* Connect your (–) meter lead to ground and move the (+) lead to touch the IC pins listed in Table 16–3. At each position move the corresponding switch segment to off, and you should read +5 Vdc; move it to on, and you should measure 0 Vdc. If all readings are reversed you have the switch in backward. If only some segments are incorrect, check around for poor soldering.

This completes the assembly of the general-purpose input card.

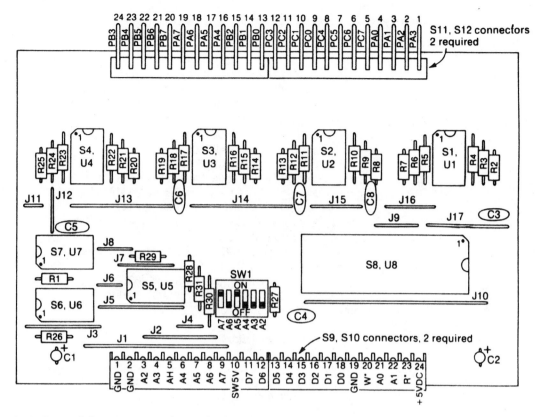

16-7 Original-design input card parts layout.

Building the Test Panel

The computer output test panel shown in Fig. 16–8, and required with the original-design I/O cards, is nearly identical to the one described in Chap. 8. The only difference being that jumpers are required to match the unique pinout arrangement associated with the original-design I/O cards. Figure 16–9 includes the schematic and parts layout and Fig. 16–10 shows the full size panel face artwork. The parts list and the assembly steps are identical to those presented in Chap. 8 except for the size of the aluminum panel and the need to install jumpers J1–J12. Card testing is identical to Chap. 8, except that you should use the test sequence defined in Table 16–5.

That completes our technology baseline by establishing all that's needed to build a universal computer interface system and to get started on its application to meet your particular interfacing desires. In the next two chapters I'll provide additional

Table 16-4. Original-Design Input Card Parts List

Qnty.	Symbol	Description
17	J1-J17	Jumpers, make from no. 24 uninsulated bus wire (Belden no. 8022)
1	R1	1.0KΩ resistor [brown-black-red]
30	R2-R31	2.2KΩ resistors [red-red-red]
7	S1-S7	14-pin DIP sockets (Digi-Key A9314)
1	S8	40-pin DIP socket (Digi-Key A9340)
1	SW1	6-segment DIP switch (Digi-Key CT2066)
2	S9,S10	12-pin Waldom side entry connectors (Digi-Key WM3309)
2	S11,S12	12-pin Waldom right angle headers (Digi-Key WM4510)
2	C1,C2	2.2μF, 35V tantalum capacitors (Jameco 33734)
6	C3-C8	100000pF, 50v ceramic disk capacitors (Digi-Key P4164)
4	U1-U4	74LS04 hex inverters (Jameco 46316)
2	U5,U6	74LS136 QUAD EXCLUSIVE-OR gates with open collector (Jameco 46586) or (JDR 74LS136)
1	U7	74LS00 quad 2-input NAND gate (Jameco 46252)
1	U8	8255 or 8255A-5 programmable peripheral interface (Jameco 52724 or 52732) or (JDR 8255-5)

Mating connector for cable

2	—	12-pin Waldom terminal housing (Digi-Key WM2110)
24	—	Crimp terminals (Digi-Key WM2301 for wire sizes 18-24 or WM2301 for wire sizes 22-26)

Author's recommendations for suppliers given in parentheses above with part numbers where applicable. Equivalent parts may be substituted. Resistors are ¼W, 5 percent and color codes are given in brackets.

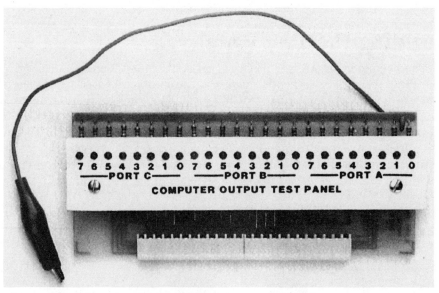

16-8 Original-design output test panel.

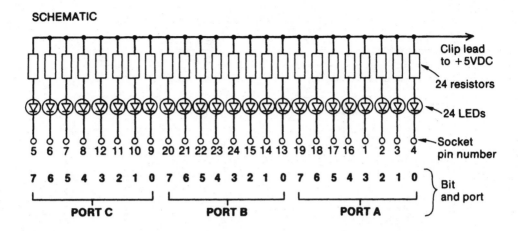

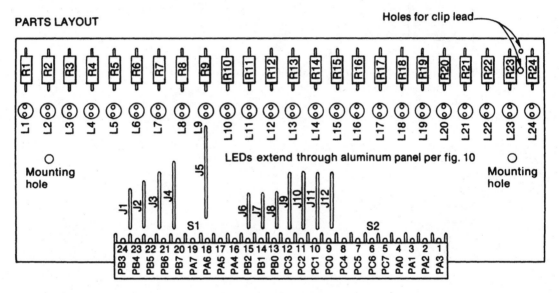

16-9 Original-design test panel schematic and parts layout.

16-10 Original-design test panel artwork.

Table 16-5. Original-Design Panel Test Sequence

CONNECTOR-PIN/LED CONTINUITY TESTS

✓	SOCKET PIN NO.	LED TURNED ON	
		PORT	BIT
	1	A	3
	2	A	2
	3	A	1
	4	A	0
	5	C	7
	6	C	6
	7	C	5
	8	C	4
	9	C	0
	10	C	1
	11	C	2
	12	C	3

CONNECTOR-PIN/LED CONTINUITY TESTS

✓	SOCKET PIN NO.	LED TURNED ON	
		PORT	BIT
	13	B	0
	14	B	1
	15	B	2
	16	A	4
	17	A	5
	18	A	6
	19	A	7
	20	B	7
	21	B	6
	22	B	5
	23	B	4
	24	B	3

assistance by presenting an example application of the UCIS by attaching it to model railroad hardware.

This detailed application example will serve to demonstrate the dynamic capabilities of UCIS with its attachment to many types of hardware devices, and the generation of real-time application-dependent software to help you better suit your particular UCIS application requirement.

17
CHAPTER

Sample Interface Application: External Hardware

In the last 16 chapters we have built and tested all the basic hardware required for a general-purpose universal computer interface system, or UCIS. This included software, up to the level needed to test the system plus a simple I/O example. Now I'll take you one step further by demonstrating an example of a real-world, real-time application of UCIS. Understanding this application should go a long way to help you adapt your UCIS to your specific desired application. Although the cards we'll discuss in this chapter might appear at first to have a specialized purpose, the technological ideas presented (as well as the circuits that result) have far-reaching implications applicable to most any interfacing project.

For illustration purposes I'll use monitoring and control of model railroad hardware as the sample interface application. We could just as well use as our example: machine tool control, robotics, greenhouse climate control, laboratory test equipment automation, or a multitude of other interfacing applications.

Using as a demonstration the interface of a computer to a model railroad, however, provides several definite advantages:

- Many people are fascinated by watching model trains operate, whether they be at a store display, an around-the-Christmas-tree setup, or a large model railroad system.
- The example is a very easily visualized application. Everyone can relate (frequently to the point of fascination) to a model train running around a track.
- Computer control adds another whole level of fascination.
- When multiple-train operation is included on the same track, the application

becomes a highly dynamic example, with the need for automatic block power routing and computer-controlled signaling to prevent collisions.

- I use a UCIS to interface a computer with my Sunset Valley Lines model railroad, so I am able personally to share with you dual areas of expertise, computers and railroading.
- The hardware hookups, although easy to visualize, are quite complex in nature; thus yielding many ideas that set the stage for other externally-attached circuits to satisfy the needs of many additional applications.

In addition to these advantages, it is also important to note that the model railroad application adds a lot of pragmatic value. For example, the University of Arizona built a UCIS-equipped model railroad system to demonstrate basic transportation principles and perform operational simulations to support a research study for a local government to establish a full-size rail transportation system.

A technical institute in the State of Maine built up a UCIS-equipped model railroad to use as a class project to teach computer interfacing. Numerous universities have done the same. Such UCIS applications make excellent school projects, and it is an effective way to provide hands-on interfacing experience. When connected to a train set, a UCIS really keeps student interest piqued.

The Intel Corporation assembled pieces of the hardware I'll show you in this chapter to support their product release of a new advanced real-time software product. They selected monitoring and control of model trains as the most effective medium for demonstrating the true dynamic capability of their product.

Since a preliminary form of UCIS originally appeared as a 16-part series in the *Model Railroader* magazine in the form of a Computer/Model Railroad Interface (C/MRI), there is a multitude of model railroads already interfacing via UCIS.

Introduction to the Application

For those who are not model railroaders, it's important that we take just a moment and introduce you to the subject of train control. Most model railroaders strive to make their layout as realistic as possible. For many, this author included, achieving realism on a model railroad includes operating the trains in a manner emulating how real railroads operate. It's important to follow prototype practice in the way the models are operated just as in the way the models are made to look realistic. The important point is that a UCIS can make running trains more realistic, easier, and more fun too.

Typically model railroad tracks are divided into isolated electrical sections, called blocks, by placing insulated rail joiners or gaps between the blocks. Manual cab control includes a control panel switch used to assign correct control power to blocks. Cab control in this typical manual form is the conventional way in which most operators run two or more trains simultaneously. With this approach the operator throws panel toggles or turns rotary switches to keep each locomotive's speed and direction controls (hereafter referred to as a cab) electrically tied to their respective trains. Such manual manipulations impose very unrealistic requirements on the operation. Switching cab power from block to block doesn't correspond to *any* procedure on a real railroad.

Obviously, real railroad engineers don't need block switches to control their locomotives. A model railroad engineer can spend as much time throwing toggles as actually running trains, especially on a small layout. Many railroaders just get used to this, but only at some cost in realism and fun. A computer can easily overcome these limitations by automatically switching cab power from block to block.

Computer Cab Control

Automatic cab control systems have been around at least since the 1950s, but prior to the C/MRI and our UCIS, they haven't been used on a lot of model railroads. Prior to computerization, they have been awkward and inflexible in implementation and limited in their capabilities. Now that we can use the UCIS to apply the flexibility and logic capacity of a personal computer to this task, computer cab control (CCC) is far easier to apply than any earlier automated system.

How does computer cab control work? Simply and almost entirely automatically. A bit of manual setup is required at the start of operation to tell the computer which cabs to start from which blocks where trains are waiting. Once initialized, CCC automatically electrically connects (assigns) blocks to each cab one step ahead of each train and disconnects blocks behind each as the last car clears. Clear blocks can be reassigned to other cabs automatically.

CCC-driven, trackside signals show when each train can enter the next block. When a train stops in a siding for a meet and its cab direction switch is set in its neutral/off position, only the block(s) the train occupies stay assigned to that cab. If an operator reverses direction, the computer automatically assigns the next block in the other direction, if it isn't occupied or already assigned to another cab. CCC follows turnout settings to assign blocks on branching routes. At the end of the run a little more manual input tells the CCC computer to disconnect the cab from the block where the train tied up, so the operator can use the same cab to run another train.

While CCC handles cab-control functions automatically, operators run the trains. Operators are free to concentrate on operating realistically, or just to watch the trains roll through the scenery. With computerized interfacing, each operator has more fun being the engineer with the computer doing the donkey work.

As with most applications in addition to the UCIS itself, there's some special hardware needed for CCC. Most basically, there's the block switching circuitry to assign one of several cabs to each block by computer command. A display showing the status of each block, particularly whether or not it is occupied and by which cab, is nice to have. I'll show how to do this using the computer's monitor and/or with a separate LED cab number display digit for each block on a track-diagram panel. Figure 17–1 illustrates the basic elements of a CCC system.

CCC and signaling are the most popular online applications of computers to model railroading, but there are at least three other applications where it's desirable to go one step further and have the computer actually run the trains:

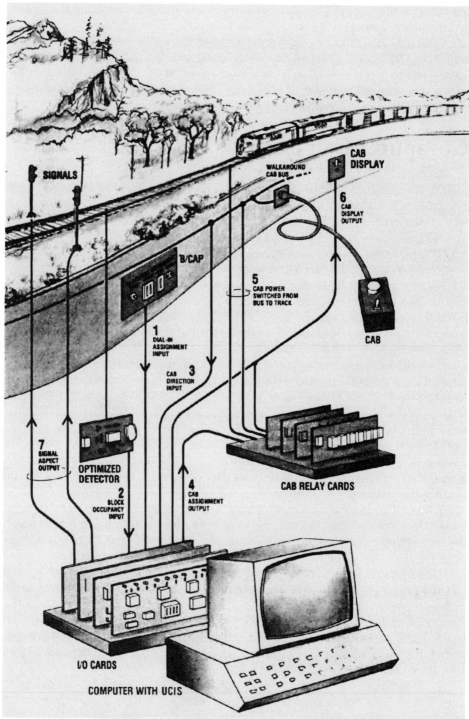

SIGNALS

WALKAROUND
CAB BUS

CAB
DISPLAY

B/CAP

5
CAB POWER
SWITCHED FROM
BUS TO TRACK

6
CAB
DISPLAY
OUTPUT

CAB

1
DIAL-IN
ASSIGNMENT
INPUT

3
CAB
DIRECTION
INPUT

7
SIGNAL
ASPECT
OUTPUT

OPTIMIZED
DETECTOR

2
BLOCK
OCCUPANCY
INPUT

4
CAB
ASSIGNMENT
OUTPUT

CAB RELAY CARDS

I/O CARDS

COMPUTER WITH UCIS

17-1 Overview of computer cab control.

- *Solo operation.* The computer could run some trains while a single operator ran one, to add realism and fun when operating alone. The operator might run a local freight, for example, or play the role of yardmaster while the computer ran the through trains automatically.
- *Display operation.* I find it's great to be able to have the computer run the railroad while I entertain guests. I can chat and answer questions while the computer orchestrates meets between trains, has passenger trains make station stops, and even employs helper locomotives to push loaded freights up a big hill.
- *Cab realism.* With the computer as an intermediary between the operator and the locomotive, the software can simulate prototype cab controls. Separately purchased, transistorized throttles do this to some extent by mimicking momentum effects and brake functions, but with a computer in the loop the sky is the limit. One could even build a full-fledged, locomotive/train simulator.

This last idea brings to light a related UCIS application area, whereby a flight simulator could be built around UCIS using realistic flight instruments, a far more realistic setup than the popular computer game.

In the closing sections of this book, I'll touch on adding throttle control extensions to CCC using the motor control DAC circuit we developed in Chap. 14. This way the computer, with a bit more software, can run the trains fully automatically.

In the remainder of this chapter, we'll look at the basic external hardware needed to connect a model railroad to the UCIS for computer cab control and computer-controlled signaling. In the next chapter I'll cover system hookup and application software. Once you have these applications under your belt, the excitement for generating your own software and applying UCIS for your own application will be wide open.

There are many different types of external hardware circuits that can be built to interface easily to UCIS. Each circuit we'll go over will expand your knowledge of computer interfacing. We'll start out looking at an especially neat, current-sensing, circuit card. Its model railroad use is for track occupation detection. This is the most fundamental of all circuits required for interfacing a computer to a model railroad. Other applications, where you desire for the computer to sense presence of current flow are also prime candidates for using this same circuit.

Optimized Detection

No matter whether you desire to apply the UCIS to automate operation of multiple trains or simply demonstrate manipulation of a few signal lights for a train looping around a Christmas tree, the fundamental requirement is for the computer to read the location of each train. By sensing current flow into each block, it's easy to have the computer determine which blocks are occupied, and thereby monitor train location.

Another scheme to determine train movement and thus occupied block location is to mount, at the entrance and exit from each block, either photosensitive detectors or magnetic reed switches. The latter are operated by a magnet placed under each locomotive. When a train enters a block, the block is set occupied, and when it leaves the block the block is set clear. The current detector, however, is easier to use and provides a better direct measure of block occupation.

Figure 17–2 shows the detector I recommend. In fact, I consider the detector so good I call it the optimized detector. My friend Paul Zank, who is an N-scale model railroader and an award-winning aerospace electronics engineer, helped in its design. Here are a few of its important properties:

- Its sensitivity is easy to adjust with a trim potentiometer and LED monitor.
- Its built-in turn-on delay of .75 s and turn-off delay of 3.5 s eliminates problems from dirty track and other causes of intermittent contact.
- Its monitor LED goes on and off before the time delays, giving instant occupancy indication to help in setting sensitivity.
- It has only two active components, one IC and one transistor, so it's easy to debug and maintain.

17-2 Optimized detector.

- Its open-collector transistor output allows easy connection to LEDs, TTL logic circuits, relays, and UCIS input cards.
- It works with conventional dc, ac, pulse power, and all forms of command control including the latest NMRA (National Model Railroad Association) DCC standard, and any sound system.
- The design handles currents from microamps up to three amps and more if you substitute higher-current diodes.
- It's a small modular unit (one per block), so it's ideal for plug-in circuit card construction. That eases system debugging and maintenance, but alternative connection methods are also provided.
- Its price is very reasonable.
- Thousands of these optimized detectors (ODs) have been placed in service around the world, and this experience shows their performance to be exceptional.

The OD's schematic is shown in Fig. 17–3, and its current capacity is determined by diodes D1 and D2. I typically recommend 3-A diodes that have a surge capability of 50 A. If more is required you can substitute 5-, 6-, or 10-A diodes.

The product of R11 and C2 determines the turn-off delay, and the product of R13 and C2 determines the turn-on delay as long as R13 is considerably smaller than R11. Thus delay times can be easily changed as desired. I enjoy the rather long 3.5-s turn-off delay, which not only completely solves the problem of intermittent contact, but also simulates the massive, slow-moving relays in prototype detection circuits.

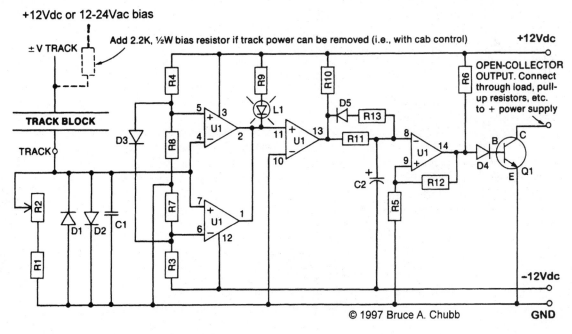

© 1997 Bruce A. Chubb

17-3 Optimized detector schematic.

The value of R6 can also be varied to select the level of detector drive capability. I selected 3.6 kΩ for reasonably high drive capability from the output transistor to handle loads as high as .2 A and still maintain a good logic low for TTL connections. For example, I use a single detector to drive parallel loads of 10 LEDs and four TTL logic gates, a total load of about .2 A, but still with a logic low under .08 Vdc. Reducing R6 to a lower value, such as 1 kΩ, would take more current from the power supply, but would allow driving output loads up to .3 A at 40 V, the ratings of the 2N3904 transistor. For values of R6 at the 3.6-kΩ level or lower, use a ½-W resistor.

Because applications require one OD per block, it's convenient, as with the general purpose I/O cards, to have a motherboard into which the detectors plug. Ready-to-assemble cards are available from JLC Enterprises, and assembled and tested cards can be ordered from ECO Works, EASEE Interfaces, and AIA. Figure 17–4 gives the parts layout for the OD, and Table 17–1 the parts list. Here's how to assemble it:

- ☐ *R1, R3–R13.* Note that R6 and R9 are ½-W resistors.
- ☐ *R2.* Install this trim potentiometer as in Fig. 17–4, pushing the three prongs all the way into the holes as they are soldered.
- ☐ *D1, D2[+].* Use needle-nose pliers to bend the heavy leads of these power diodes at right angles so they drop into the holes. The banded ends must face in opposite directions as in Fig. 17–4. Slip a ¹⁄₁₆-in spacer between the card and the diodes as they are soldered, then remove the spacer. The space helps ventilate the diodes and protects the card.
- ☐ *D3–D5[+].*
- ☐ *S1[+].*
- ☐ *U1[+].*
- ☐ *S2.* (Optional, for use with detector motherboard. Alternate holes, with wider spacing for hardwiring, are provided for track and common connections.)
- ☐ *Q1[+].*
- ☐ *C1.*
- ☐ *C2[+].*
- ☐ *L1 [+].* Note orientation of flat side and positive hole (longer lead) in Fig. 17–4. With needle-nose pliers, hold the leads securely next to the housing and bend at right angles as in the Fig. 17–4 detail. The LED sticks out over the edge of the card so it can be seen when the detectors are plugged into their motherboard.
- ☐ *Cleanup and inspection.*

If detectors are assembled with the S2 connectors, it's important also to build up one or more of the detector motherboards in Fig. 17–5. Each motherboard accommodates 12 detectors; if more motherboards are needed they can be connected in parallel, with each of the three bus traces wired from one board to the next.

Figure 17–5 includes the parts list and shows the parts layout for the motherboard. Here are the assembly steps:

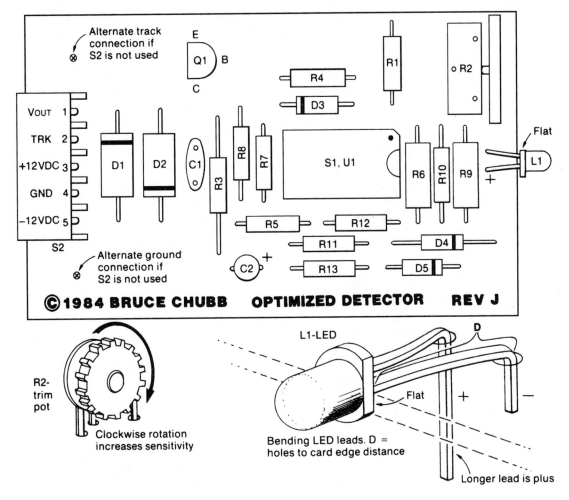

17-4 Parts layout for optimized detector.

☐ *Terminal screws.* Install 27 terminal screws with brass nuts soldered to the circuit traces: +12 Vdc, COMMON, −12 Vdc, 12 track connections, and 12 detector outputs.

☐ *S1–S12.*

☐ *Adjacent trace test.* Using a VOM set on R × 100, check the resistance between the COMMON terminal and the +12- and −12-V terminals. It should be infinite, an open circuit. Then clip one lead to the +12-V terminal and touch the other to each of the track terminal screws for each card slot: each should also read open circuit. Then for each card slot, touch one lead to the track terminal and the other to the output terminal. Again, the test should read open circuit. A reading close to 0 Ω for any of these indicates a solder

Table 17-1. Optimized Detector Parts List

Qnty.	Symbol	Description
1	R1	10Ωresistor [brown-black-black]
3	R3-R5	10KΩ resistors [brown-black-orange]
1	R6	3.6KΩ ½ W resistor [orange-blue-red]
2	R7,R8	2.2KΩ resistors [red-red-red]
1	R9	2.2KΩ ½ W resistor [red-red-red]
1	R10	10KΩ resistor [brown-black-orange]
1	R11	2.2MΩ resistor [red-red-green]
1	R12	220KΩ resistor [red-red-yellow]
1	R13	330KΩ resistor [orange-orange-yellow]
1	R2	10KΩ potentiometer (Jameco 94713) or (Mouser 320-1510-10K)
2	D1,D2	For regular DC or AC track power select from:
		3A, 50V diodes (Digi-Key 1N5400GICT)
		6A, 50V diodes (Digi-Key GI750CT)
		For command control, e.g. DCC or Railcommand, select from:
		3A, 50V fast recovery diodes (Digi-Key GI850CT)
		5A, 50V fast recovery diodes (Digi-Key GI820CT)
3	D3-D5	1A, 100V diodes (Digi-Key 1N4002GICT)
1	S1	14-pin DIP socket (Digi-Key A9314)
1	S2	5-pin Waldom side entry connector (Digi-Key WM3302)
1	C1	100000pF, 50V ceramic disk capacitor (Digi-Key P4164))
1	C2	1.5µF, 35V tantalum capacitor (Digi-Key P2060)
1	Q1	2N3904 small signal transistor (Jameco 38359) or
		(Digi-Key 2N3904)
1	L1	Red diffused size T1 LED (Digi-Key P363)
1	U1	LM339N quad voltage comparator (Jameco 23851)

Author's recommendations for suppliers given in parentheses above with part numbers where applicable. Equivalent parts may be substituted. Resistors are ¼W, 5 percent unless otherwise noted and color codes are given in brackets.

bridge between adjacent pads somewhere along the two bus lines under test. Locate it, remove it, and retest to be sure.

□ *Cleanup and inspection.*

For mounting, use six ¼-in long standoffs as for the UBEC in Chap. 6. The ODs need a ±12-Vdc power supply. Actually, any voltage between 5 and 15 would work, but the nearer the high end the better, and the positive and negative values must be the same. Each detector draws about 10 mA (not counting external current on the open-collector output connection), so a 1.5-A supply handles up to 150 detectors.

Figure 17–6 shows the schematic for the supply. The parts list is given in Table 17–2. It's similar to the +5-Vdc supply in Chap. 8, except that regulators V1 and V2 must be insulated from each other. Because the currents are lower the heat sinks could be smaller, but I like the reliability of an overdesigned supply with extra-large heat sinks.

Figure 17–7 shows how to connect optimized detectors on a layout with cab control. Each block has a bias resistor to provide a current path to activate the detector

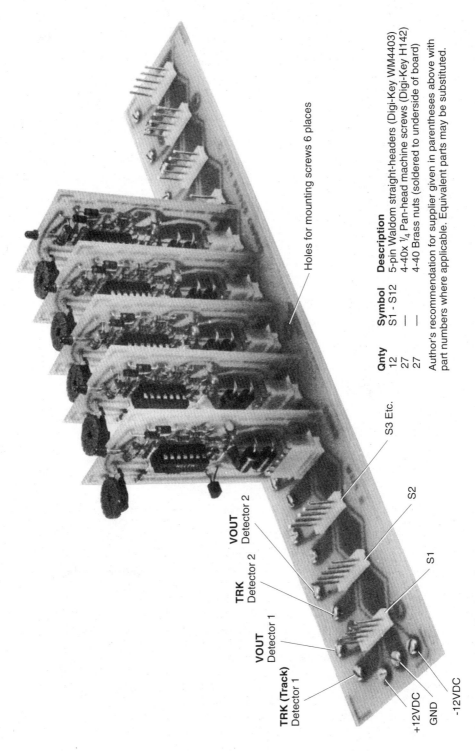

Holes for mounting screws 6 places

Qnty	Symbol	Description
12	S1 - S12	5-pin Waldom straight-headers (Digi-Key WM4403)
27	—	4-40x ¼ Pan-head machine screws (Digi-Key H142)
27	—	4-40 Brass nuts (soldered to underside of board)

Author's recommendation for supplier given in parentheses above with part numbers where applicable. Equivalent parts may be substituted.

S3 Etc.

S2

S1

VOUT
Detector 2

TRK
Detector 2

VOUT
Detector 1

TRK (Track)
Detector 1

+12VDC

GND

-12VDC

17-5 Detector motherboard.

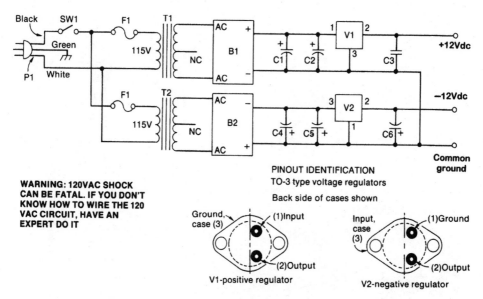

17-6 ±12-Vdc detector power supply schematic.

Table 17-2. ±12-Vdc Detector Power Supply Parts List

Qnty.	Symbol	Description
1	P1	Three-conductor power cord w/plug (Jameco 37997)
1	—	Strain relief (Radio Shack 278-1636)
2	F1	Panel-mount fuse holder (Jameco 18702)
2	—	2A slow-blow fuse (Jameco 25646)
1	SW1	Toggle switch (Jameco 76241)
2	T1,T2	18V, 1A power transformer (Jameco 105531)
2	B1, B2	6A, 200V Bridge rectifier (Digi-Key KBL02)
2	C1, C4	4,700µF, 25V electrolytic capacitors (Digi-Key P6354)
1	C2	.22µF, 35V tantalum capacitor (Jameco 33507)
1	C3	100000pF, 50V ceramic disk capacitor (Digi-Key P4164)
1	C5	2.2µF, 35V tantalum capacitor (Jameco 33734)
1	C6	1µF, 35V tantalum capacitor (Jameco 33662)
1	V1	Positive 12V, 1.5A voltage regulator (Mouser 511-L7812CT)
1	V2	Negative 12V, 1.5A voltage regulator (Mouser 511-L7912CT)
2	—	TO-3 mounting kit with socket (Mouser 534-4732)
2	H1,H2	Fin-type heat sink for V1 and V2 (Jameco 16512)

Author's recommendations for suppliers given in parentheses above with part numbers where applicable. Equivalent parts may be substituted.

if the block is occupied but the selector switch is set to off and/or the assigned cab is off. Each block so occupied will draw about 10 mA from the +12-Vdc supply.

The typical turn-on sequence with the ODs is that they come up in the occupied state, and after the 3.5-s delay the unoccupied blocks turn off. However if you have a significant number of detectors and they are located quite remote from the ± detector supply, you might find that the detectors tend to hang up in the occupied state, or sometimes their LEDs have a dim glow when a block is unoccupied. The problem is easily solved by adding two 2.2-μF tantalum capacitors to each OD motherboard. One connects between the +12-Vdc terminal and ground, and the other between the −12-Vdc terminal and ground. On the latter the + lead of the capacitor needs to be the ground connection.

With certain cabs it's possible to find that a particular throttle setting in one direction precisely cancels out the bias current, causing an occupied block to show up as unoccupied. If this happens the problem is easily circumvented by driving all the bias resistors with 12–24 Vac instead of the steady +12 Vdc. I've illustrated this alternate hookup in Fig. 17–8, which also shows how to set up good ground wiring, following the practice we established in Chap. 13, with separate ground wires for different functions all tied to one common point.

For use with command control (where a digital receiver is placed into each locomotive for totally independent control), track power is always present, so the bias resistors (and, obviously, the selector switches) can be eliminated. A DPDT reversing switch is still required for each reversing block, but with the UCIS a DPDT relay can be used to enable the computer to switch the reversing block automatically.

When optimized detectors are installed, turn the sensitivity-adjustment pot R2 on each one up as high (clockwise) as it will go. If the LED lights, showing the block occupied when it is not, back off the pot until the LED goes out.

Powered locomotives or lighted cars entering a block should activate the detector and turn on the LED. To detect unlighted cars, a resistive path is added across at least one wheelset on each car. Commercial metal wheelsets with built-in resistance, in both HO and N scale, are available from Jay Bee, P.O. Box 7031, Villa Park, IL 60181 and from Logic Rail Technologies, 21175 Tomball Parkway, Houston, TX 77070. To do it on your own, there are at least two ways: silver-based resistance paint, or a small resistor across a pair of metal wheels.

Peter Thorne describes the resistance- and conductive-paint method in his book *Practical Electronic Projects for Model Railroaders*, available from Kalmbach Publishing Co., P.O. Box 21027, Waukesha, WI 53187. It can be used with plastic wheels and/or axles, as well as metal ones. Peter recommends a resistance of about 5 kΩ. If too low, the resistance paint can get hot enough to damage plastic at full track voltage. He also points out that conductive paint eventually wears off plastic wheels.

I prefer to use a 12-kΩ, ⅛-W resistor on one wheelset of each unlighted car, as in Fig. 17–9. For this approach to work the wheels must be metal. The resistor has a definite value, and wear can't cause any loss of detection. At 12 kΩ per car and about 350 such cars on a layout, the total current drain on the whole system is still less than ½ A.

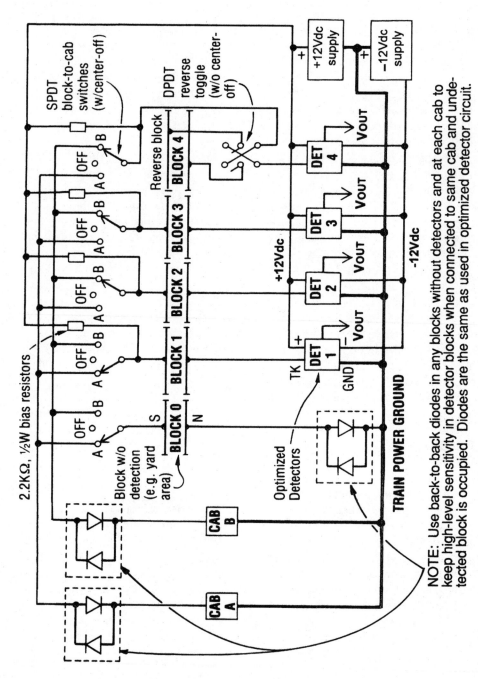

17-7 Detector wiring with conventional cab control.

NOTE: Use back-to-back diodes in any blocks without detectors and at each cab to keep high-level sensitivity in detector blocks when connected to same cab and undetected block is occupied. Diodes are the same as used in optimized detector circuit.

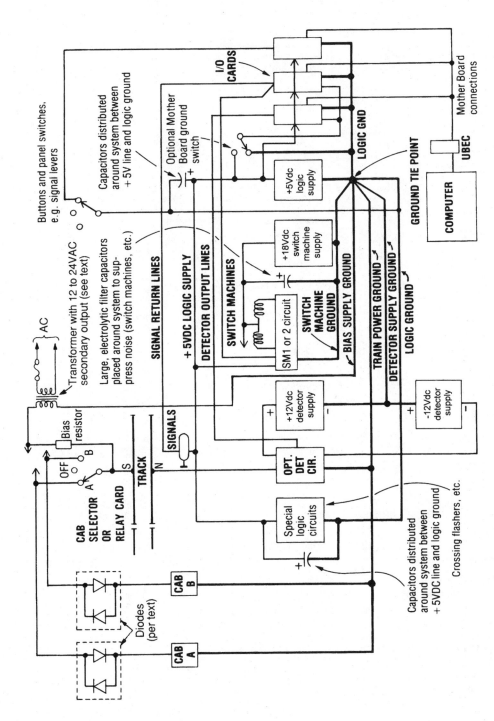

17-8 Good ground wiring.

12KΩ, ⅛W resistor (plastic-axle location)

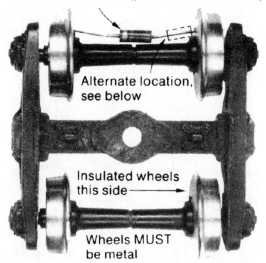

Alternate location, see below

Insulated wheels this side ——→

Wheels MUST be metal

Axle may be metal or plastic. Locate resistor near insulated wheel on metal axles to avoid short from lead to axle

17-9 Detecting unlighted cars.

I drill a #78 (.016-in) hole in each wheel, poke the lead wires through the holes, cut them off, and then bend the ends over against the wheel faces to secure the resistor. The leads fit tightly in the holes and make good contact.

I adjust the ODs for single-car protection in each block by temporarily clipping a 120-kΩ resistor across the track and setting the sensitivity pot so that the LED just lights. That way the 12-kΩ car resistor activates the detector with a 10-to-1 margin of safety!

Figure 17–10 demonstrates the sensitivity range of the optimized detector as a linear function of the potentiometer position. With the sensitivity pot set to maximum (fully clockwise) the detector triggers with 1 MΩ or less across the track. Setting the detector to minimum sensitivity requires 1 kΩ across the track before the detector activates, so the 10-kΩ pot provides a 1000-to-1 linear range in sensitivity adjustment!

Connecting devices to the detector output is just like connecting them to a UCIS output card, as presented in Chap. 13 and summarized in Fig. 17–10.

Switch Machine Control

Track switches, called turnouts by model railroaders, are used to route trains over divergent routes. Turnout position, when electrically operated, is typically controlled by a twin-coil solenoid, frequently referred to as a switch machine.

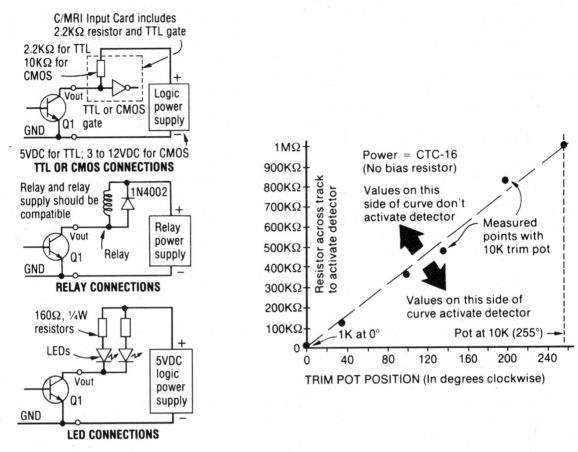

17-10 Detector outputs and linear control of sensitivity.

Two of the booster circuits shown in Fig. 13–5 could be used to drive the twin-solenoid switch machine, one for each coil, but it wouldn't be a good idea. If an error in the control program allowed a switch machine output line to be low for more than the .25 s or so needed to actuate the machine, or set both lines to a given machine low at the same time, the twin coil solenoid could easily burn out. Even with perfect software, when initializing the system the 8255 can come up with random outputs, which causes problems.

I'll show two circuits that can be used to drive twin-coil machines from the UCIS safely. The same circuit can be used for controlling any other two relay-type devices for different applications. Ready-to-assemble cards are available for both circuits from JLC Enterprises and assembled-and-tested from ECO Works, EASEE Interfaces, and AIA.

The first, circuit SM1 in Figs. 17–11 and 17–12, is easy and direct. Two UCIS output lines drive the coils via two booster circuits and an individual capacitor-discharge (CD) power supply. The boosters are like the one in Fig. 13–5, but with

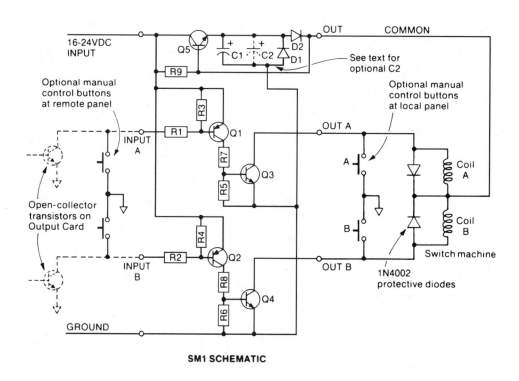

SM1 SCHEMATIC

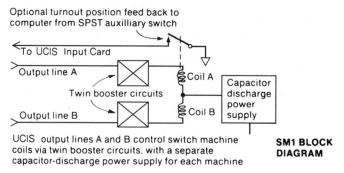

UCIS output lines A and B control switch machine
coils via twin booster circuits, with a separate
capacitor-discharge power supply for each machine

**SM1 BLOCK
DIAGRAM**

17-11 The SM1, dual-output switch machine control schematic.

resistor values selected so that everything, including the CD supply, can use the same 16–24-Vdc power. Each SM1 circuit can be mounted close to the machine it controls, with all SM1s fed by a central filtered supply. Table 17–3 is the parts list.

Using a separate CD supply for each machine protects the coils, because the capacitor stores just enough energy to throw the machine once, and the supply circuit won't let it recharge until the controlling signals from the UCIS return to high state. That makes an accidental burnout impossible, and makes the circuit very immune to electrical noise. Here are the steps for building the SM1:

Table 17-3. SM1 Parts List

Qnty.	Symbol	Description
7	—	4-40 x ¼" pan-head machine screws (Digi-Key H142)
7	—	4-40 brass nuts
4	R1-R4	4.7KΩresistors [yellow-violet-red]
2	R5,R6	1.0KΩ resistors [brown-black-red]
2	R7,R8	470Ω 1W resistors [yellow-violet-brown]
1	R9	470Ω 2W resistor [yellow-violet-brown]
2	D1,D2	1A, 100V diodes (Digi-Key 1N4002GICT)
1	C1	4700µF, 25V electrolytic capacitor (Mouser 140-XAL25V4700)
1	C2*	Optional (see text and note below)
2	Q1,Q2	2N2907 or PN2907 PNP small signal transistors (Jameco 28644)
2	Q3,Q4	2N6387 NPN power Darlington transistors (Mouser 511-2N6387)
1	Q5	2N3055 NPN power transistor (Jameco 38308)
1	L1	Red diffused size T1 LED (Digi-Key P363)
1	U1	LM339N quad voltage comparator (Jameco 23851)

Author's recommendations for suppliers given in parentheses above with part numbers where applicable. Equivalent parts may be substituted. Resistors are ¼W, 5 percent unless otherwise specified and color codes are given in brackets.

*Start with the 4700µF. If that throws your switch machine harder than necessary, substitute a smaller capacitor as listed below. For more power, add C2. Try the 1000µF value first, and the 2200µF if that's still not enough power.

C1 2200µF, 25V Mouser 140-XAL25V2200   C2 1000µF, 50V Mouser 140XRL50V1000
C1 3300µF, 25V Mouser 140-XAL25V2200   C2 2200µF, 35V Mouser 140XRL35V2200

Larger voltage values for C2 selected to obtain needed lead spacing = .3" (7.5mm) to fit board

☐ *Terminal screws.* Insert seven 4-40 screws from the top side, tighten brass
   nuts on the bottom, and solder the nuts to the traces.
☐ *R1–R9.*
☐ *D1, D2[+].* Orient the banded ends as shown in Fig. 17–12.
☐ *C1, C2[+].* C2 is optional; see the parts list.
☐ *Q1, Q2[+].*
☐ *Q3, Q4 [+].* Bend the center (collector) leads back and orient them as
   shown in Fig. 17–12, with the metal tabs toward the terminal screws. Push
   in until the wide parts of the leads touch the card, solder, and trim.
☐ *Q5[+].* Orient as shown in Fig. 17–12, and secure with 4-40 screws and
   brass nuts. Solder and trim the two leads, then solder the nuts to the traces.
☐ *Cleanup and inspection.*

Circuit SM2, Figs. 17–13 and 17–14, uses just one UCIS output line to control the two coils, taking advantage of the fact that a two-coil machine (like the turnout it operates) is a binary device. Its two states are its normal and reverse positions. I've labeled the single input to the logic driver X* (active low), so we can use X* = 0 for normal position and X* = 1 for reverse. Table 17–4 is the parts list.

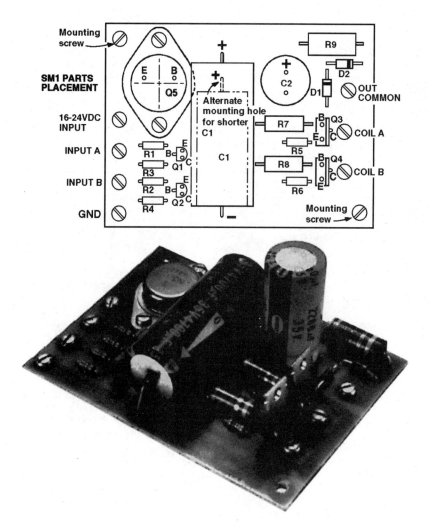

17-12 SM1 parts layout and completed board.

Outputs from pins 3 and 8 of the 74LS00 NAND gates drive the coils via booster circuits. The 74LS00 output is low only if both inputs are high. One of the inputs for each gate comes from the 555 timer IC, which when triggered puts out a high pulse for about .25 s. The other two gate inputs are the X* input and its inversion, or X. Thus a low input energizes one coil, a high input the other, and in either case only during the .25-s period of the 555's pulse output.

The 74LS86 exclusive-OR IC, along with R1 and C1, triggers the 555, which requires a negative-going narrow-pulse input. The 74LS86 output goes low only when its two inputs are identical (00 or 11). Because X* and X can't be identical, by definition, it might seem that the 74LS86 output would always be a logic high.

However, R1 and C1 cause a small delay in the X-path every time X* changes state, so the X signal arrives at the 74LS86 a bit later than the X*. That gives a 1 μs

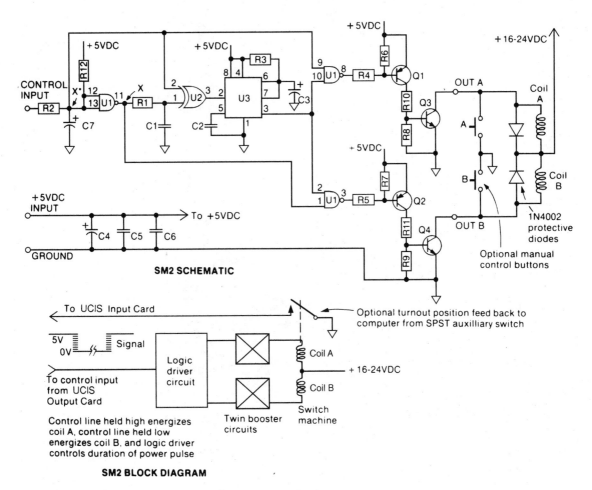

SM2 SCHEMATIC

SM2 BLOCK DIAGRAM

To UCIS Input Card

Optional turnout position feed back to computer from SPST auxilliary switch

5V / 0V Signal

Logic driver circuit

Twin booster circuits

Coil A

Coil B

+ 16-24VDC

Switch machine

To control input from UCIS Output Card

Control line held high energizes coil A, control line held low energizes coil B, and logic driver controls duration of power pulse

17-13 SM2, dual-output switch machine control schematic.

logic 0 trigger-pulse input to the 555, which in turn energizes the desired coil for the desired .25-s duration. To vary the length of the coil pulse, change the value of R2; decreasing R2 shortens the pulse proportionately.

The SM2 isn't as noise-immune as the SM1, but R2 and C7 do help. Usually a noise pulse just causes the machine to try to move the way it's already thrown. Here are the steps for building the SM2:

    ☐ *Terminal screws.* Five 4-40 screws and brass nuts.
    ☐ *J1, J2.*
    ☐ *R1–R12.*
    ☐ *C1, C2, C5, and C6.*
    ☐ *C3, C4, and C7[+].*
    ☐ *S1–S3[+].*
    ☐ *Q1, Q2[+].*

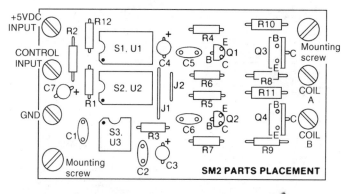

17-14 SM2 parts layout and completed card.

☐ *Q3, Q4[+].*
☐ *U1–U3[+].*
☐ *Cleanup and inspection.*

Local, manual control of machines is easy to add with pushbuttons to ground as shown in Figs. 17–11 and 17–13.

The use of low-current stall motors is gaining popularity and gradually replacing the twin-coil machine. The stall motor is easier to install with less adjustment hassle, and its slow movement is easier on turnouts. These motor-driven machines are simple to control using the switch machine card (SMC). Figure 17–15 shows the circuit's block diagram and schematic for driving one stall motor. Terminals A and B connect to the stall motor, and power input from a filtered dc supply is connected to the +12–16-Vdc and GND terminal screws. A single-control input line is connected to the IN terminal screw. Zener diode Z1 and R2 set the offset voltage at pins 9 and 12 of U1 at about +4.3 Vdc. With the IN line open-circuited, the addition of D1 and R1 pull up the voltage on pins 10 and 13 of U1 to one diode drop above the Z1 voltage, or to about +4.9 Vdc. This condition causes pin 8 of U1 to swing to saturated positive output, the supply voltage minus about 3.5 Vdc, while pin 14 approaches ground;

Table 17-4. SM2 Parts List

Qnty.	Symbol	Description
5	—	4-40 x ¼" pan-head machine screws (Digi-Key H142)
5	—	4-40 brass nuts
2	J1,J2	Jumpers, make from no. 24 uninsulated bus wire (Belden 8022 or equivalent)
2	R1-R2	47Ω resistors [yellow-violet-black]
1	R3	200KΩ resistor [red-black-yellow]
2	R4,R5	470Ω resistors [yellow-violet-brown]
4	R6-R9	1.0KΩ resistors [brown-black-red]
2	R10,R11	47Ω ½ W resistors [yellow-violet-black]
1	R12	2.2KΩ resistor [red-red-red]
1	C1	22000pF, 50V ceramic disk capacitor (Digi-Key P4160)
1	C2	10000pF, 50V ceramic disk capacitor (Digi-Key P4161)
2	C3,C4	2.2μF, 35V tantalum capacitors (Jameco 33735)
2	C5,C6	100000pF, 50V ceramic disk capacitors (Digi-Key P4164)
1	C7	1μF, 35V tantalum capacitor (Jameco 33662)
2	S1,S2	14-pin IC sockets (Digi-Key A9314)
1	S3	8-pin IC socket (Digi-Key A9308)
2	Q1,Q2	2N2907 or PN2907 PNP small signal transistors (Jameco 28644)
2	Q3,Q4	2N6387 NPN power Darlington transistors (Mouser 511-2N6387)
1	U1	74LS00 quad 2-input NAND gate (Jameco 46252)
1	U2	74LS86 quad 2-input EXCLUSIVE-OR gate (Jameco 48098)
1	U3	LM555CN timer (Jameco 27422)

Author's recommendations for suppliers given in parentheses above with part numbers where applicable. Equivalent parts may be substituted. Resistors are ¼W, 5 percent unless otherwise specified and color codes are given in brackets.

this drives the switch motor to stall in one direction. Grounding the IN control terminal causes pins 8 and 14 of U1 to change state, thereby driving the motor to stall in the opposite direction. The addition of R3 provides a margin of safety in preventing the burnout of Z1 in case you accidentally connect the IN terminal to a positive voltage source.

Because the IN terminal voltage switches between +4.9 Vdc and ground, it is TTL compatible and therefore can be also fed into an input card to report the position of manually controlled turnouts. Using a DPDT center-off toggle switch as shown in the SMC block diagram makes it easy to provide both manual and automatic control of each stall motor.

The outputs of the U1 are rated at 20 mA, sufficiently greater than the requirement of most low-current stall motors (which are typically in the 10–16-mA range). The most popular brand is the Tortoise, available from Circuitron, P.O. Box 322, Riverside, IL 60546-0322. It has a 16-mA maximum stall current when operated at 12 Vdc.

Figure 17–16 includes the parts layout and Table 17–5 gives the parts list for the SMC. Ready-to-assemble SMCs are available from JLC Enterprises, and assembled and tested cards can be ordered from ECO Works and AIA.

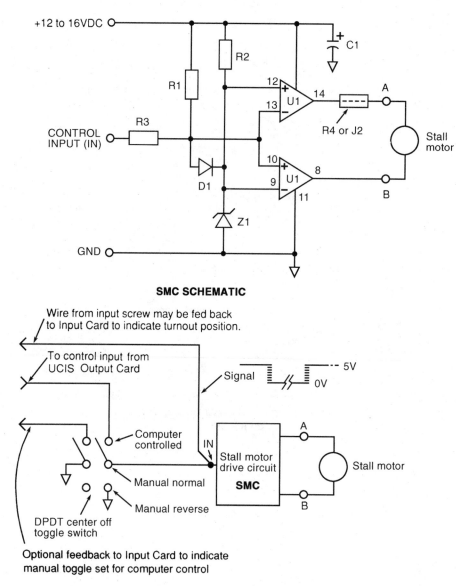

SMC SCHEMATIC

Wire from input screw may be fed back
to Input Card to indicate turnout position.

To control input from
UCIS Output Card

Signal

5V

0V

Computer
controlled

IN

Stall motor
drive circuit

SMC

A

Stall motor

B

Manual normal

Manual reverse

DPDT center off
toggle switch

Optional feedback to Input Card to indicate
manual toggle set for computer control

SMC BLOCK DIAGRAM

17-15 SMC, stall motor switch machine control schematic.

Eleven circuits can be built on each card, and because all are identical I've repeated the parts nomenclature from circuit to circuit. The parts list quantities are for
a single motor. The assembly steps are as follows:

☐ *Terminal screws.* Use the existing holes to solder wires directly from the
power supply to the IN terminal pads, or enlarge the holes with a #33 drill

Table 17-5. SMC Parts List

Qnty.	Symbol	Description
3	—	4-40x ¼" pan-head machine screws (Digi-Key H142)
3	—	4-40 brass nuts
2	J1,J2*	Jumpers, make from no. 24 uninsulated bus wire (Belden no. 8022)
2	R1,R2	1.0KΩ resistors [brown-black-red]
1	R3	100Ω resistor [brown-black-brown]
1	R4	Optional resistor used in place of J2 (see text)
1	D1	1A, 100V diode (Digi-Key 1N4002GICT)
1	Z1	4.3V, 1W zener diode (Digi-Key 1N4731ACT)
1	C1	1µF, 35V tantalum capacitor (Jameco 33662)
1	S1	14-pin DIP socket (Digi-Key A9314)
1	U1	LM324N low power quad operational amplifier (Jameco 23683)

Author's recommendations for suppliers given in parentheses above with part numbers where applicable. Equivalent parts may be substituted. Resistors are ¼W, 5 percent unless otherwise noted and color codes are given in brackets.

for 4-40 terminal screws and brass nuts for soldering to the terminal pads.

☐ *J1.* Note that this jumper lies under S1.

☐ *R1–R4.* Note that R4 is optional to allow current limiting in the motor, and if it is not used it must be replaced with jumper J2.

☐ *C1[+].*

☐ *Z1 and D1[+].*

☐ *S1[+]*

☐ *U1[+]*

☐ *Cleanup and inspection.*

Using the Tortoise, it is also desirable to place a .1µF disk capacitor across the output screw terminals, or cut and jumpered into the board. This prevents a small bouncing action when the motor comes up against the stop in each direction.

As further examples of control, Fig. 17–17 shows how to use UCIS-controlled relays to drive switch motors that are not the stall-type, or for stall motors that might draw more than 20 mA. Where both manual and computer control are desired, set the turnout's normal position as the open-circuit, switch-off state with the relay not energized, so the center-off position of the toggle can be the normal setting under manual control.

Whichever method is used to control the turnout position, an auxiliary SPST contact on the switch machine can be used to feed back the turnout position. The feedback reaches the computer as an input through a UCIS input card, as I'll illustrate in the next chapter. If position is not fed back to the computer, the software must include a power-up initialization procedure that throws each turnout to a known position. Otherwise the machine position can get out of step with the computer memory. If any independent manual controls are used, as with most interface setups, position feedback is mandatory. Such important characteristics are something to look for in all interface designs.

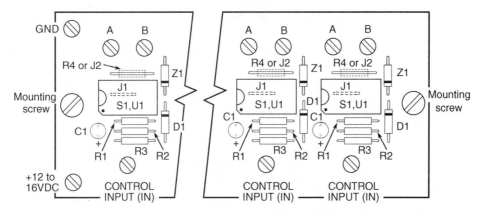

SMC PARTS PLACEMENT

17-16 SMC parts layout and completed card.

Block Power Switching

The biggest challenge in CCC is the block power switching itself. Current levels can be high, considering short circuits from derailments or wheels across turnout gaps, long passenger trains with lighted cars, multiple powered units, and especially the larger scales.

There's a variety of power-switching devices to choose from, including stepping switches, power optoisolators, solid-state photovoltaic relays (PVRs), gate-turnoff devices (GTOs), and triacs. Still, it's hard to beat the old familiar mechanical relay.

Relays are easy to understand, to wire, and to debug. They can provide years of dependable, trouble-free operation. Relays are also a lot cheaper than the newest solid-state devices, the PVRs and GTOs, even though the high-current relays we need are fairly expensive even as surplus. New solid-state devices will be developed and the prices of existing ones will come down, but right now relays are still the best choice for CCC power switching.

Figures 17–18 through 17–20 and Table 17–6 shows a Cab Relay Card (CRC) along with its parts layout and parts list. Each card connects one of up to seven cabs with one block using single-pole relays. Three BCD logic lines, labeled 1, 2, and 4, come to the CRC as outputs from a UCIS output card to determine which cab is to be selected. The three lines are decoded as in the Fig. 17–19 logic table by the 7445 BCD-to-one-of-10 decoder driver, to select which of the seven cab relays is activated

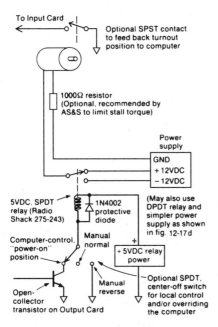

To Input Card

Optional SPST contact
to feed back turnout
position to computer

1000Ω resistor
(Optional, recommended by
AS&S to limit stall torque)

Power
supply

GND
+12VDC
−12VDC

5VDC, SPDT
relay (Radio
Shack 275-243)

1N4002
protective
diode

(May also use
DPDT relay and
simpler power
supply as shown
in fig. 12-17d

Computer-control,
"power-on"
position

Manual
normal

+5VDC relay
power

Open-
collector
transistor on Output Card

Manual
reverse

Optional SPDT,
center-off switch
for local control
and/or overriding
the computer

a. DISPLAY MOTOR SWITCH MACHINES
(American Switch and Signal: Rebco)

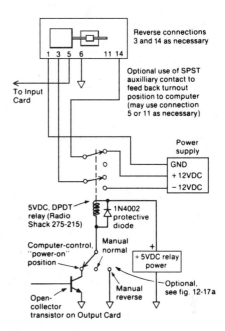

Reverse connections
3 and 14 as necessary

1 3 5 6   11 14

To Input
Card

Optional use of SPST
auxilliary contact to
feed back turnout
position to computer
(may use connection
5 or 11 as necessary)

Power
supply

GND
+12VDC
−12VDC

5VDC, DPDT
relay (Radio
Shack 275-215)

1N4002
protective
diode

Computer-control,
"power-on"
position

Manual
normal

+5VDC relay
power

Open-
collector
transistor on Output Card

Manual
reverse

Optional,
see fig. 12-17a

b. JACK-SCREW SWITCH MACHINE
(GB Electronics Turnout Motor)

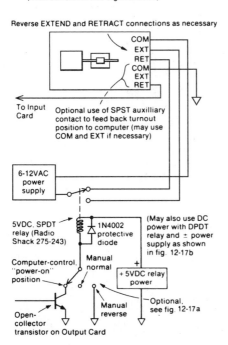

Reverse EXTEND and RETRACT connections as necessary

COM
EXT
RET
COM
EXT
RET

To Input
Card

Optional use of SPST auxilliary
contact to feed back turnout
position to computer (may use
COM and EXT if necessary)

6-12VAC
power
supply

5VDC, SPDT
relay (Radio
Shack 275-243)

1N4002
protective
diode

(May also use DC
power with DPDT
relay and ± power
supply as shown
in fig. 12-17b

Computer-control,
"power-on"
position

Manual
normal

+5VDC relay
power

Open-
collector
transistor on Output Card

Manual
reverse

Optional,
see fig. 12-17a

c. JACK-SCREW SWITCH MACHINE
(Mann-Made Point Drive)

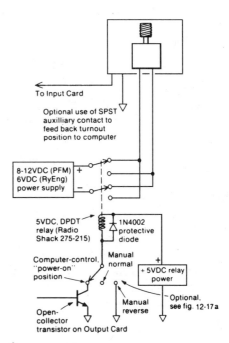

To Input Card

Optional use of SPST
auxilliary contact to
feed back turnout
position to computer

8-12VDC (PFM)
6VDC (RyEng)
power supply

+

−

5VDC, DPDT
relay (Radio
Shack 275-215)

1N4002
protective
diode

Computer-control,
"power-on"
position

Manual
normal

+5VDC relay
power

Open-
collector
transistor on Output Card

Manual
reverse

Optional,
see fig. 12-17a

d. WORM-DRIVE SWITCH MACHINE
(PFM Slow Action; Railway Engineering Rotor Motor)

17-17 Use of relays for switch motor control.

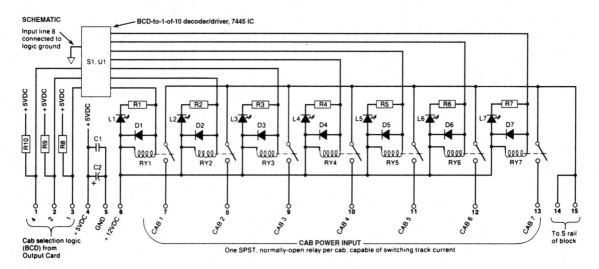

17-18 Cab relay card schematic.

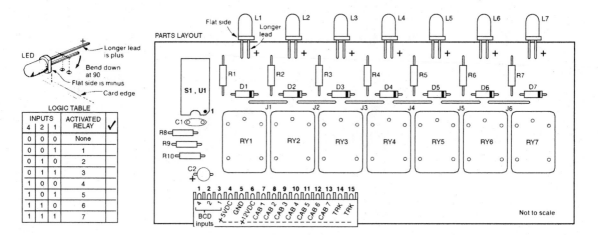

17-19 CRC parts layout.

(closed). The LEDs light to indicate which relay is closed, and there's a protective diode across each of the relay coils. Ready-to-assemble circuit cards are available from JLC Enterprises, and assembled and tested cards are available from ECO Works, EASEE Interfaces, and AIA.

For applications of less than seven cabs, or where desired to keep the initial cost down, assemble depopulated CRCs with just the parts needed to handle a minimum number of cabs and add more cabs later by installing additional parts. Parts in this category are marked by an asterisk in the quantity column of the parts list. One each of these parts are required for each cab. Install them first in the cab 1 position and work up. For assembly refer to Fig. 17–19 and follow these steps:

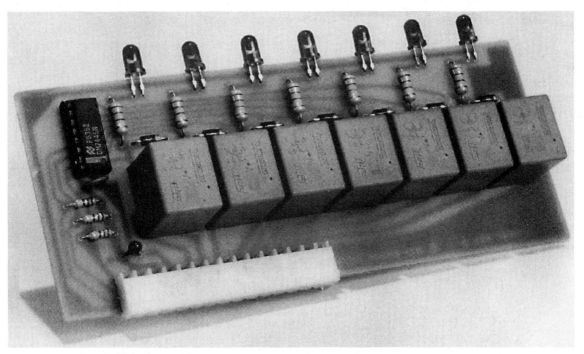

17-20 Completed cab relay card.

☐ *J1–J6.* Use #20 wire rather than the lighter wire used on previous cards; the heavier wire is better to handle track current. All six jumpers are required even if you are not installing all seven relays.

☐ *R1–R10.*

☐ *D1–D7[+].*

☐ *S1[+].*

☐ *S2.*

☐ *U1[+].*

☐ *RY1–RY7.* Insert each relay, making sure that all five pins pass through the holes and that the relay case is held firmly against the card while you are soldering.

☐ *C1.*

☐ *C2[+].*

☐ *L1-L7[+].* Note the orientation of the flat side and the + hole (longer lead) in Fig. 17–19. Hold the leads securely next to the housing with needle-nose pliers, and bend them at right angles so the LEDs stick out over the edge of the card (so they are visible when the CRCs are plugged into their motherboard). Then insert, solder, and trim the leads.

☐ *Cleanup and inspection.*

Table 17-6. CRC Parts List

Qnty.	Symbol	Description
6	J1-J6	Jumpers, make from no. 20 (to carry track current) uninsulated bus wire (Belden no. 8020 or equivalent)
7*	R1-R7	510Ω ½ W resistors [green-brown-brown]
3	R8-R10	2.2KΩ resistors [red-red-red]
7*	D1-D7	1A, 100V diodes (Digi-Key 1N4002GICT)
1	S1	16-pin DIP socket (Digi-Key A9316)
1	S2	15-pin Waldom side entry connector, make from one 12-pin (Digi-Key WM3309) and one 3-pin (Digi-Key WM3300)
1	U1	7445 BCD-to-decimal decoder/driver (Jameco 50403)
7*	RY1-RY7	SPDT relay, 12Vdc coil with contact rated currents as: 5A = (Digi-Key Z721 or Z731) or 10A = (Digi-Key Z777)
1	C1	100000pF ceramic disk capacitor (Digi-Key P4164)
1	C2	1μF, 35V tantalum capacitor (Jameco 33662)
7*	L1-L7	Red diffused T1¾ size LED (Digi-Key P300)
1	U1	7445 BCD-to-decimal decoder/driver (Jameco 50403)

Author's recommendations for suppliers given in parentheses above with part numbers where applicable. Equivalent parts may be substituted. Resistors are ¼W, 5 percent unless otherwise specified and color codes are given in brackets.

*Select quantity desired for number of desired cabs

We'll hold off testing the card until we have the CRC motherboard assembled that accommodates 12 CRCs as shown in Fig. 17–21. If more are needed they can be connected together with inputs wired in parallel. Refer to Table 17–7 for a parts list.

Resistors R1–R12 are bias resistors that provide a current path to let the detectors read an occupied block even when the track power is turned off. Here are the motherboard assembly steps:

☐ *Terminal screws.* Install the 59 terminal screws and brass nuts, and solder the nuts to the circuit traces.
☐ *S1-S12.*
☐ *R1-R12.* Install these parts with the leads toward S1–S12 a bit longer than the others, as shown in Fig. 17–21, to allow clearance for the CRC connectors.
☐ *Adjacent trace test.* Set the VOM to R × 100 and check the resistance between each of the adjacent traces. It should be infinite, an open circuit. A reading close to 0 Ω for any of these indicates a solder bridge between adjacent pads or traces. Locate and remove any solder bridges, and retest the card to be sure.
☐ *Cleanup and inspection.*

For mounting, use six ¼-in-long standoffs as for the UBEC in Chap. 6.

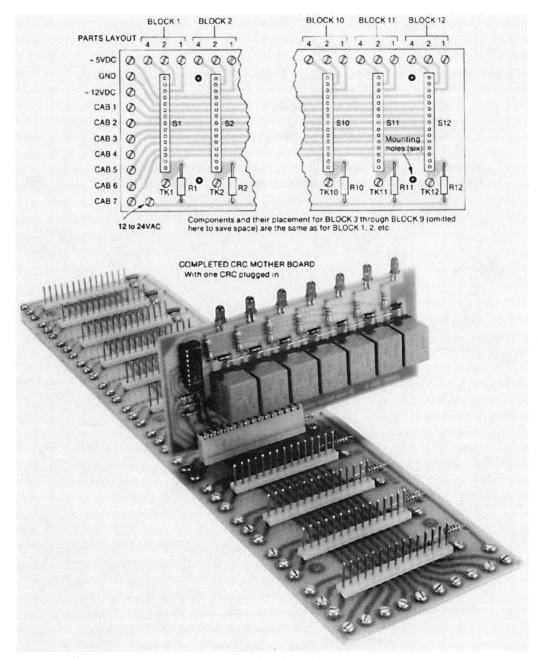

17-21 CRC motherboard layout and completed card.

Table 17-7. CRC Motherboard Parts List

Qnty.	Symbol	Description
59	—	4-40 x ¼" pan-head machine screws (Digi-Key H142)
59	—	4-40 brass nuts
12	S1-S12	15-pin Waldom headers, make from 12 each 12-pin (Digi-Key WM4410) and 12 each 3-pin cut from (Digi-Key WM4410)
12	R1-R12	2.2KΩ ½W resistors [red-red-red]

Author's recommendations for suppliers given in parentheses above with part numbers where applicable. Equivalent parts may be substituted. Resistors are 5 percent ½W and color codes are given in brackets.

The relays need a +12-Vdc power supply. Combined relay and LED current for each occupied block is about 45 mA, and assuming that only about a third of the blocks are ever assigned at any one time, a 1-A supply as shown in Fig. 17–22 would handle up to about 70 blocks. Build it like the +5-Vdc supply except, of course, use the +12-Vdc regulator.

Run a heavy (#16 or larger) ground line from the 12-Vdc supply ground terminal to the common-ground single tie point as I described in Chap. 13. To connect supplies to the cab motherboard, run #16 or larger wire from the ground-tie point to motherboard terminal screw number 2, and #18 or larger wires from +5 Vdc to screw 1 and from +12 Vdc to screw 3.

The following tests check that each CRC functions correctly. I'll assume that the CRCs have the full complement of seven relays each. If they have fewer, ignore the tests for those not installed.

☐ *Plug in the card.* Place the CRC to be tested in the first slot (block 1) on the cab motherboard, taking care that all socket pins go into the proper holes, then turn on the 5- and 12-Vdc supplies. The cab 7 LED should light. If not, check that power from both supplies is reaching the CRC, and that U1 is inserted correctly.

☐ *Zero (no) cabs assigned.* Using clip leads, ground the 1, 2, and 4 terminals for block 1. All the LEDs should be off. Set the VOM on R × 100 and connect one lead to the block 1 track terminal TK1, then touch cab input screws 4 through 10 (cabs 1–7) with the other. All should read an open circuit, showing no cabs connected to the block. If a cab tests connected, there most likely is a solder bridge; find and correct it before going on.

☐ *Cabs 1 through 7.* Rearrange three clip leads to ground block 1 logic terminals 1, 2, and 4 according to the logic table in Fig. 17–19. Attach a lead where the table shows a 0 (logic 0 = ground) and remove any lead where the table shows a 1. For each case the indicated relay should close and the LED light. Use a VOM to check continuity between the TK1 terminal and the terminal for the activated cab and to see that there is no continuity to the terminal of any unactivated cab. Check off the corresponding box in the

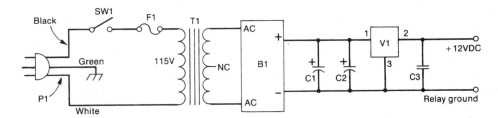

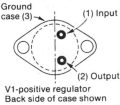

Ground case (3)
(1) Input
(2) Output

V1-positive regulator
Back side of case shown

PINOUT IDENTIFICATION
TO-3 type voltage regulator

+12VDC RELAY POWER SUPPLY PARTS LIST (In order of assembly)

Qnty.	Symbol	Description
1	P1	Three-conductor power cord w/plug (Jameco 37997)
1	—	Strain relief (Radio Shack 278-1636)
1	F1	Panel-mount fuse holder (Jameco 18702)
1	—	2A slow-blow fuse (Jameco 25646)
1	SW1	Toggle switch (Jameco 76241)
1	T1	18V, 1A Power transformer (Jameco 105531)
1	B1	6A, 200V Bridge rectifier (Digi-Key GBPC602)
1	C1	10,000µF, 25Vdc Filter capacitor(s) (Digi-Key P6480)
1	V1	7812K 12Vdc 1A regulator (Jameco 5131867)
1	—	TO-3 mounting kit with socket (Mouser 534-4732)
1	C2	1µF, 35V tantalum capacitor (Jameco 33662)
1	C3	100000pF, 50V disk capacitor (Digi-Key P4164)
1	H1	Fin-type heat sink for V1 (Jameco 16512)

Author's recommendations for suppliers given in parentheses above with part numbers where applicable. Equivalent parts may be substituted.

17-22 12-Vdc relay power supply.

table in Fig. 17–19 as each test is passed. If you find problems, look for a faulty solder joint, a part left out, or a solder bridge. Correct and continue until all eight conditions are passed, and repeat for each CRC.

Block/Cab Assignment Panel

A CCC-equipped system also needs one or more control stations for initial cab assignments and reassignments. As with all interfacing initialization, this can be accomplished using the computer keyboard, but mixing keyboard input into a tight-loop, real-time program as needed for CCC slows down the processing. This kind of control can also require interrupt processing, complicating the program. Besides, engineers can be anywhere around the system when they need to make or change assignments, and the keyboard might not be handy.

Figure 17–23 shows a Block/Cab Assignment Panel (B/CAP) that attaches direct to UCIS. Input commands are achieved via three binary-coded decimal (BCD) push-button thumbwheel switches (two to identify the block and one to identify the cab) along with two pushbuttons to activate the assignment and drop functions. Like panels and similar hookups can be applied to many applications where you desire direct operator control interface to UCIS. This panel is for input only, but I'll soon show you how to add direct display output as well.

To assign a block to a cab, the operator dials in the block and cab numbers desired, then presses the ASSIGN button. To cancel cab assignment to a block, the operator would dial the block number and press the DROP button. The cab setting doesn't matter in the drop case—any cab assigned to the block dialed will be dropped.

The output of BCD switches can go directly into the UCIS, and thumbwheel types let one enter more block numbers than possible with readily available rotary switches. Two thumb-wheel switches can select up to 100 blocks (0 through 99), and three up to 1000. The single switch for cab input handles to nine cabs, with 0 used for no cab, or off. Multiple B/CAPs can be installed so they are convenient wherever trains start and end their runs. Assemble each one as follows:

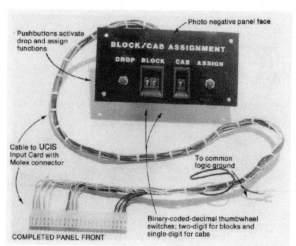
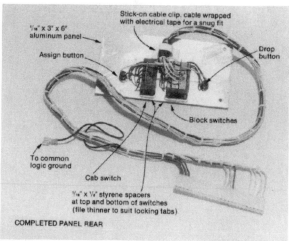

17-23 Block/cab assignment panel.

☐ *Pushbutton thumbwheel switches.* Snap together the two block number switches with one pair of end plates, and the cab number switch, also with a pair of end plates, as shown in Fig. 17–24.

☐ *Panel plate.* Lay the printed panel face from Fig. 17–25 over the aluminum and prick-punch through the paper to mark the metal for drilling. Drill the eight corner holes with a #7 bit, being careful to keep the drill inside the outline of the switch openings. Drill a ½-in hole for file clearance in the center of each switch opening, then file the openings out to their outlines, filing the drilled-out corners square to fit the switches. Drill the button and mounting-screw holes and sand the surfaces smooth, then check the fit of all parts and screws.

☐ *Panel face.* See a commercial printer who does photolithography work to have a one-to-one photo negative made of the printed panel face using Kodak graphic art reproduction reversal film, specifying that the emulsion be on the back side. Spray the back side of the negative any color(s) desired, and allow it several days to dry. Then spray the painted side of the negative with spray adhesive. Attach the negative to the panel, being careful to align it with the holes, then use a pointed knife from the front side to cut out the

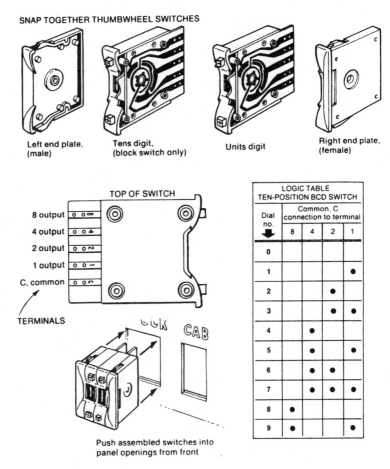

SNAP TOGETHER THUMBWHEEL SWITCHES

Left end plate, (male)   Tens digit, (block switch only)   Units digit   Right end plate, (female)

TOP OF SWITCH

8 output
4 output
2 output
1 output
C, common

TERMINALS

Push assembled switches into panel openings from front

LOGIC TABLE TEN-POSITION BCD SWITCH			
Dial no.	Common, C connection to terminal		
	8	4	2	1
0				
1				●
2			●	
3			●	●
4		●		
5		●		●
6		●	●	
7		●	●	●
8	●			
9	●			●

17-24 B/CAP assembly.

film at each hole, and to trim away excess film around the panel edge.

A simpler approach is to use the paper panel face directly from Fig. 17–25, or paint the aluminum and apply your own lettering. I prefer the film negatives. They look great, and the lettering and colors can't be scratched or worn off. I demonstrate this method for facing any type of panel in the commercially available videotape *The Basics of Model Railroad Wiring*, available from Kalmbach Publishing Co., P.O. Box 21027, Waukesha, WI 53187.

☐ *Assembly.* To install the switches, push them into position from the front side of the panel so the locking tabs snap into place. Then mount the two pushbuttons.

☐ *Wiring.* Connect a common logic ground lead to the C position of each switch and to one side of each button. Then connect 13 logic line wires, four to each block-number switch, three to the cab-number switch, and one each to the buttons. We won't test the B/CAP until it's connected to the UCIS, when we can test it easily with software.

Directional Input

In addition to B/CAP input, the CCC computer needs input for the direction of travel to be set on each cab. This could come from direction-sensitive, occupancy detectors, or by direct feedback from each cab's direction switch. For dc trains, the output of a direction-sensitive detector is reliable as long as sufficient voltage is applied to the occupied block. However, when the throttle is off, or nearly so, a direction-sensitive detector can't do the job.

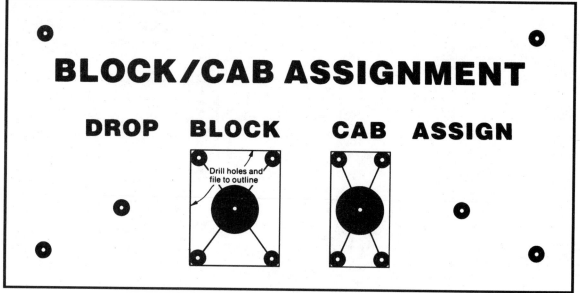

17-25 B/CAP panel.

Also, directional input from detectors takes two lines (wires) per detector to the computer, compared to just one for occupancy. Typically there are many more blocks than cabs, so monitoring direction by direction-switch input takes less wiring and requires fewer UCIS input cards.

Finally, in order to control return loops and other turning tracks automatically, CCC needs direction input from reverse *blocks* before those blocks are occupied. For that the computer must read the auxiliary direction switch (or its equivalent) that controls a reverse block and not a detector.

To get direction-switch input from cabs, simply replace their DPDT direction switches with 3PDT versions. Figure 17–26 shows how to wire them. For commercial packs where it's not practical to make modifications, add separate switches. If walka-round throttles on cables are used, the cables likely don't have extra wires for a 3PDT direction switch. Rather than change the cables use the existing DPDT switch to drive 3PST relays, as Fig. 17–26 also shows.

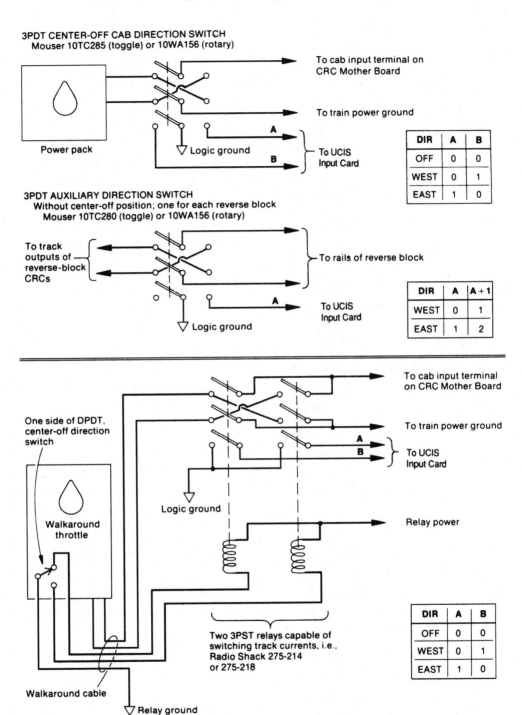

3PDT CENTER-OFF CAB DIRECTION SWITCH
Mouser 10TC285 (toggle) or 10WA156 (rotary)

To cab input terminal on CRC Mother Board

To train power ground

Power pack

Logic ground

A

B

To UCIS Input Card

DIR	A	B
OFF	0	0
WEST	0	1
EAST	1	0

3PDT AUXILIARY DIRECTION SWITCH
Without center-off position; one for each reverse block
Mouser 10TC280 (toggle) or 10WA156 (rotary)

To track outputs of reverse-block CRCs

To rails of reverse block

A

To UCIS Input Card

Logic ground

DIR	A	A+1
WEST	0	1
EAST	1	2

One side of DPDT, center-off direction switch

To cab input terminal on CRC Mother Board

To train power ground

A

B

To UCIS Input Card

Logic ground

Walkaround throttle

Relay power

Two 3PST relays capable of switching track currents, i.e., Radio Shack 275-214 or 275-218

Walkaround cable

Relay ground

DIR	A	B
OFF	0	0
WEST	0	1
EAST	1	0

17-26 Direction switch input for CCC.

The UCIS connections are identical from either a switch or a relay, with two lines to an input card for each cab. One line low indicates that the cab is set in one direction, and the other low indicates the other direction. Neither line low indicates that the cab is in the center off position, not set for either direction. A center off position is a nicety for CCC, but it can also function with a single direction input from each cab.

Cab Display Card

The LEDs on the CRC indicate which cab is assigned to a block. They're handy for tests and when working under the layout. However, it's also handy to have a cab assignment display where it can be seen when assigning and dropping cabs. In the next chapter I'll show you how to use software to put such a system status display on the computer monitor. Placing numeric displays on a track diagram panel or along the edge of the layout also can be a very effective option.

Figure 17–27 shows such an optional cab display card (CDC). The LED digit (L4) displays the cab assignment for the block, and the LED below that (L1) indicates block occupancy. The LEDs to the left and right of the digit in the photo (L2 and L3) indicate the direction set on the assigned cab. Similar displays come in handy for many interfacing projects, such as displaying machine tool/table position and direction of travel, or test result readouts.

Ready-to-assemble CDC circuit cards are available from JLC Enterprises, and assembled and tested cards can be ordered from ECO Works, EASEE Interfaces, and AIA. One CDC is required for each CCC block where one desires the optional display. Figure 17–27 also includes the parts layout, and parts list. Figure 17–28 provides a handy mounting pattern for the CDC.

Here's how to assemble the CDC:

- ☐ *R1–R13.*
- ☐ *S1, S2[+].*
- ☐ *U1[+].*
- ☐ *Spacers.* Install the spacers as shown and secure them with 4-40 nuts.
- ☐ *Panel mounting template.* The template also serves as an LED assembly jig. Lay the printed pattern, Fig. 17–28, over the aluminum and prick punch through the paper to mark the drill centers. Drill the mounting holes and check the fit to the CDC by screwing to the spacers. Once this fits, drill and file the opening for the LED digit as done for the BCD switches on the B/CAP, but make the center hole $3/8$ in. Then drill holes for the occupancy and direction LEDs.
- ☐ *L4[+].* Install L4 in its socket, making sure its orientation is correct with the decimal point as shown in Fig. 17–27.
- ☐ *L1–L3, positioning[+].* Insert the positive leads in the positive holes as indicated on the parts layout. DO NOT solder these parts in place at this time
- ☐ *Template assembly.* Mount the circuit card on the template as in Fig. 17–27, carefully working each LED into its hole in the template.
- ☐ *L1–L3, soldering.* One at a time, hold each LED in position and solder it.

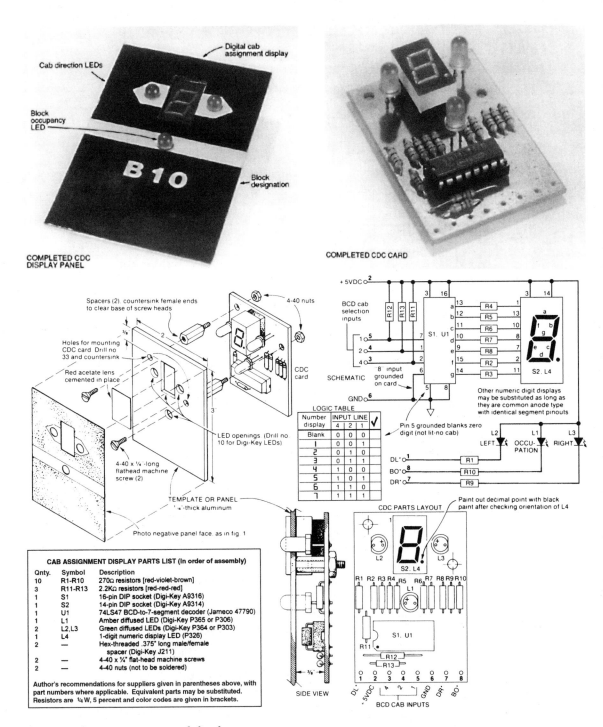

17-27 Cab assignment panel display.

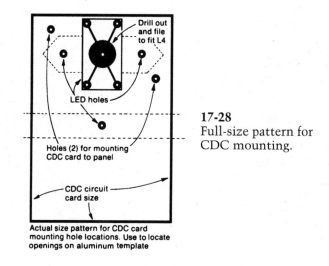

17-28
Full-size pattern for
CDC mounting.

Actual size pattern for CDC card
mounting hole locations. Use to locate
openings on aluminum template

Table 17-8. B/CAP Parts List

Qnty.   Description

3   10-position BCD pushbutton thumbwheel switches (Jameco 26788)
2   Thumbwheel switch end plates - pair (Jameco 26761)
1   Stick on cable clip (Radio-Shack 278-1640)
2   Normally-open SPST pushbuttons (Jameco 26622)

**Author's recommendations for suppliers given in parentheses above, with
part numbers where applicable.   Equivalent parts may be substituted.**

☐ *Cleanup and inspection.*
☐ *Card test.* Temporarily solder a ground test lead to card input hole 6 and a
+5-Vdc test lead to hole 2. Connect these to the 5-Vdc supply and turn on
the power. The LED digit should light up with a numeral 7. If not, check for
improper insertion of L4 or U1, or a solder bridge. Now check the digits
from 7 down to 0, following the logic test diagram in Fig. 17–27 to ground
the input lines 1, 2, and 4 temporarily so as to check the digits from 7 down
to 0. Once these pass, ground input hole 1 to check the left direction LED,
hole 7 to check the right LED, and hole 8 to test the occupation LED. If any
of these fail, check for a solder bridge or a reversed or faulty LED.

When mounting the CDCs directly to a panel as shown in Fig. 17–27, it's best to
secure the screws and spacers to the panel with cyanoacrylate adhesive, otherwise
known as superglue, so they won't come loose when taking the nuts off to remove
the card.

Cut a lens of transparent red acetate to fit over the face of L4. This makes the unlit segments darker so the displayed digit is easier to read. Red acetate is available in office supply stores that stock materials for overhead projector transparencies.

That's all the specialized hardware needed for CCC. I hope the numerous and varied circuits and cards I've covered have given you ideas for matching your own interfacing requirements to the UCIS I/O lines.

18
CHAPTER

Sample Interfacing Application: System Hookup and Software

In the last chapter, as an example of building external hardware for interfacing with UCIS, I covered the general-purpose hardware required for connecting our interface to a model railroad. In this chapter we'll make the actual connection assignments and then develop sample application software to drive the system. Even if your desired UCIS application isn't model railroading, the technological details I'll explain about this real-world hookup and the development of its application software program will go a long way to help you program the specific application you desire.

Example Layout

Figure 18–1 shows a straight-line schematic track plan for the layout I'll use as the example for hooking up Computer Cab Control via UCIS. The layout has 12 controlled blocks, numbered 1 through 12. It also has an assortment of representative features: a terminal yard, a passing siding, a reversing loop, a division-point or staging yard, and spurs off the signaled route. Even if you are a model railroader, your layout probably won't be the same as my example; but the basic steps I'll describe for this example apply to any system. In fact, the same basic techniques apply to any monitoring/control application requirement. Making computerized block assignments is just like the assignment of a given control panel switch to a particular remote flow valve in a processing plant. Likewise, turnout control is synonymous with switching conveyor flow in a material-handling system.

For this example seven turnouts, TU(0) through TU(6), control train routing through the 12 powered blocks. For simplicity, I'm assuming the turnouts are not controlled through the UCIS; they could be either manually or remotely controlled

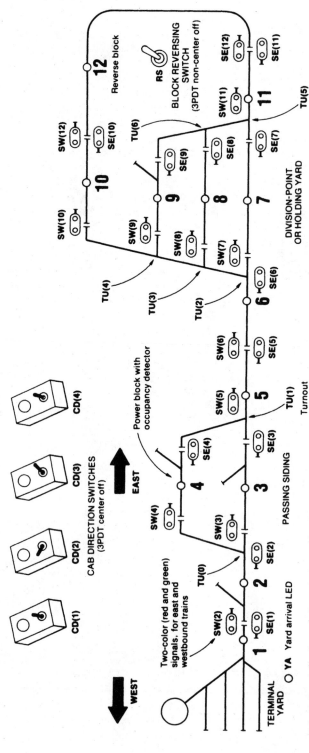

18-1 Straight-line schematic track plan.

from panel switches; or, with slight modification to the application software we'll develop, they could be computer controlled as well.

Our software will also drive trackside signals that tell operators (or other parts of the software, if the software itself is running a train) not to run into a block controlled by another cab. In general each block has two signals—one at the east end and one at the west. The exception is block 1, a stub terminal yard with no west-end signal.

For easier signal software, we'll set up two variables; *SE( )* for signals east, controlling eastward traffic, and *SW( )* for signals west, controlling westward traffic. The signals at the east and west ends of block 2 are thus *SE(2)* and *SW(2)*. As you'll soon see, signal-control output is almost a by-product of the CCC software.

For the sake of example, I'll assume the layout has four road cabs, though CCC using the hardware we developed in Chap. 12 could handle up to seven just as easily. The cab direction switches are labeled CD(1) through CD(4). For block 12, the reverse-loop block, the auxiliary reversing switch is labeled RS. For more than one, use RS with parenthetical notation: RS(1), RS(2), and so forth.

I/O Tables

Tables 18–1 and 18–2 show I/O tables for the Fig. 18–1 layout made out on the worksheet forms introduced in Chap. 15. Cards 1 and 2 are input cards, and cards 3 through 6 are output cards all connected to the hardware derived in Chap. 17. This example uses the original-design I/O cards. Everything works just the same if you are using the new I/O cards, with the only difference being the pinout numbers for the I/O lines.

The line locations are mostly arbitrary for any application, but it's handy to group related or similar lines. When it takes more than one line (more than one bit) to represent a variable, the lines should be contiguous on the same port. An example of this is the input cab-number thumbwheel bits 2 through 4 on card 1, port A.

I likewise grouped the two block-number thumbwheels, four bits each, to fill card 1, port B. Grouping them is not only logical but expedient in programming, because the program can skip past the complete port-and-bit decoding if its decoding of port A shows that neither the assign nor drop buttons were pushed. To use more than one B/CAP, give each its own pair of ports on separate input cards.

Similarly, I grouped the turnout position inputs to fill a single port, even though that wastes one line. The advantage is that a simple comparison of the complete input byte with that of the previous iteration will show if any turnout positions have changed. If none have, subsequent operations involving turnout position input can be skipped.

The center-off cab direction switches need two bits each. The lines for each switch go together, and the group of four cabs neatly fills one port. Direction switch RS for block 12 is not a center-off type. One line can sense two positions, so I tucked it in back on card 1, port A, bit 5.

We have only four basic types of UCIS output for the example layout:

- Cab Number, *CN( )*, to drive the cab relay cards and the digital readouts of the cab display cards.
- Direction of travel, *DT( )*, to drive the direction LEDs on the cab display card panels.

Table 18-1. Table Worksheets

C/MRI I/O WORKSHEET

CARD No. _1_ INPUT ~~OR OUTPUT~~

PORT	BIT	PIN	DESCRIPTION OF FUNCTION PERFORMED
	0	4	ASSIGN PUSHBUTTON — AS
	1	3	DROP PUSHBUTTON — DP
	2	2	1 ⎫
A	3	1	2 ⎬ CAB NUMBER THUMBWHEEL
	4	16	4 ⎭ INPUT — IC
	5	17	BLOCK REVERSE SWITCH— RS
	6	18	⎫ SPARE
	7	19	⎭
	0	13	1 ⎫
	1	14	2 ⎬ BLOCK NUMBER LOW ORDER
	2	15	4 ⎬ THUMBWHEEL DIGIT – BL
B	3	24	8 ⎭
	4	23	1 ⎫
	5	22	2 ⎬ BLOCK NUMBER HIGH ORDER
	6	21	4 ⎬ THUMBWHEEL DIGIT – BH
	7	20	8 ⎭
	0	9	TURNOUT 0 POSITION — TU(0)
	1	10	TURNOUT 1 POSITION — TU(1)
	2	11	TURNOUT 2 POSITION — TU(2)
C	3	12	TURNOUT 3 POSITION — TU(3)
	4	8	TURNOUT 4 POSITION — TU(4)
	5	7	TURNOUT 5 POSITION — TU(5)
	6	6	TURNOUT 6 POSITION — TU(6)
	7	5	SPARE TURNOUT

C/MRI I/O WORKSHEET

CARD No. _2_ INPUT ~~OR OUTPUT~~

PORT	BIT	PIN	DESCRIPTION OF FUNCTION PERFORMED
	0	4	BLOCK 1 OCCUPATION — BO(1)
	1	3	BLOCK 2 OCCUPATION — BO(2)
	2	2	BLOCK 3 OCCUPATION — BO(3)
A	3	1	BLOCK 4 OCCUPATION — BO(4)
	4	16	BLOCK 5 OCCUPATION — BO(5)
	5	17	BLOCK 6 OCCUPATION — BO(6)
	6	18	BLOCK 7 OCCUPATION — BO(7)
	7	19	BLOCK 8 OCCUPATION — BO(8)
	0	13	BLOCK 9 OCCUPATION — BO(9)
	1	14	BLOCK 10 OCCUPATION — BO(10)
	2	15	BLOCK 11 OCCUPATION — BO(11)
B	3	24	BLOCK 12 OCCUPATION — BO(12)
	4	23	SPURS IN BLOCK 2 — SP(2)
	5	22	SPURS IN BLOCK 3 — SP(3)
	6	21	SPURS IN BLOCK 4 — SP(4)
	7	20	SPURS IN BLOCK 9 — SP(9)
	0	9	⎫ CAB 1 DIRECTION SWITCH CD(1)
	1	10	⎭
	2	11	⎫ CAB 2 DIRECTION SWITCH CD(2)
C	3	12	⎭
	4	8	⎫ CAB 3 DIRECTION SWITCH CD(3)
	5	7	⎭
	6	6	⎫ CAB 4 DIRECTION SWITCH CD(4)
	7	5	⎭

- Signals east and west, *SE( )* and *SW( )* as explained earlier.
- A single line to drive a yard-arrival indicator, *YA,* an LED to tell the yardmaster that a mainline train is approaching his yard.

Each cab-number output needs three contiguous bits to handle our four-cab example, and in fact the same three bits are enough for up to seven cabs. Grouping two cab numbers per port leaves two bits left over in each port for signal lines. Following this pattern, cards 3 and 4 take care of all the cab assignment numbers for the 12 blocks. Card 5 fills out the rest of the signal lines and the yard-arrival LED. I've used card 6 to handle the direction-of-travel LED lines, two each per block.

System Wiring

Figure 18–2 illustrates the general system wiring, with emphasis on the hookups to the UCIS Input/Output cards. Because the wiring is repetitive, I've only shown examples of each kind of device.

Table 18-2. Table Worksheets

C/MRI I/O WORKSHEET

CARD No. 3 ~~INPUT OR~~ OUTPUT

PORT	BIT	PIN	DESCRIPTION OF FUNCTION PERFORMED
A	0	4	CAB NUMBER BLOCK 1 — CN(1)
	1	3	
	2	2	
	3	1	CAB NUMBER BLOCK 2 — CN(2)
	4	16	
	5	17	
	6	18	SIGNAL EAST BLOCK 1 — SE(1)
	7	19	SIGNAL EAST BLOCK 2 — SE(2)
B	0	13	CAB NUMBER BLOCK 3 — CN(3)
	1	14	
	2	15	
	3	24	CAB NUMBER BLOCK 4 — CN(4)
	4	23	
	5	22	
	6	21	SIGNAL EAST BLOCK 3 — SE(3)
	7	20	SIGNAL EAST BLOCK 4 — SE(4)
C	0	9	CAB NUMBER BLOCK 5 — CN(5)
	1	10	
	2	11	
	3	12	CAB NUMBER BLOCK 6 — CN(6)
	4	8	
	5	7	
	6	6	SIGNAL EAST BLOCK 5 — SE(5)
	7	5	SIGNAL EAST BLOCK 6 — SE(6)

C/MRI I/O WORKSHEET

CARD No. 4 ~~INPUT OR~~ OUTPUT

PORT	BIT	PIN	DESCRIPTION OF FUNCTION PERFORMED
A	0	4	CAB NUMBER BLOCK 7 — CN(7)
	1	3	
	2	2	
	3	1	CAB NUMBER BLOCK 8 — CN(8)
	4	16	
	5	17	
	6	18	SIGNAL EAST BLOCK 7 — SE(7)
	7	19	SIGNAL EAST BLOCK 8 — SE(8)
B	0	13	CAB NUMBER BLOCK 9 — CN(9)
	1	14	
	2	15	
	3	24	CAB NUMBER BLOCK 10 — CN(10)
	4	23	
	5	22	
	6	21	SIGNAL EAST BLOCK 9 — SE(9)
	7	20	SIGNAL EAST BLOCK 10 — SE(10)
C	0	9	CAB NUMBER BLOCK 11 — CN(11)
	1	10	
	2	11	
	3	12	CAB NUMBER BLOCK 12 — CN(12)
	4	8	
	5	7	
	6	6	SIGNAL EAST BLOCK 11 — SE(11)
	7	5	SIGNAL EAST BLOCK 12 — SE(12)

C/MRI I/O WORKSHEET

CARD No. 5 ~~INPUT OR~~ OUTPUT

PORT	BIT	PIN	DESCRIPTION OF FUNCTION PERFORMED
A	0	4	SIGNAL WEST BLOCK 2 — SW(2)
	1	3	SIGNAL WEST BLOCK 3 — SW(3)
	2	2	SIGNAL WEST BLOCK 4 — SW(4)
	3	1	SIGNAL WEST BLOCK 5 — SW(5)
	4	16	SIGNAL WEST BLOCK 6 — SW(6)
	5	17	SIGNAL WEST BLOCK 7 — SW(7)
	6	18	SIGNAL WEST BLOCK 8 — SW(8)
	7	19	SIGNAL WEST BLOCK 9 — SW(9)
B	0	13	SIGNAL WEST BLOCK 10 — SW(10)
	1	14	SIGNAL WEST BLOCK 11 — SW(11)
	2	15	SIGNAL WEST BLOCK 12 — SW(12)
	3	24	YARD ARRIVAL LED — YA
	4	23	LOOP TIMER LED — X0
	5	22	
	6	21	
	7	20	
C	0	9	SPARE
	1	10	
	2	11	
	3	12	
	4	8	
	5	7	
	6	6	
	7	5	

C/MRI I/O WORKSHEET

CARD No. 6 ~~INPUT OR~~ OUTPUT

PORT	BIT	PIN	DESCRIPTION OF FUNCTION PERFORMED
A	0	4	DIRECTION TRAVEL BLOCK 1 — DT(1)
	1	3	
	2	2	DIRECTION TRAVEL BLOCK 2 — DT(2)
	3	1	
	4	16	DIRECTION TRAVEL BLOCK 3 — DT(3)
	5	17	
	6	18	DIRECTION TRAVEL BLOCK 4 — DT(4)
	7	19	
B	0	13	DIRECTION TRAVEL BLOCK 5 — DT(5)
	1	14	
	2	15	DIRECTION TRAVEL BLOCK 6 — DT(6)
	3	24	
	4	23	DIRECTION TRAVEL BLOCK 7 — DT(7)
	5	22	
	6	21	DIRECTION TRAVEL BLOCK 8 — DT(8)
	7	20	
C	0	9	DIRECTION TRAVEL BLOCK 9 — DT(9)
	1	10	
	2	11	DIRECTION TRAVEL BLOCK 10 — DT(10)
	3	12	
	4	8	DIRECTION TRAVEL BLOCK 11 — DT(11)
	5	7	
	6	6	DIRECTION TRAVEL BLOCK 12 — DT(12)
	7	5	

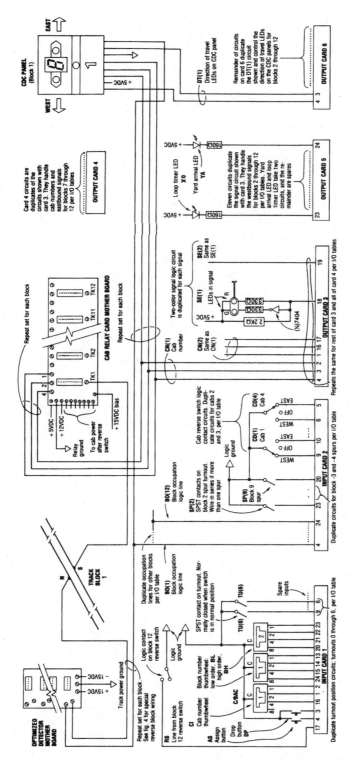

18-2 Layout wiring.

As described in Chaps. 13 and 17, good ground wiring is essential, so for every application make sure you keep your logic grounds separate from your power grounds, that all the ground lines tie together at only one central point close to the power supplies and UCIS, and that heavy wire is used for all ground lines; use at least #18 even for the smallest system, and #16 up through #10 or greater for larger systems.

Figure 18–2 also shows a sample circuit to decode line 18, card 3, for signal SE(1)—all other signal lines are identical. The 7404 is a hex inverter, so one IC can handle six signals. A low output turns on the red LED, and high turns on the green.

The only exception to the repetitiveness of the layout wiring is block 12, the reverse block. It needs special attention as shown in Fig. 18–3, a schematic of CCC wiring for a reverse block (12) with regular blocks (1 and 2) included for comparison. Notice that each reverse block requires two cab relay cards to switch the two power lines from each cab ahead of the cabs' reversing switches. The two CRCs for each reversing block have their 1, 2, and 4 logic lines wired in parallel, so they'll switch simultaneously to the same cab-assignment state on command from the UCIS.

For a system that has three or more reversing blocks, it would be handiest to use separate CRC motherboards (or cut-off portions of the CMB cards) for each set of cab power leads. For only one or two reverse blocks, it is more cost effective to hardwire the reverse-block CRCs instead of plugging them into CRC motherboards.

There are two ways to connect an optimized detector to a reverse block. The Fig. 18–4 main schematic shows the Cadillac approach, which retains the detector's full 1-MΩ sensitivity but requires a separate ±12-Vdc power supply for each reverse block and an optoisolator at the detector output. The separate supply powers just the one detector and its optoisolator, so it only needs a low current output. The Fig. 11–2 supply works great.

The bottom of Fig. 18–3 shows a way to connect the detector without a separate power supply, by moving the optoisolator ahead of the detector to read the track current directly. An ac-input optoisolator is required because the polarity of the track current will change. Wire it with a bridge rectifier in parallel and a 10-Ω resistor in series to protect the optoisolator against high track currents; otherwise, any track short circuit would exceed the 60 mA rating of this part.

The disadvantages of this method are significantly lower detection sensitivity and slight jumps in train speed entering and leaving the reverse block. The front-end optoisolator reduces the optimized detector's sensitivity from 1 MΩ to about 30 kΩ, and the added diode drop in the bridge rectifier will make the track voltage about .8 V less in the reverse block than in regular blocks.

Application Software

CCC is an excellent online real-time demonstration of UCIS capability. The computer and the railroad are connected through the UCIS to interact as the trains run. Figure 18–4 outlines the CCC program. Once initialized, the computer loops through its set of instructions over and over (the real-time loop), for as long as the railroad is operated.

In its initialization step the program defines variables and constants, initializes the UCIS by sending command bytes to the 8255s, and so on. Once into the real-time

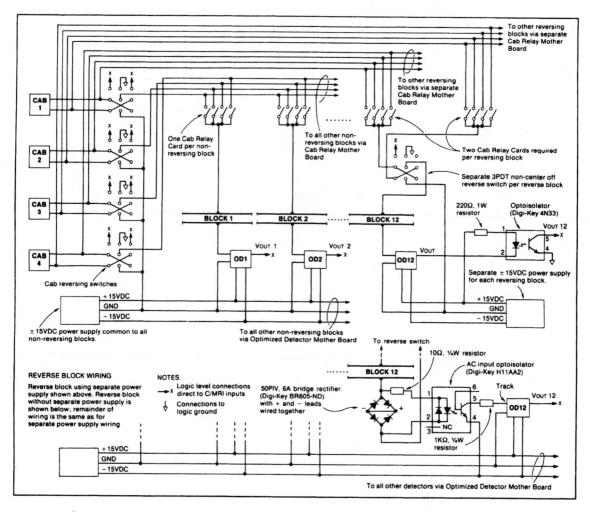

18-3 Reverse block wiring schematic.

loop, the program first reads the input cards for the status of the railroad, checking block/cab assignment panels, turnout positions, block occupancy, and cab direction switch settings.

From these inputs the program determines whether any blocks need to be assigned or dropped and calculates the settings for the signals and cab displays. Once the outputs are calculated, the program updates the monitor display (if used) and writes the outputs. Then it branches back to reread the inputs and begin the loop over again.

The loop time is typically a fraction of a second. As panel inputs change and trains move from block to block, new cab assignments are made, block assignments no longer needed are dropped, and the signals change automatically. I'll now explain the highlights of each program segment.

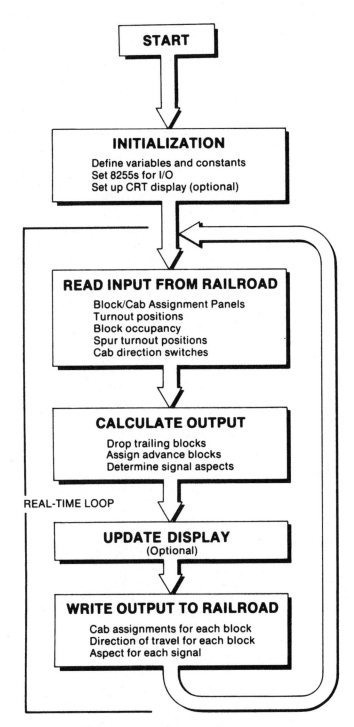

18-4 Computer cab control program structure.

Initialization

The main program for CCC with two-color signaling on our example layout is listed in Fig. 18–5. This version assumes an IBEC or UBEC application with memory-mapped I/O and a UCIS starting address of 640K using an IBM PC, x86, or Pentium. It should be straightforward by now to convert this to port I/O. The example uses Microsoft BASIC and I've avoided exotic codes, so most of the statements should work on any computer with few changes.

Because this is our first real-world, application-operational UCIS program, I'll walk you through it, pointing out significant additional technology beyond that provided in the simple introductory example I/O software covered in Chap. 15. Figure 18–6 defines the important variables for reference. Don't try to understand all of them now; they'll be clear once you understand the program shown in Fig. 18–5.

Lines 40 through 70 dimension, or define the size of, each array variable. For example, variable *BO* is used to store the block occupancy status for each block. There are 12 blocks, so *BO* is dimensioned to BO(12). In use, *BO(7) = 0* will mean that block 7 is clear and *BO(7) = 1* that it's occupied. The string variables, those followed by a $, store alphanumeric characters for the optional monitor display subroutines.

Lines 90 and 100 set up the special constants used to unpack and pack the input and output bytes received and transmitted over the UCIS following the same procedure we set up in Chap. 15.

Lines 120 through 240 attach two-letter symbols to frequently-used constants. As an example, for the program 1 means westbound and 2 eastbound, but lines 190 and 200 let us use *WB* and *EB* so we don't have to remember the numbers. Also, line 240 makes the program more general-purpose by defining variable *MB* for the maximum block number. If a layout has 40 blocks, simply set *MB* = 40 and all other relationships that depend on *MB* are automatically set for 40 blocks. The only other required change is the size of the DIM statement.

Lines 260 and 270 are layout-dependent and define the constant next-east or next-west blocks, those that don't change with turnout position. Look at the track schematic, and they should be obvious. The next block east of 1 is 2, so *NE(1) = 2*; likewise *NE(5) = 6*.

The next blocks east of 2, 3, and 4 aren't constant and so aren't defined here. For example, *NE(2)* is 3 if turnout 0, *TU(0)*, is normal, and 4 if *TU(0)* is reversed. The program decides this after reading turnout positions in the real-time loop. Also note that *NW(1)* is NB for no block, because the whole terminal yard is block 1; and there is no block farther west. These statements are layout-dependent and must be tailored to fit the actual track plan.

Lines 290 through 310 define the input and output control bytes for the I/O cards, and the UCIS starting address of 640K via DEF SEG = 40960. In line 330 the GOTO statement branches to a subroutine to initialize a monitor status display. I'll come back to that later.

Lines 350 through 400 set the 8255s on the first two cards for input and on cards 3 through 6 for output. Line 420 sets S1, the saved turnout-position byte, to an unlikely value. The program won't have to unpack the turnout-position byte on every iteration, but this forces it to do so on the first one.

```
10 PRINT "COMPUTER CAB CONTROL -EXAMPLE PROGRAM"
20 REM**Define variable types and array sizes**
30 DEFINT A-Z
40 DIM BO(12), CN(12), DT(12), BS(12), CS(12), DS(12)
50 DIM NE(120, NW(12), SP(12), SE(12), SW(12), ES(12)
60 DIM WS(12), BE(12), BW(12), SS(12), TU(6), TS(6), CD(4)
70 DIM CC(4), DD$(2), DC$(4), DO$(4), DR$(1)
80 REM**Define constants for packing/unpacking bits**
90 B0 = 1: B1 = 2: B2 = 4: B3 = 8: B4 = 16: B5 = 32: B6 = 64: B7 = 128
100 W1 = 1: W2 = 3: W3 = 7: W4 = 15: W5 = 31: W6 = 63: W7 = 127
110 REM**Define general constants**
120 CL = 0      'Clear
130 OC = 1      'Occupied
140 RD = 1      'Red
150 GN = 0      'Green
160 NB = 99     'No next block
170 ND = 0      'No direction
180 NC = 0      'No cab assignied
190 WB = 1      'Westbound
200 EB = 2      'Eastbound
210 TN = 1      'Turnout normal position
220 TR = 0      'Turnout reverse position
230 PP = 1      'Pushbutton pressed
240 MB = 12     'Maximum block number
250 REM**Define next east/west constant blocks**
260 NE(1) = 2: NE(5) = 6: NE(10) = 12: NE(11) = 12: NE(12) = 11
270 NW(1) = NB: NW(2) = 1: NW(6) = 5: NW(12) = 10
280 REM**Define I/O interface constants**
290 CI = 155    '8255 Control byte for input
300 CO = 128    '8255 Control byte for output
310 DEF SEG = 40960  (Assumes Memory-mapped I/O at 640KB)
320 REM**Initialize moniator status display - optional**
330 GOTO 3020   'REMark out if not using monitor display
340 REM**Transmit control bytes to set up I/O ports
350 POKE 3, CI   'card 1 input
360 POKE 7, CI   'card 2 input
370 POKE 11, CO 'card 3 output
380 POKE 15, CO 'card 4 output
390 POKE 19, CO 'card 5 output
400 POKE 23, CO 'card 6 output
410 REM**Force initial iteration thru turnout logic**
420 S1 = 99'Set turnout save byte to off-beat value
430 REM**Initialize count variable for loop timing**
440 X = 0
450 REM**Read inputs and separate out bit areas**
460 GOTO 1010 'Line for keyboard test to bypass inputs
470 IB = PEEK(0)       'Card 1 port A
480 AS = IB AND W1
490 DR = IB \ B1 AND W1
500 IC = IB \ B2 AND W3
510 RS = IB \ B5 AND W1
520 REM**Read port B later only if ASign/DRop pressed**
530 T1 = PEEK(2)       'Card 1 port C
540 IB = PEEK(4)       'Card 2 port A
550 BO(1) = IB AND W1
560 BO(2) = IB \ B1 AND W1
570 BO(3) = IB \ B2 AND W1
580 BO(4) = IB \ B3 AND W1
590 BO(5) = IB \ B4 AND W1
600 BO(6) = IB \ B5 AND W1
610 BO(7) = IB \ B6 AND W1
620 BO(8) = IB \ B7 AND W1
630 IB = PEEK(5)       'Card 2 port B
640 BO(9) = IB AND W1
650 BO(10) = IB \ B1 AND W1
660 BO(11) = IB \ B2 AND W1
670 BO(12) = IB \ B3 AND W1
680 SP(2) = IB \ B4 AND W1
690 SP(3) = IB \ B5 AND W1
700 SP(4) = IB \ B6 AND W1
710 SP(9) = IB \ B7 AND W1
720 IB = PEEK(6)       'Card 2 port C
730 CD(1) = IB AND W2
740 CD(2) = IB \ B2 AND W2
750 CD(3) = IB \ B4 AND W2
760 CD(4) = IB \ B6 AND W2
770 REM**If assign button pressed perform assign**
780 IF AS<>PP GOTO 850
790 IB = PEEK(1)       'Card 1 port B
800 BL = IB AND W4
810 BH = IB \ B4 AND W4
820 BI = BL + (10 * BH)
830 CN(BI) = IC
840 REM**If drop button pressed perform drop**
850 IF DR<>PP THEN GOTO 920
860 IB = PEEK(1)
870 BL = IB AND W4
880 BH = IB \ B4 AND W4
890 BI = BL + (10 * BH)
900 CN(BI) = NC
910 REM**If turnout changed decode/update next blocks**
920 IF T1 = SI THEN GOTO 1150
930 S1 = T1
940 TU(0) = T1 AND W1
950 TU(1) = T1 \ B1 AND W1
960 TU(2) = T1 \ B2 AND W1
970 TU(3) = T1 \ B3 AND W1
980 TU(4) = T1 \ B4 AND W1
990 TU(5) = T1 \ B5 AND W1
1000 TU(6) = T1 \ B6 AND W1
1010 IF TU(0) = TN THEN NE(2) = 3: NW(4) = NB: GOTO 1030
1020 NE(2) = 4: NW(3) = NB: NW(4) = 2
1030 IF TU(1) = TN THEN NW(5) = 3: NW(3) = 5: NE(4) = NB: GOTO 1050
1040 NW(5) = 4: NE(3) = NB: NE(4) = 5
1050 NW(7) = NB: NW(8) = NB: NW(9) = NB: NW(10) = NB
1060 NE(7) = NB: NE(8) = N: NE(9) = NB
1070 IF TU(2) = TN THEN NE(6) = 7: NW(7) = 6: GOTO 1110
1080 IF TU(3) = TR THEN NE(6) = 8: NW(8) = 6: GOTO 1110
1090 IF TU(4) = TR THEN NE(6) = 9: NW(9) = 6: GOTO 1110
1100 NE(6) = 10: NW(10) = 6
1110 IF TU(5) = TN THEN NW(11) = 7: NE(7) = 11: GOTO 1150
1120 IF TU(6) = TR THEN NW(11) = 8: NE(8) = 11: GOTO 1150
1130 NW(11) = 9: NE(9) = 11
1140 REM**Set direction of travel for assigned blocks**
1150 FOR N = 1 TO MB
1160 DT(N) = ND
1170 IF CN(N) > NC THEN DT(N) = CD(CN(N))
1180 NEXT N
1190 DT(12) = RS + 1
1200 REM**Drop block if not valid advance block**
1205 FOR N = 1 TO MB
1210 IF BO(N) = OC THEN GOTO 1260
1215 IF CN(N) = NC THEN GOTO 1260
1220 IF DT(N) = ND THEN GOTO 1255
1225 IF DT(N)=WB THEN AB=NE(N): GOTO 1235
1230 AB = NW(N)
```

18-5 Computer cab control main program using an IBEC or UBEC.

```
1235 IF AB = NB THEN GOTO 1255
1240 IF BO(AB) = CL THEN GOTO 1255
1245 IF CN(AB) <> CN(N) THEN GOTO 1255
1250 GOTO 12160
1255 CN(N) = NC
1260 NEXT N
1280 REM**Assign available advance block to desired cab**
1285 FOR N = 1 TO MB
1290 IF BO(N) = CL THEN GOTO 1370
1295 IF CN(N) = NC THEN GOTO 1370
1300 IF DT(N) = ND THEN GOTO 1370
1310 IF DT(N) = WB THEN AB = NW(N): GOTO 1330
1320 AB = NE(N)
1330 IF AB = NB THEN GOTO 1370
1340 IF BO(AB) = OC THEN GOTO 1370
1350 IF CN(AB) > NC THEN GOTO 1370
1360 CN(AB) = CN(N)
1370 NEXT N
1380 REM**Compute signal indications**
1390 FOR N = 1 TO MB
1400 SE(N) = RD: SW(N) = RD: IE = NE(N): IW = NW(N)
1410 IF IE = NB THEN GOTO 1490
1420 IF BO(IE) = OC THEN GOTO 1490
1430 IF SP(IE) = TR THEN GOTO 1490
1440 IF BO(N) = OC THEN GOTO 1470
1450 IF CN(IE) > NC THEN GOTO 1490
1460 GOTO 1480
1470 IF CN(IE) <> CN(N) THEN GOTO 1490
1480 SE(N) = GN
1490 IF IW = NB THEN GOTO 1570
1500 IF BO(IW) = OC THEN GOTO 1570
1510 IF SP(IW) = TR THEN GOTO 1570
1520 IF BO(N) = OC THEN GOTO 1550
1530 IF CN(IW) > NC THEN GOTO 1570
1540 GOTO 1560
1550 IF CN(IW) <> CN(N) THEN GOTO 1570
1560 SW(N) = GN
1570 NEXT N
1580 REM**Set reverse block signals red if polarity wrong**
1590 IF DT(10) <> DT(12) THEN SE(10) = RD: SW(12) = RD
1600 IF DT(11) = ND THEN SE(11) = RD: SE(12) = RD: GOTO 1630
1610 IF DT(11) = DT(12) THEN SE(11) = RD: SE(12) = RD
1620 REM**Set yard approach LED**
1630 YA = 0: IF DT(2) = WB THEN YA = 1
1640 REM**Update monitor display - optional**
1650 GOTO 3510   'REMark out if don't want display
1660 REM**Set XOut bit for timing**
1670 XO = X AND W1
1680 REM**Write outputs to railroad**
1690 OB = 0        'Card 3 port A
1700 OB = (7 - CN(1)) OR OB
1710 OB = (7 - CN(2)) * B3 OR OB
1720 OB = SE(1) * B6 OR OB
1730 OB = SE(2) * B7 OR OB
1740 POKE 8, OB
1750 OB = 0        'Card 3 port B
1760 OB = (7 - CN(3)) OR OB
1770 OB = (7 - CN(4)) * B3 OR OB
1780 OB = SE(3) * B6 OR OB
1790 OB = SE(4) * B7 OR OB
1800 POKE 9, B
1810 OB = 0        'Card 3 port C
1820 OB = (7 - CN(5)) OR OB
1830 OB = (7 - CN(6)) * B3 OR OB
1840 OB = SE(5) * B6 OR OB
1850 OB = SE(6) * B7 OR OB
1860 POKE 10, B
1870 OB = 0        'Card 4 port A
1880 OB = (7 - CN(7)) OR OB
1890 OB = (7 - CN(8)) * B3 OR OB
1900 OB = SE(7) * B6 OR OB
1910 OB = SE(8) * B7 OR OB
1920 POKE 12, OB
1930 OB = 0        'Card 4 port B
1940 OB = (7 - CN(9)) OR OB
1950 OB = (7 - CN(10)) * B3 OR OB
1960 OB = SE(9) * B6 OR OB
1970 OB = SE(10) * B7 OR OB
1980 POKE 13, OB
1990 OB = 0        'Card 4 port C
2000 OB = (7 - CN(11)) OR OB
2010 OB = (7 - CN(12)) * B3 OR OB
2020 OB = SE(11) * B6 OR OB
2030 OB = SE(12) * B7 OR OB
2040 POKE 14, OB
2050 OB = 0        'Card 5 port A
2060 OB = SW(2) * OB
2070 OB = SW(3) * B1 OR OB
2080 OB = SW(4) * B2 OR OB
2090 OB = SW(5) * B3 OR OB
2100 OB = SW(6) * B4 OR OB
2110 OB = SW(7) * B5 OR OB
2120 OB = SW(8) * B6 OR OB
2130 OB = SW(9) * B7 OR OB
2140 POKE 16, OB
2150 OB = 0        'Card 5 port B
2160 OB = SW(10) OR OB
2170 OB = SW(11) * B1 OR OB
2180 OB = SW(12) * B2 OR OB
2190 OB = YA * B3 OR OB
2200 OB = XO * B4 OR OB
2210 POKE 17, OB
2220 OB = 0        'Card 6 port A
2230 OB = DT(1) OR OB
2240 OB = DT(2) * B2 OR OB
2250 OB = DT(3) * B4 OR OB
2260 OB = DT(4) * B6 OR OB
2270 POKE 20, OB
2280 OB = 0        'Card 6 port B
2290 OB = DT(5) OR OB
2300 OB = DT(6) * B2 OR OB
2310 OB = DT(7) * B4 OR OB
2320 OB = DT(8) * B6 OR OB
2330 POKE 21, OB
2340 OB = 0        'Card 6 port C
2350 OB = DT(9) OR OB
2360 OB = DT(10) * B2 OR OB
2370 OB = DT(11) * B4 OR OB
2380 OB = DT(12) * B6 OR OB
2390 POKE 22, OB
2400 REM**Increment loop counter
2410 X = X + 1: IF X > 31000 THEN X = 0
2420 GOSUB 4020   'FOR TEST ONLY
2430 REM**Return to begining of loop**
2440 GOTO 450
```

18-5 *Continued.*

Line 440 sets variable *X* to 0 (zero), to be incremented by 1 each time through the real-time loop. This can be printed on the monitor to show that the program is running and to permit timing of the loops.

Input

The real-time loop starts at line 450 with the reading of inputs and separation of values for assignment to variables. Line 460 is a temporary instruction to skip reading the UCIS inputs so the program can be tested with simulated input from the keyboard. For normal operation using UCIS input, omit line 460.

Line 470 reads the input byte from port A of card 1 and stores it at location *IB*. After each PEEK we unpack the bits once for each variable read from that port.

As defined in the I/O tables, card 1 port A includes four variables: assign button *AN*, drop button *DR*, input from cab thumbwheel setting *IC*, and block 12 auxiliary reverse switch *RS*. All take one bit each except *IC*, which takes three.

The pattern for each unpacking statement is identical. The variable to be defined is calculated by first dividing the input byte by a bit-position constant, B0 through B7, depending on the position of the variable's least significant bit. This effectively shifts the byte to the right by the required number of bits so that the LSB is in the B0 position.

That result is ANDed with a second constant, W1 through W5, depending on the variable's width in bits. The AND operation zeros out or masks all but the bits defining the variable in question.

For example, the I/O table shows that *AN* is one bit wide at position 0, so line 480 is written as *AN* = *IB*/B0 AND W1. Variable *IC* is three bits wide with its LSB at position 2, so line 500 is *IC* = *IB*/B2 AND W3. Once all the variables from one port are defined, the next port is handled with another PEEK to read in its input byte.

Lines 530 through 1000 follow the same process. A PEEK brings in the port data followed by repeated executions of the same basic unpacking instruction format to separate the input bits defining the respective variables.

For any application it's important to group your input lines to cut the cycle time of your real-time loop. For example, input and decoding of card 1, port B is completely skipped unless the reading of port A shows that either the assign or drop button is pressed.

I purposely grouped all the turnout position inputs into a single port. Line 530 reads the port and stores the input byte as variable *T1*. In line 920 the present *T1* is compared with the last iteration's *T1* value, stored as *S1*. If no turnouts have been thrown, *T1* = *S1* and the program branches to line 1150, bypassing the unpacking of the individual bits and the calculation of the next blocks which depend on turnout position. If more than eight CCC routing turnouts exist, the program can skip each group of eight by using variables *T2* and *S2* for the second group, followed by *T3* and *S3* for the third group, and so on.

It is important to go back and see how the thumbwheel switches are handled. Line 780 invokes a branch to skip to line 850 if *AN* is not equal to *PP* (*AN*<>*PP*), the variable defined in line 230 for indicating whether a pushbutton is pressed. *AN* = *PP* if the button is pressed, and line 790 reads card 1 port B. Line 800 sets *BL* to the block low-order thumbwheel switch value (units digit), and 810 sets *BH* to the block

BO(N): Array dimensioned to the maximum block number used to define the current occupation status of the Nth block; e.g. BO(7)= 1 for block 7 occupied and BO(7)=∅ for block 7 clear.

BS(N): Array dimensioned to the maximum block number used to Save the last iteration of the block occupation array BO(N).

N: Array index (pointer) used for subscripted variable notation.

CN(N): Array dimensioned to the maximum block number used to define the current cab assignment(if any) of the Nth block; e.g. CN(5)=∅ for no cab assigned to block 5 and CN(5)=3 for cab number 3 assigned.

CS(N): Array dimensioned to the maximum block number used to Save the last iteration of the CN(N) array.

DT(N): Array dimensioned to the maximum block number used to define when a cab is in the Nth block the direction of travel(if any); e.g. DT(7)=∅ if the cab assigned to block 7 is set for no direction or if there is no cab assigned to block 7, DT(7)=1 if the cab assigned to block 7 is set for Westbound, DT(7)=2 if the cab assigned to block 7 is set for Eastbound.

DS(N): Array dimensioned to the maximum block number used to Save the last iteration of the DT(N) array.

NE(N): Array dimensioned to the maximum block number used to define the current Next Eastward block number connected to the Nth block; e.g. NE(1)=2 defines that the next eastward block for block 1 is block 2. Block number is set to NB (where NB=No Block=99) if there is no next block connected.

BE(N): Array dimensioned to the maximum block number used to save the last iteration of the NE(N) array.

NW(N): Array dimensioned to the maximum block number used to define the current Next Westward block number connected to the Nth block; e.g. NW(6)=5 defines that the next westward block for block 6 is block 5. Block number is set to NB (where NB=No Block=99) if there is no next block connected.

BW(N): Array dimensioned to the maximum block number used to save the last iteration of the NW(N) array.

SP(N): Array dimensioned to the maximum block number used to define the current status of any spurs left open in the Nth block; e.g. SP(4)=∅ indicates all spurs(if any) are set for normal position in block 4 and SP(4)=1 indicates that one or more spurs in block 4 are in the reverse(unsafe position).

SS(N): Array dimensioned to the maximum block number used to Save the last iteration of the SP(N) array.

SE(N): Array dimensioned to the maximum block number used to define the current state of the Signal at the East end of the Nth block; e.g. SE(12)=∅ defines a red signal and SE(12)=1 defines a green signal.

ES(N): Array dimensioned to the maximum block number used to save the last iteration of the SE(N) array.

SW(N): Array dimensioned to the maximum block number used to define the current state of the Signal at the West end of the Nth block; e.g. SW(12)=∅ defines a red signal and SW(12)=1 defines a green signal.

EW(N): Array dimensioned to the maximum block number used to save the last iteration of the SW(N) array.

TU(N): Array dimensioned to the maximum turnout number used to define the current position status of route selection turnouts; e.g. TU(∅)=1 defines that turnout ∅ is set for the turnout normal route and TU(∅)=∅ defines that turnout ∅ is set for the turnout reversed route.

TS(N): Array dimensioned to the maximum turnout number used to Save the last iteration of the TU(N) array.

CD(N): Array dimensioned to the maximum cab number defining Cab Direction i.e. position of the Nth cabs reverse switch; e.g.: CD(3)=∅ if cab 3 reverse switch is set at center off(no direction), CD(3)=1 if cab 3 reverse switch is set for westbound, CD(3)=2 if cab 3 reverse switch is set for eastbound.

CC(N): Array dimensioned to the maximum block number used to Save the last iteration of the CD(N) array.

DD$(2): Array used to define the Displayed Direction used on the CRT for direction of travel; i.e.: DD$(∅)=" " (a blank) for no direction, DD$(1)="<" for westbound, DD$(∅)=">" for eastbound.

DO$(1): Array used to define the Displayed Occupation used on the CRT: DO$(∅)=" " (a blank) for clear, DO$(1)="1" for occupied.

DC$(4): Array used to define the Displayed Cab number used on the CRT: DC$(∅)=" " (a blank) for no cab. DO$(1)="1", DO$(2)="2", DO$(3)="3" and DO$(4)="4".

DR$(1): Array used to define the Displayed Red signal used on the CRT: DR$(∅)=" " (a blank) for green, DR$(1)="R" for red.

B∅, B1, B2, B3, B4, B5, B6, and B7: Constants used to define the low order Bit position for each variable to be unpacked or packed in an input or output byte.

18-6 Table of variables.

W1, W2, W3, W4, W5, W6, W7: Constants used to define the number of bits Wide taken up by a variable in the input or output byte.

YA: Yard Arrival indication used to inform the yardmaster that a train has entered single track heading for the yard: YA=1 indicates train arriving, YA=∅ indicates train not arriving,

X$: String variable used to temporarily store the Next East or Next West block numbers used on the CRT.

FNDC$(X,Y): Direct Cursor function used to direct the cursor to the point on the CRT screen defined by the argument input coordinates where X=line number and Y=column number.

CHR$(I): Intrinsic function used to return a string whose element has the ASCII code corresponding to the input argument I.

LEN(X$): Intrinsic function used to return the length of the string X$ with non-printing characters and blanks counted.

CI: Variable set to the Control Input byte for the 8255; CI=155.

CO: Variable set to the Control Output byte for the 8255; CO=128.

SA: User defined variable to be set to the C/MRI's Starting Address e.g. SA=1∅24 for the Sunset Valley.

IB: Variable used for temporary storage of the Input Byte read in from an input port.

OB: Variable used for temporary storage of the Output Byte being formulated to send to an output port.

AS: Variable used to define the status of the ASsign pushbutton on the cab/block initialization panel; AS=1 if pushbutton pressed and AS=∅ if pushbutton not pressed.

DR: Variable used to define the status of the DRop pushbutton on the cab/block initialization panel; DR=1 if pushbutton pressed and DR=∅ if pushbutton not pressed.

IC: Variable used to define the Input Cab thumbwheel switch position: IC=∅ defines no cab number selected, IC=1 through 7 defines the cab number input selected.

BH: Defines the input Block High order thumbwheel switch setting(∅ thru 9).

BL: Defines the input Block Low order thumbwheel switch setting(∅ thru 9).

BI: Defines the Block Input selected on the cab/block initialization panel as calculated by BI=BL+(1∅*BH).

RS: Defines the position the the block 12 reverse switch (w/o center off): RS=∅ for eastbound and RS=1 for westbound.

T1: Defines the 1st(and only for this example) Turnout input byte status.

S1: Used to Save the 1st(and only for this example) Turnout input byte for comparison to see if a turnout has changed status.

CB: Working variable used to temporarily define a specific CaB number entry in CN( ).

AB: Working variable used to temporarily define the Advance Block and Approach Block numbers.

IE: Working variable used to temporarily define the next East block number.

IW: Working variable used to temporarily define the next West block number.

X: Loop counter used for program timing.

XO: Lowest order bit of X used for C/MRI display Output for timing program iteration rate.

CL: Variable set to constant=∅ for CLear.

OC: Variable set to constant=1 for OCcupied.

RD: Variable set to constant=1 for ReD.

GN: Variable set to constant=∅ for GreeN.

NB: Variable set to constant=99 for No next Block.

ND: Variable set to constant=∅ No Direction.

NC: Variable set to constant=∅ for No Cab assigned.

WB: Variable set to constant=1 for WestBound.

EB: Variable set to constant=2 for EastBound (East is even).

TN: Variable set to constant=1 for Turnout Normal.

TR: Variable set to constant=∅ for Turnout Reversed.

PP: Variable set to constant=1 for Pushbutton Pressed.

MB: Variable set to maximum block number (MB=12 for this example).

VA$: String variable used to enter the two-alpha-character code when using simulated input.

NI: Variable used to enter the index to the array position to be altered when using simulated input.

VA: Variable used to enter the numeric value for parameter to be altered when using simulated input.

high-order value (tens digit). The total block number input is then calculated as *BI* = *BL* + (10*BH) in line 820.

Because the assign button is pressed, line 830 sets the cab number in block *BI* to the number read from thumbwheel switch *IC*. Lines 850 through 900 handle the drop case in the same manner except that *CN(BI)* is set to 0, NC for no cab.

Lines 1010 through 1130 are layout-dependent and define those next blocks east and west that vary with turnouts. In line 1010, for example, if turnout 0 is set to turnout normal, TN, then *NE(2)* = 3, *NW(3)* = 2, and *NW(4)* = NB, and the program branches around line 1020. If turnout 0 is reversed, not normal, line 1020 is executed to set *NE(2)* = 4, *NW(3)* = NB and *NW(4)* = 2. This all follows from the track plan, and lines 1030 and 1040 do the same thing for turnout 1.

The three-track yard, blocks 7, 8, and 9, and branch-into-block-10 require a bit more logic. Because at most times *NE* and *NW* for blocks 7, 8, and 9 and *NW* for block 10 will be no block, these are set to NB in lines 1050 and 1060. Then lines 1070 through 1130 reset whichever next blocks become valid when turnouts are thrown.

Lines 1150 through 1180 set the direction of travel *DT* for each block. First 1160 sets *DT(N)* for no direction, ND. Then 1170 resets *DT* for each block that has a cab assigned (*CN(N)* greater than 0), equal to the direction set on the assigned cab. Line 1190 takes care of reverse block 12, where *DT(12)* is determined by the block 12 auxiliary reversing switch, not the cab direction switch.

Block/Cab Assignment

It's best to have the program automatically drop any assigned blocks that are no longer needed, such as trailing blocks that have become unoccupied, then search for and assign clear blocks ahead of moving trains. That way any blocks that were dropped are immediately available for reassignment. For 12 blocks or 120, the 12 code statements in lines 1205 through 1260 will do the job.

To help you understand this code, Fig. 18–7 shows the logic flow for dropping cab assignments. Rectangular boxes in this diagram represent calculations performed by the computer. Diamonds represent decision points such as IF statements. The statement numbers let you compare the program listing to the chart.

To find cab assignments that can be dropped, the program loops through each block from 1 to MB. If line 1210 finds the *N*th block occupied or 1215 finds that it has no cab assigned, no action is taken and the program branches to line 1260 to increment *N* for testing the next block. If at 1220 it finds that there is no direction of travel setting, the cab assignment is dropped via a branch to 1255.

For each clear block with a cab assigned and a direction of travel identified, program lines 1225 and 1230 look for an approach block AB in the direction from which the train is approaching, opposite the direction of travel. If there is no approach block, or if it is also clear, or if it is assigned to another cab, the program again branches to program line 1255 to drop the assignment for the *N*th block. If it gets past all of the IF statements for a given block, the program again reaches line 1260 and the assignment is retained.

Lines 1285 through 1370 automatically assign cabs to the blocks ahead of each moving train. Again, these lines stay the same for any layout! The program takes another look at each block. For each one it finds occupied, and with both a cab

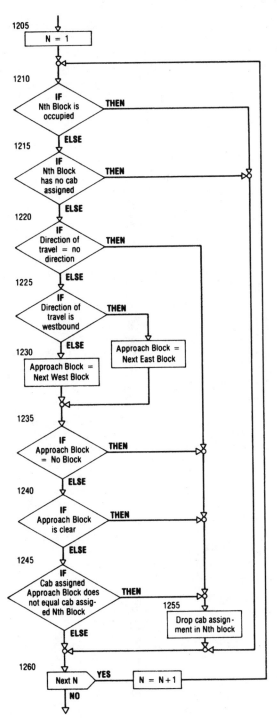

18-7
Logic flow for dropping cab assignments.

assignment and a direction of travel, the program looks for an advance block, also called AB, the next one in the direction of travel. If there is an AB, and if it is both clear and unassigned, the program makes the assignment (Fig. 18–8).

Any time an IF condition is true the program makes no assignment, but instead branches to line 1370 to increment N for testing the next block. If it gets past all the IF conditions for a given block, the program reaches line 1360 and sets the cab number for the advance block equal to the cab number for the Nth block.

Signaling

Once all the block occupancy, turnout position, and cab assignment data are in the computer, it takes only a few added steps to operate trackside signals on a CCC layout. I'll explain software to drive a simple two-color signal system to make CCC operation easier. If desired, it only takes the addition of a few statements to convert the program to drive three-color signaling.

Lines 1390 through 1570 perform all the general logic for the signals, and again these 19 lines are the same for any layout. The software loops through each block and initially sets the signals at the east and west ends to red. Also, to save execution time by minimizing use of subscripted array variables, working variables IE and IW are temporarily set equal to the next-east and next-west block numbers.

Lines 1410 through 1480 then determine if it is safe to change the east end signal from red to green. Figure 18–9 shows the logic flow. If the next block east is no block, if it is occupied, or if it has a spur turnout reversed, the signal guarding the block is kept red by having the program branch to line 1490.

If these tests all fall through, the program reaches line 1440 to check occupancy of the Nth block. If occupied, 1440 branches to line 1470, an important check of whether the cab assigned to the Nth block is the same one running the approaching train. If the cab numbers are not equal, a branch to 1490 keeps the signal red. If the assigned cabs are identical, 1480 changes the signal at the east end of the block to green.

If, back in 1440, the Nth block is unoccupied, line 1450 is executed; if the next block east has a cab assigned, a branch to line 1490 keeps the signal red. If there is no cab assigned to the next block, line 1480 is reached and the signal at the east end of the block changes to green. Lines 1490 through 1560 are just like 1420–1480, except that they govern the signal at the west end of the block.

That's the standard signal logic, but reverse-loop block 12 is a special case. How nice it would be to have the computer interlock the block 12 auxiliary reversing switch with the appropriate cab's direction switch, and then set the signals so that a train wasn't permitted to enter or leave block 12 unless the track polarities were correct!

With the UCIS, all sorts of nice things are possible just by adding a little software. Lines 1590, 1600, and 1610 operate according to the reverse-loop-polarity logic shown in Fig. 18–10 to control the signals both in and out of block 12. Obviously, these lines would have to be changed to reflect whichever block or blocks are reversing blocks on any particular railroad.

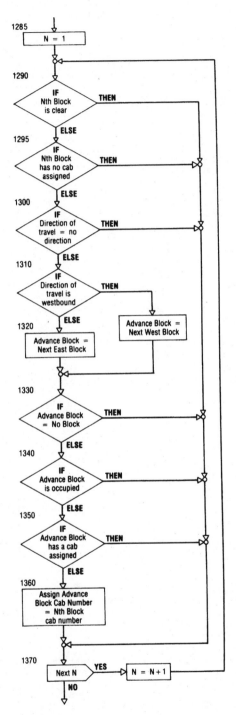

18-8
Logic flow for assigning cabs
to blocks.

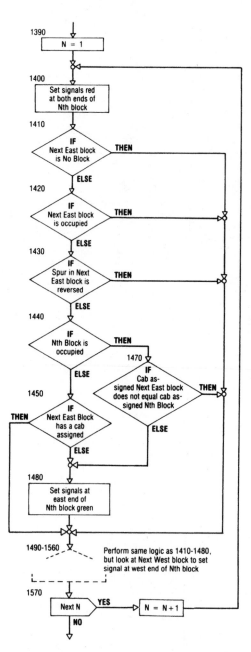

18-9
Logic flow for two-color trackside signals.

Line 1630 sets the yard approach LED variable *YA* so that the LED lights to tell the yardmaster a train is approaching. This line initializes *YA* to off and then, if the direction of travel in block 2 is westbound, resets *YA* to on.

Line 1650 is a GOTO instruction for use of a monitor status display as part of a real-time loop. Typically you'll find that you will want such a feature for every application, at least in the front-end integration phase of a program development. Read-

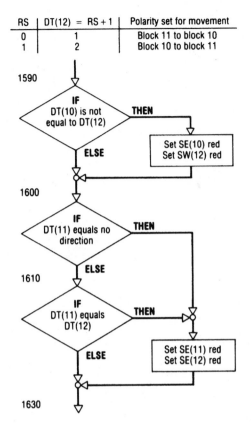

RS	DT(12) = RS + 1	Polarity set for movement
0	1	Block 11 to block 10
1	2	Block 10 to block 11

18-10
Logic flow for reversing block signals.

ing in hardware lines and immediately seeing the results displayed on a monitor is great for system testing and debugging. I'll cover the actual subroutine later, but if the feature isn't wanted, simply leave out the line.

Line 1670 comes in very handy, for any application, to give you a hardware indication for timing your software. Variable X increments by 1 each time through the real-time loop, so that the LSB of X alternates between 0 and 1. By setting $X0 = X$ AND $W1$, $X0$ takes on the value of the LSB of X. $X0$ can be sent as output to an LED (pin 23 on card 5 in Fig. 18–2 or Table 18–2 for this example), and the diode will change state, flash on or off, each time the program loops. How long the LED stays on or off shows you the iteration time of your real-time loop. Your loop might well be too fast to see the LED blink. If so, then set up a loop counter and change the LED state only after executing a number of times through the loop.

Output

The last part of our main program sends the calculated variables as output to the external hardware. As for handling inputs, we'll use repetitions of a standard procedure to pack variables into bytes and send output via each port a byte at a time.

Taking card 3, port A as an example, the first step is to initialize output byte OB to zero. The bit-packing instruction format is OB = the name of the variable to be packed multiplied by the position constant of the variable's LSB, then ORed with the previous value of *OB*. This instruction format is repeated once for each variable to go into the byte.

The packing of the cab number variables illustrates a point that might well be very important for your application, handling inverted logic I/O. This phenomenon is illustrated by line 1700, which at first glance may appear more complicated than you might expect. It packs variable *CN(1)*, the cab number assigned to block 1. The input lines on CRC and CDC cards are active high, positive logic, while the output card output transistors are active low; when they are turned on, they connect their output lines to ground, 0 V. This negative logic is best for most UCIS loads, but it has to be reversed for the *CN( )* outputs. The correct values are obtained by outputting 7 − *CN(1)*, instead of *CN(1)* itself. This effectively changes 1 s to 0 s and 0 s to 1 s.

Following that format, line 1700 packs the cab number for block 1 starting at bit position 0, and line 1710 the cab number for block 2 starting at bit position 3. Cab numbers 0–4 or 0–7 take 3 bits each. Lines 1720 and 1730 pack the settings for the east-end signals for blocks 1 and 2 into bit positions 6 and 7.

Once all the bits are packed, a POKE (or OUT) instruction sends the output byte to the appropriate port of an output card. The remainder of the output lines follow the same procedure. Only the variable to be output is changed along with the bit position and card/port address, all following the I/O tables.

To close off the program, line 2410 increments the loop counter, and resets it to 0 when it exceeds 31000. Line 2420 is only for using the simulated-input test subroutine I'll cover later. The last main program statement is line 2440. It completes the real-time loop with a GOTO branch back to line 450, to repeat the loop.

As discussed in Chap. 15, with the Apple II series of machines using Applesoft BASIC and Apple's Integer BASIC you need to use different unpacking and packing procedures. See the enclosed disk for a copy of this example program modified to include these special Apple procedures.

Monitor Status Display

Once the main real-time loop program is written, it's easy to add a monitor display to present the variable status for any particular application. As an example for this application we'll display such variables as block occupancy, cab assignments, and directions on the computer's monitor. The display can be simple or complex, up to full use of color graphics.

Graphics capabilities vary from computer to computer, and graphics commands are anything but standard. The display would also have to be tailored for each track plan. Although many UCIS users make use of color graphics displays for an abundance of applications, their use is not within the scope of this book. I can, however, show you a subroutine for a general-purpose tabular status display. That way you can do the same for your application, with the only requirement that you change the variable names and format.

Figure 18–11 shows typical screen output in a tabular monitor display tailored for the example layout. The top row includes the program title and "ITERATION =," followed by the current value of *X,* the loop counter variable. The next row is the display title, and the next lists block numbers 1–12.

```
COMPUTER CAB CONTROL-EXAMPLE PROGRAM-ITERATION = 35
BLOCK/CAB STATUS DISPLAY
BN   1   2   3   4   5   6   7   8   9   10  11  12
BO       1   1   1                       1
CN       3   3   2   3                   1       1
DT       >   >   <   >                   >       >
SE   R   R       R           R   R           R   R
SW   R   R   R   R   R   R       R   R   R       R
NE   2   3   5   99  6   7   99  99  11  12  12  11
NW   99  1   2   99  3   5   6   99  99  99  9   10
SP               1
TU   0   1   2   3   4   5   6
                         1
CD   1   2   3   4
     >   <   >
ENTER VA$,NI,VA?
```

18-11 Monitor display.

The fixed parts (shown as bold characters in Fig. 18–11) of succeeding rows contain the symbols for block occupation, cab number assigned, direction of travel, signal east, signal west, next-east block, next-west block, and spur position. This is followed by fixed title lines for turnouts 0–6 and cab direction switches 1–4.

The last line isn't part of the display subroutine, but is the data input request line for the simulation test subroutine I'll cover later. I've included it in Fig. 18–11 so you can see how it appears in conjunction with the status display.

The status printouts, nonbold characters in Fig. 18–11, are updated as required each time the program completes an iteration of the real-time loop. The block occupancy indicators ("1"s in Fig. 18–11) can be any graphic symbol in a computer's repertoire. A solid rectangle, for example, would avoid confusion with cab 1.

The cab number line shows assignments by number, with a blank for no assignment. Greater-than and less-than symbols serve as arrows to show direction of travel, with blanks for no direction. Each signal is displayed as a blank if green and as an R if red.

The next blocks east and west are indicated by number, with 99 used for no block. This next-block display is handy for debugging, but once the main program's next-block logic is working this display isn't of much use, and it can be deleted.

For the spur and route turnout status I chose a blank if normal and a 1 if reversed. The cab direction display is like the direction of travel for each block.

In Fig. 18–11, we see that cab 3 is running an eastbound train in blocks 2 and 3. The computer has already assigned block 5, which is clear, to cab 3, and signal SE(3) is green to let cab 3's train enter block 5. Cab 3's train has cleared the trailing block, so block 1's cab assignment has automatically been dropped. That leaves the block available for reassignment.

Cab 2 has a train switching in block 4, the passing siding, with the spur off the siding in reverse position. Note that the signals at both ends of the passing track are red, restricting cab 2's train to the siding as it works the local industries.

Cab 1 has a train in block 10, moving toward reverse-loop block 12. The track polarities are correct, as shown by the direction arrows as well as by signal SE(10) being green for entry into block 12.

Once cab 1's train enters block 12, the computer will automatically assign cab 1 to block 11. When his train is entirely in block 12, the cab 1 operator will reverse the cab direction switch to westbound, allowing the CCC program to clear signal SE(12). That shows the cab 1 operator that his train can enter block 11. Turnout 5 is reversed, routing cab 1's train into yard block 9.

Monitor Display Subroutines

You can add as many extra features as you like to your application program, and typically such moves are very tempting. As a counterpoint it's very important to keep the execution time of the real-time loop as short as possible, because that is the index of how fast the computer reacts to changes in the external hardware. A good design requirement is to have the monitor display program update only things that have changed since the last iteration, and let anything that hasn't changed stay the same rather than being written over.

For comparison's sake, the enclosed disk contains example monitor display subroutines written using several different techniques. This way you can look over different approaches and use the features that work best for you and your computer. Be sure to read the DISPLYME file, which will provide you with text information about the different display programs.

Simulating UCIS Input

A handy way to check the execution of any real-time program is to simulate the UCIS input with the computer keyboard. One can enter variable changes that would occur in normal operation and see the program response on the monitor. Figure 18–12 lists the sample subroutine I used to debug the CCC main program using the keyboard to simulate railroad input.

```
4000   REM**SUBROUTINE TO SIMULATE INPUTS...
4010     LOCATE 16, 1 '....FROM KEYBOARD
4020     PRINT "ENTER VA$, NI, VA"
4030     INPUT VA$, NI, VA
4040     IF VA$ = "EN" THEN GOTO 4120
4050     IF VA$ = "BO" THEN BO(NI) = VA
4060     IF VA$ = "CD" THEN CD(NI) = VA
4070     IF VA$ = "TU" THEN TU(NI) = VA
4080     IF VA$ = "CN" THEN CN(NI) = VA
4090     IF VA$ = "SP" THEN SP(NI) = VA
4100     IF VA$ = "RS" THEN RS(NI) = VA
4110     GOTO 4020
4120     RETURN
```

18-12 Optional keyboard-input test.

Line 4010 directs the cursor to row 16, column 1 and line 4020 prints the prompt message "ENTER VA$,NI,VA." Again note the semicolon used after each PRINT command to suppress the automatic carriage-return/line-feed. Following the INPUT command in line 4030 the program waits for the operator to enter variable changes.

Simply key in any of the two-character codes in lines 4050 through 4100 (such as BO for block occupancy), followed by a comma and the array index number (such as 1 for block 1), followed by another comma and the desired new value to give the variable (0, clear, or 1, occupied), followed by a carriage return. The entry format is fixed, but one can loop through as many variable changes as desired.

When done changing variables for a given iteration, enter EN for end followed by any two "dummy" numbers, and line 4040 branches to line 4120. That turns control back to the main program at the line just after the GOSUB 4020 line that invoked the test subroutine.

One can use keyboard input to move imaginary trains around the layout, assign or drop cabs, set turnouts, and change direction switches, then watch as the monitor displays the results of the actions in automatic block/cab assignments, changing signals, and so on. The same process can be used no matter what your application. Remember, however, that to use this test subroutine you must include the GOTO statement in the main program (line 460 in this example) to bypass the normal UCIS input so it won't get mixed up with keyboard input.

Program Modification Using USIC

The program listing in Fig. 18–5, as modified to work with the serial USIC, is included on the 3.5-in disk enclosed with this second edition. All the core portions of the program remain nearly identical, so the same text explanations apply.

That completes our baseline CCC example, but before we close off this chapter and the book, it's important to introduce another important advantage resulting from interfacing a computer to your external hardware: automated diagnostics.

Automated Diagnostics

Through the UCIS, your computer has the power to check out your external hardware. For example, I've used such an automated diagnostic system on my model railroad ever since I installed my first computer in 1980, and it has saved me hundreds of hours of testing and debugging. Without the computer I could easily spend an hour or two running a train while throwing all combinations of panel switches to make sure that everything worked. Even so, there would be many situations left unchecked. If I did find a problem, say a signal LED that wouldn't light, it would still take some time to track down the exact fault. The logic paths ran through hundreds of relay contacts with over 300 relays, and checkout was a very time-consuming process. Now all the logic is performed by software, wiring is much, much simpler, and the computer itself checks out the system.

With the UCIS it's possible to make a very complete check of an entire layout's electrical system in less than ten minutes. By complete I mean checking every panel switch, pushbutton, LED, trackside signal, occupancy detector, and switch machine for both control and position feedback. For CCC one can check all CBAC panel inputs, cab display cards, and cab relay cards.

The test includes each monitored/controlled device plus the wiring from each device to the I/O cards, the I/O cards themselves, the motherboard, the IBEC, (or the UBEC/CBAC, or the USIC), and all the DIP switches, and so on. That's a thorough test, and all in 10 min—indeed a lot of power!

Digital simplicity, documentation, and plug-in card replacement make UCIS diagnosis an easy way to isolate and repair faults. When I find a problem with the diagnostic program, I usually have the defect located and repaired in less than a tenth of the time it took before the computer. Because the UCIS and its field devices are almost an all-solid-state system, there are considerably fewer failures in the first place. It's a much better system that gives me more time to enjoy other facets of life.

Figure 18–13 shows sample monitor displays from a typical layout's diagnostic program. The actual display uses color. For sake of explanation, parts that don't change I've shown with a light gray screen, and those that are updated are shown on white. I won't detail the code because once you understand programming your UCIS, adding test capability is simple.

Screen A is the test menu that appears when the test diagnostic program is activated. It lists all the tests that can be performed. The operator simply keys in any of the two-character codes followed by a RETURN, and the program branches to that test.

The programming for this branching follows the procedure used for simulating UCIS input in Fig. 18–12. Entering a CTRL-C command at any point ends the test in process, and then RUN returns operation to the test menu. From the menu, entering

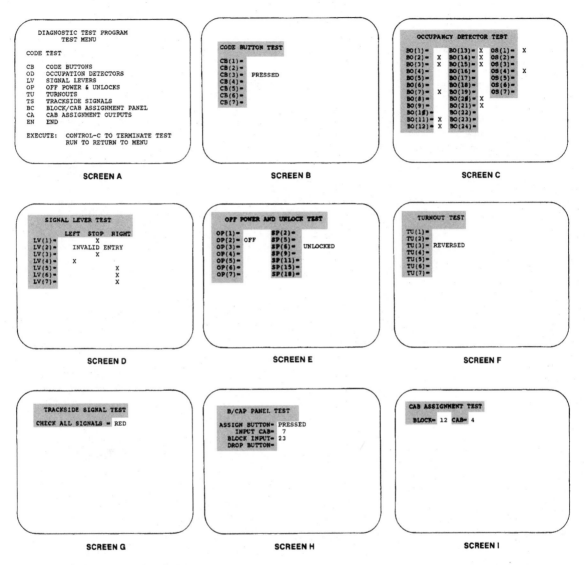

18-13 Sample diagnostic displays.

EN for end terminates the diagnostic program and returns to the operating system so that you are in position to invoke any other program.

In screens B through I, the constant or standing part of each display is shown against a gray-screen background, and the updated part on white. Screen B is the display for code button testing. (Code button is simply a term for a pushbutton interfaced to a computer as part of a railroad's centralized traffic control, or CTC, panel.) The word PRESSED appears next to the symbol for any button the computer detects as being pressed, in this case CB(3), but otherwise this part of the display is blank.

By pushing each button in sequence and watching the monitor, it is easy to see that each button is working properly, including all wiring back to the computer. If the panel for an application is remote from the monitor, it's easy to have the diagnostic program blink a panel LED or beep an annunciator horn as each button is pressed, in addition to updating the monitor. This makes testing even faster, as the operator doesn't have to look at the monitor unless a problem is detected.

Screen C is the display for the occupancy detector test: the Xs show which blocks or sections are read as occupied. By running a train, or (faster but not as much fun) by touching a resistor across each block and OS section, the operator verifies the operation of each detector and its wiring. With optimized detectors, use a resistor corresponding to the desired value for the maximum sensitivity adjustment, 30 kΩ for example, to check the potentiometer setting of each detector. Again, it's also handy to have the horn beep for each new detection.

Screen D is the test display for CTC panel signal levers, a three-position rotary switch. The Xs mark the read-in position of each lever; moving each lever to left, stop, and right tests all its connections. The diagnostic program also drives the signal-lever panel LEDs (via the SDC card described in Chap. 13) to follow the switch position as read. This provides checking out the lever and all its wiring into the software and back out through the SDC to the panel LEDs just by flipping the switch!

The program prints INVALID ENTRY when both lines for any lever are read as pulled to ground, as for LV(2) in this example. This appears to show that the switch is left and right simultaneously and indicates either a faulty switch or, more likely, a line shorted to ground.

Screen E is the display used to check the off-power and unlock circuits, again part of a CTC panel. Flipping each toggle on and off in sequence checks out all actions from the toggle out to some IC logic cards and back through the UCIS to the computer.

Screen F is for the turnout test, which can be tailored to fit any particular application. If the turnouts aren't computer controlled, just display the state of each turnout-position feedback line. For CTC-controlled turnouts the program can display the lever position, throw the turnout, and set the panel LEDs. Another useful test is to have the program progressively throw each turnout back and forth automatically, as the operator watches for proper switch point operation and spring tension.

The trackside signal test is on screen G. The monitor display is simple, but the test is powerful. The software first initializes all signal LEDs to red, and when a RETURN is entered all reds go out and all yellows go on, then another RETURN turns on all greens. Another RETURN sets all signals dark.

For a quick overall lamp test, one could have the program turn on all signal LEDs simultaneously, so that a quick walk around the layout could check all signal LEDs. Or the program can automatically, without waiting for carriage return input, cycle through all dark, all red, all yellow, and all green. This would permit watching for correct, independent operation of each signal line without manual intervention.

Screen H is an example of a B/CAP test display for a CCC railroad. The two buttons are handled as before, and the numbers read from the thumbwheel switches are displayed. If the B/CAPs aren't near the monitor, it's best to have a friend help with this test or build in a parallel form of audio/visual display.

Screen I is a test display for CCC cab assignment. The program initializes all blocks to no cab assigned, cab number 0. For each block, 1 through MB, the cab assignment is switched from 0 through 7 and back to 0 sequentially, and the CRC LEDs indicate that each cab relay is being activated.

Automated Operation

From a computer hobbyist's point of view, having the computer run the train might be the main objective for the system hookup. Even for most model railroaders, who like to play the role of locomotive engineer and run their own trains while the computer automatically does the block power switching and signal control, automated operation can be a positive factor. Remember, at the beginning of Chap. 17, I gave three good reasons why most hobbyists might want the computer to run the train.

The DAC card can be used to control almost any throttle used for manual cab control and connected to any layout. One can also use the CCT card presented in Chap. 14. In either case the throttles can be used for computer cab control (CCC) by feeding the throttles' output through CRC cards.

An alternative to CCC is to use one computer-controlled throttle for each block to provide computer block control, or CBC. CBC's separate throttles for each block eliminate block power switching and give great control flexibility. A train can be operated manually with a cab consisting simply of a pot and a switch to generate speed and direction input to be read by an ADC card and a general-purpose input card, respectively. Or use a hexadecimal rotary switch (Fig. 13–10) as a speed control; its output can be read directly by an input card.

With CBC the software keeps track of which blocks are being occupied by which trains. With that data the software calculates and transmits settings for each of the separate CBC block throttles. Wherever an operator runs a train over the layout, the train responds to its cab through the CBC software. The CRC cards limit CCC to seven cabs, but with CBC it's possible to have up to $N-1$ trains in simultaneous independent operation on a railroad with N blocks. If desired the trains can be running entirely automatically via software directly driving the CCT cards.

CBC might sound costly, but by using the CCT card covered in Chap. 14 the overall system can be very cost-effective. CBC requires just one set of power connections per card, and multiple cards can be powered in parallel. The CCT card can be cut along the dashed lines on the trace side to make four individual CCTs to install close to the blocks they control. If ODs are used in a CBC system, omit jumper J4 on each throttle and connect the detector as in Fig. 18–14.

Figure 18–14 also shows how the CCT can be used as a conventional (non-CBC) throttle to establish a fully automated cab using CCTs as an extension to CCC. This

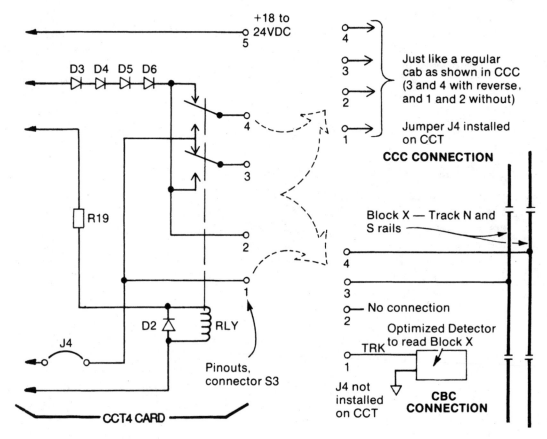

18-14 CCT hookups for CBC.

application uses the four output terminals on each S3. Terminals 1 and 2 are before the reversing relay, while 3 and 4 are after it, for CCC use as shown in Fig. 18–4. Jumper J4 is installed, and a separate 18–24-Vdc supply is required for each throttle. To isolate these, cut away the two outside circuit traces between each throttle at the dashed arrows on the card. You will also need to include the optoisolator circuit shown in Fig. 18–15 between the UCIS output lines and the CCT input lines.

That completes our application example. You should now have the capability to put almost everything under software control for any type of application, whatever your desires. I hope you've learned much from this book and that you have found building your own interface both enjoyable and rewarding.

If you've followed along but haven't yet jumped in to build your first card, now is the time. Working one step at a time, the steps are meaningful and easy to understand, and before you realize it you'll have a working system. Happy computer-interfacing!

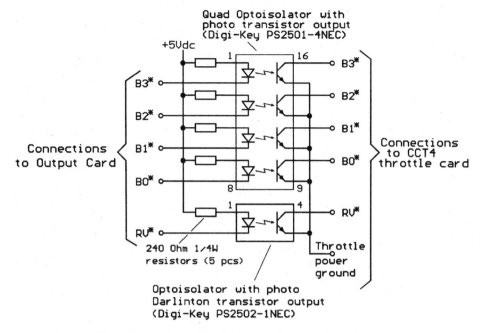

Note: Circuit not required when using analog throttle input or when using digital input in CBC configuration.

18-15 Optoisolator inputs to CCT.

Appendixes

Appendix A
Circuit Board Order Blank
and Circuit Artwork

JLC Enterprises Inc.
P.O. Box 88187
GRAND RAPIDS, MI 49518 USA
Telephone (616) 243-4184

Yes, we are making circuit boards available for the Bruce Chubb designed Computer Interface System. To order simply fill out the order form provided and send a check or money order to the above address. We are not equipped to accept credit card or C.O.D. orders.

All the JLC circuit cards are fabricated from epoxy-glass stock with copper laminate. All holes are drilled and all circuit traces are solder coated for easy parts assembly by the User. Where double sided boards are provided, all feed through holes are plated-through with copper plus solder coated. Where edge tab connectors are required, they are gold plated for increased contact reliability. The IBEC, UBEC, USIC and RS422 boards have the parts nomenclature printed on the component side for ease in assembly and system testing. The IBEC board has solder masking on both sides.

The JLC boards are top quality and guaranteed against any defects in workmanship as delivered. If for any reason you are not satisfied with the JLC boards, return the unused board(s) within 90 days for a full refund. Quantity discounts are available as follows:

Order size	Discount Percentage
$ 0.00 - $249.99	0%
$ 250.00 - $499.99	10%
$ 500.00 - $999.99	15%
$1000.00 and over	20%

The quoted prices are for the fabricated boards only, that is no components are provided by JLC Enterprises, Inc. These are to be purchased separately, by the User, following the parts list information provided in Bruce Chubb's description of the interface assembly. All components are readily available and a summary list of the recommended suppliers is presented in Appendix B.

Fully assembled and tested UCIS boards are available from ECO Works, P.O. Box 9361, Wyoming MI 49509-0361, telephone 616-243-2893 and from Automation In Action (AIA), 1313 16 Mile Rd., Kent City MI 49330, telephone 616-887-0817. Complete "turn-key" ready-to-apply UCIS systems are available from EASEE Interfaces, P.O. Box 92260, Lakeland FL 33804, telephone 941-858-8702. Please send ECO Works , Automation In Action and EASEE Interfaces a business size SSAE to receive a copy of their current flyer material and price list. An alternative "best way" to reach AIA is via their world wide web site HTTP://WWW.ISERV.NET/~AIA. As the supplier status and part numbers, etc.. tend to change over the years, always feel free to contact JLC Enterprises for the latest available information including the pending release of additional interfacing cards.

With best regards from: **JLC Enterprises, Inc.**

ORDER FORM FOR BRUCE CHUBB DESIGNED
UNIVERSAL COMPUTER INTERFACE CARDS

FUNCTION DESCRIPTION	CARD NAME	UNIT PRICE	QTY. ORDER	EXTEND PRICE
Build Your Own Universal Computer Interface, Second Edition book by Dr. Bruce A. Chubb, including 3.5"software disk		$34.95		
Internal bus extender card for IBM PCs up through Pentium	IBEC	42.00		
Mounting bracket for connecting IBEC to computer	BK1	4.50		
Universal serial interface card	USIC	35.00		
RS-232 to RS-422 conversion card	RS422	20.00		
Universal bus extender card (use with adapters listed below)	UBEC**	24.00		
Adapter card for IBM PCs through Pentium	IBPC	32.00		
Adapter card for Apple II family	APPII	24.00		
Adapter card for TRS color computer family	TRSCC	24.00		
Adapter card for Commodore 64, 64c and 128 family	C64	24.00		
Adapter card for VIC-20	V20	24.00		
I/O Mother board (13 slots for holding I/O cards)	IOMB	38.00		
Digital output card (supports 24 output lines)	DOUT	26.00		
Digital input card (supports 24 input lines)	DIN	24.00		
Digital output test card (for connections to DOUT card)	DOTEST	16.00		
Digital to analog converter (3 channels of analog output)	DAC3	30.00		
Analog to digital converter (3 channel of analog input)	ADC3	34.00		
Original-design output card (supports 24 output lines)	COUT24	24.00		
Original-design input card (supports 24 input lines)	CIN24	22.00		
Original-design test card (for connections to COUT24)	OUTEST	16.00		
Signal decoder card	SDC20	12.00		
Computer control throttle (4 per card)	CCT4	28.00		
Pulse power control for throttle	PPC	9.00		
Optimized detector (OD)	OD	6.00		
OD mother board (12 slots for connecting OD cards)	ODMB	20.00		
Dual coil switch machine driver, dual input lines	SM1	8.00		
Dual coil switch machine driver, single input line	SM2	8.00		
Switch motor driver, controls 11 stall motors per card	SMC	22.00		
Cab relay card (up to 7 relays)	CRC	10.00		
Cab mother board (12 slots for connecting to CRCs)	CMB	34.00		
Cab display card (drives numeric digit LED plus3 other LEDs)	CDC	6.00		
Date: _____		Subtotal		
Shipment to be made to:		Subtract quantity discount		
		Total price after discount		
		MI residents add sales tax		
		Add shipping & handling: USA and Canada $6.00 All other foreign $16.00		
		Total Enclosed		

Remit payment as check or money order in US$ to: JLC Enterprises   **Special sale price
P.O. Box 88187
Grand Rapids, MI 49508 USA

All printed board circuits in this Appendix have been reduced to 65% of the original.

© 1996 BRUCE CHUBB

UBEC REV F

Universal Bus Extender Card (UBEC)
Printed circuit board patterns for the UCIS are copyrighted by Bruce A. Chubb. They may be locally reproduced as an aid to your own personal construction, but distributing the photocopies or any circuit cards made from these diagrams to others is prohibited. Ready to assemble pre-fabricated boards are available from JLC Enterprises, P.O. Box 88187, Grand Rapids, MI 49518-0817, U.S.A.

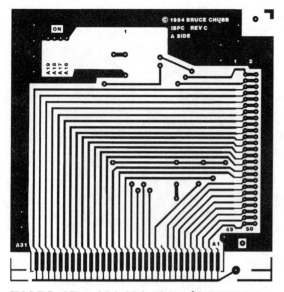

IBM PC, AT or 286, 386, 486 and PENTIUM

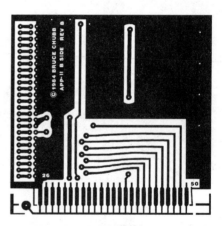

Apple II, II+, IIe, IIGS, and Compatibles

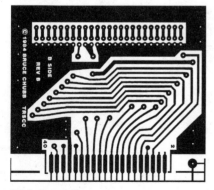
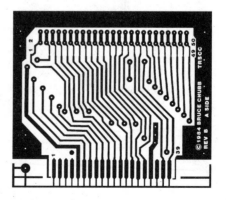

TRS Color Computers

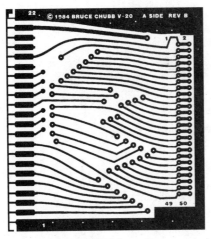
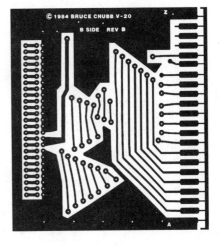

Commodore VIC-20

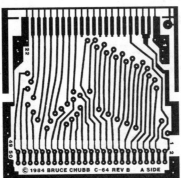
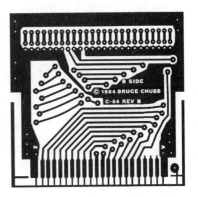

Commodore 64, 64C, and 128

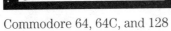

Output Test Card (for connections to DOUT)

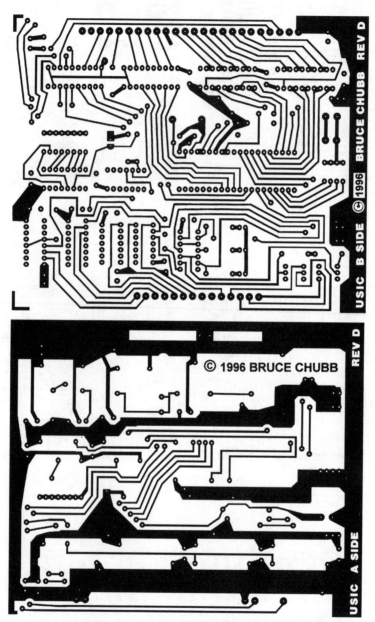

Universal Serial Interface Card (USIC)

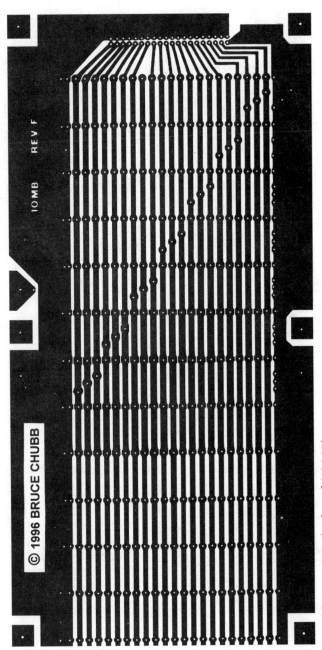

Input/Output Motherboard (IOMB)

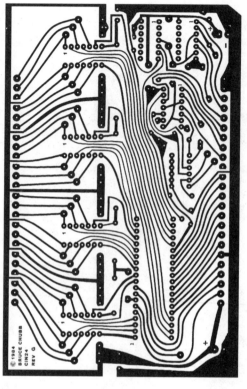

Input Card (CIN24)

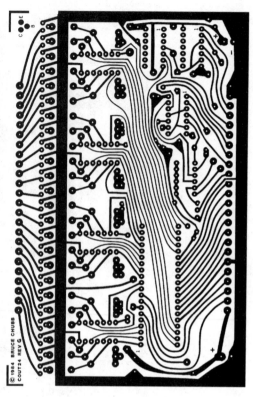

Output Card (COUT24)

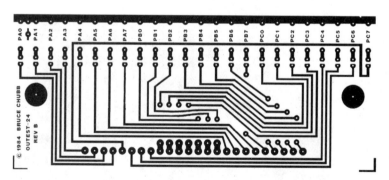

Output Test Card (for connections to COUT24)

Appendix B
Electronic Parts Ordering: Address Summary

Author-recommended sources for electronic parts. Free catalog provided upon request.

Digi-Key Corporation
P.O. Box 677
Thief River Falls, MN 56701
1-800-344-4539

Jameco Electronics
1355 Shoreway Rd.
Belmount, CA 94002
1-800-831-4242

JDR Microdevices
1224 S. Bascom Ave.
San Jose, CA 95128
1-800-538-5000

Mouser Electronics
11433 Woodside Ave.
Santee, CA 92071
1-800-346-6873

Source for all ready-to-assemble circuit cards presented in-this book. See Appendix A for card flyer sheet description and order form. Additional forms available upon request.

JLC Enterprises
P.O. Box 88187
Grand Rapids, MI 49518-0187
616-243-4184

Sources for assembled and tested UCIS circuit boards

ECO Works
P.O. Box 9361
Wyoming MI 49509-0361
616-243-2893

Automation In Action
1313 16 Mile Rd.
Kent City MI 49330
616-887-0817

EASEE Interfaces
801 Bryson Loop
Lakeland FL 33809
941-858-8702

Source for USIC programmed MC68701S Microcomputer Unit chip and associated support software.

Chesapeake Computer Group Inc.
3903 17th Street
Chesapeake Beach, MD 20732
301-855-8430

Sources for surplus electronic parts. Free catalog provided upon request.

All Electronics Corp.
P.O. Box 567
Van Nuys, CA 91408
1-800-826-5432

Fair Radio Sales
P.O. Box 1105
Lima, Ohio 45802
419-223-2196

Index

Index

About the Author

Dr. Bruce Chubb has over 35 years of engineering design, research, and management expertise, including extensive work developing digital computers, computer systems, and large real-time software packages for avionic flight management systems. He has served as vice president of Research and Development for Smiths Industries SLI Avionic Systems Corporation (formerly Lear Siegler), a large producer of digital computer systems for aircraft, spacecraft, and ground vehicles. He also has research experience working with government laboratories. He has gained extensive experience through consulting work in the applications of the Universal Computer Interface System (UCIS).

Dr. Chubb is also one of the world's leading model railroaders. His famous HO scale Sunset Valley System is currently being expanded to 2800 square feet with over 5000 feet of track. Each of the 128 locomotives has an internally integrated digital receiver that is connected to Dr. Chubb's UCIS.

SOFTWARE AND INFORMATION LICENSE

The software and information on this diskette (collectively referred to as the "Product") are the property of The McGraw-Hill Companies, Inc. ("McGraw-Hill") and are protected by both United States copyright law and international copyright treaty provision. You must treat this Product just like a book, except that you may copy it into a computer to be used and you may make archival copies of the Products for the sole purpose of backing up our software and protecting your investment from loss.

By saying "just like a book," McGraw-Hill means, for example, that the Product may be used by any number of people and may be freely moved from one computer location to another, so long as there is no possibility of the Product (or any part of the Product) being used at one location or on one computer while it is being used at another. Just as a book cannot be read by two different people in two different places at the same time, neither can the Product be used by two different people in two different places at the same time (unless, of course, McGraw-Hill's rights are being violated).

McGraw-Hill reserves the right to alter or modify the contents of the Product at any time.

This agreement is effective until terminated. The Agreement will terminate automatically without notice if you fail to comply with any provisions of this Agreement. In the event of termination by reason of your breach, you will destroy or erase all copies of the Product installed on any computer system or made for backup purposes and shall expunge the Product from your data storage facilities.

LIMITED WARRANTY

McGraw-Hill warrants the physical diskette(s) enclosed herein to be free of defects in materials and workmanship for a period of sixty days from the purchase date. If McGraw-Hill receives written notification within the warranty period of defects in materials or workmanship, and such notification is determined by McGraw-Hill to be correct, McGraw-Hill will replace the defective diskette(s). Send request to:

Customer Service
McGraw-Hill
Gahanna Industrial Park
860 Taylor Station Road
Blacklick, OH 43004-9615

The entire and exclusive liability and remedy for breach of this Limited Warranty shall be limited to replacement of defective diskette(s) and shall not include or extend to any claim for or right to cover any other damages, including but not limited to, loss of profit, data, or use of the software, or special, incidental, or consequential damages or other similar claims, even if McGraw-Hill has been specifically advised as to the possibility of such damages. In no event will McGraw-Hill's liability for any damages to you or any other person ever exceed the lower of suggested list price or actual price paid for the license to use the Product, regardless of any form of the claim.

THE McGRAW-HILL COMPANIES, INC. SPECIFICALLY DISCLAIMS ALL OTHER WARRANTIES, EXPRESS OR IMPLIED, INCLUDING BUT NOT LIMITED TO, ANY IMPLIED WARRANTY OF MERCHANTABILITY OR FITNESS FOR A PARTICULAR PURPOSE. Specifically, McGraw-Hill makes no representation or warranty that the Product is fit for any particular purpose and any implied warranty of merchantability is limited to the sixty day duration of the Limited Warranty covering the physical diskette(s) only (and not the software or in-formation) and is otherwise expressly and specifically disclaimed.

This Limited Warranty gives you specific legal rights; you may have others which may vary from state to state. Some states do not allow the exclusion of incidental or consequential damages, or the limitation on how long an implied warranty lasts, so some of the above may not apply to you.

This Agreement constitutes the entire agreement between the parties relating to use of the Product. The terms of any purchase order shall have no effect on the terms of this Agreement. Failure of McGraw-Hill to insist at any time on strict compliance with this Agreement shall not constitute a waiver of any rights under this Agreement. This Agreement shall be construed and governed in accordance with the laws of New York. If any provision of this Agreement is held to be contrary to law, that provision will be enforced to the maximum extent permissible and the remaining provisions will remain in force and effect.